Society of Earth Scientists Series

Series Editor

Satish C. Tripathi, Lucknow, India

The Society of Earth Scientists Series aims to publish selected conference proceedings, monographs, edited topical books/text books by leading scientists and experts in the field of geophysics, geology, atmospheric and environmental science, meteorology and oceanography as Special Publications of The Society of Earth Scientists. The objective is to highlight recent multidisciplinary scientific research and to strengthen the scientific literature related to Earth Sciences. Quality scientific contributions from all across the Globe are invited for publication under this series. Series Editor: Dr. Satish C. Tripathi.

More information about this series at http://www.springer.com/series/8785

Dhananjai K. Pandey · M. Ravichandran ·
Nisha Nair
Editors

Dynamics of the Earth System: Evolution, Processes and Interactions

Contributions from Scientific Ocean Drilling

Springer

Editors
Dhananjai K. Pandey
Geoscience Division
Earth System Sciences Organisation
(ESSO), National Centre for Polar
and Ocean Research (NCPOR)
Vasco da Gama, Goa, India

M. Ravichandran
Earth System Sciences Organisation
(ESSO), National Centre for Polar
and Ocean Research (NCPOR)
Vasco da Gama, Goa, India

Nisha Nair
Earth System Sciences Organisation
(ESSO), National Centre for Polar
and Ocean Research (NCPOR)
Vasco da Gama, Goa, India

ISSN 2194-9204 ISSN 2194-9212 (electronic)
Society of Earth Scientists Series
ISBN 978-3-030-40658-5 ISBN 978-3-030-40659-2 (eBook)
https://doi.org/10.1007/978-3-030-40659-2

This Springer imprint is published by the registered company Springer Nature Switzerland AG
The registered company address is: Gewerbestrasse 11, 6330 Cham, Switzerland

Series Editor Foreword

The Earth's ocean surface has a direct bearing on the global geological processes; therefore, understanding the ocean is as significant as the land. Our knowledge about plate tectonics, ocean circulation, climate change, monsoonal system, deep life, etc., is greatly strengthened by challenging ocean deep drilling programmes. India is an important partner in this programme. The geoscientific results coming out of some of these expeditions have been edited by the team of National Centre for Polar and Ocean Research (NCPOR), Goa, India, enhancing our awareness about the evolving Earth. I congratulate editors and all the contributors for bring out this valuable information to geoscientific community.

The mega-event of 36th International Geological Congress 2020 in India opened new chapter on the geology of India. On such an occasion, Society of Earth Scientists Series by Springer decided to bring out 36th IGC Commemorative Volumes on various recent geological and geophysical studies of India. As such geoscientists were requested to prepare comprehensive account as monographs or edited volumes. I am personally thankful to all the editors and authors for timely submission of high-quality manuscripts for inviting interest of global community of geoscientists.

Lucknow, India

Satish C. Tripathi
Series Editor

Foreword

I am delighted to see this scientific monograph entitled *Dynamics of the Earth System: Evolution, Processes and Interactions* based on scientific contributions to the ocean drilling by Indian scientists in recent years. This collection also commemorates a decade-long association of India to the International Ocean Discovery Program (IODP) consortium.

India became an associate member to IODP consortium in 2009 through a MoU between Ministry of Earth Sciences (MoES) and National Science Foundation (NSF), USA. This MoU provides Indian scientists and researchers a unique opportunity to carry out cutting-edge geoscientific research using sediment and rock sampling. The National Centre for Polar and Ocean Research (NCPOR), Goa, an autonomous institute of MoES, has been designated as the nodal agency to act as IODP India and the Programme Management Office (PMO). This MoU caters to the requirements of Indian scientists and researchers to participate in various IODP expeditions around the world on-board D/V JOIDES Resolution. Since 2009 to till date, about 50 Indian scientists have had an exclusive opportunity to get hands-on experience with the cutting-edge science of deep ocean drilling.

During initial phases (2009–2011), IODP India organised several national and international workshops and brainstorming meetings to develop new ocean drilling proposals. Long-term nurturing of scientific proposals led to the major breakthrough in terms of five dedicated IODP expeditions in the Arabian Sea and Bay of Bengal during 2015–2016. One of these expeditions (IODP 355) was jointly led by an Indian scientist for the first time. These dedicated expeditions provide immense information about long-term evolution of Indian monsoon with special emphasis on their links with the Himalayan tectonics and geohazards in the Andaman–Sumatra seismogenic zone. Indian scientists who took part in these expeditions have produced a large number of peer-reviewed high-impact scientific publications. I am glad to see a snapshot of their scientific contributions in form of this volume and wish them all the best for many more path-breaking discoveries in the near future.

Dr. M. Rajeevan
Secretary
Ministry of Earth Sciences
Government of India
New Delhi, India
e-mail: secretary@moes.gov.in

Preface

Our planet Earth is an evolving dynamic system. It comprises several subsystems or subspheres, which closely interact among each other to produce a complex and continuously changing system. Processes operating in the Earth system vary extensively on spatial as well as temporal scales. More than two-thirds of our planet Earth is covered by vast oceans. Global oceans constitute an ideal domain where potential processes, interactions and their effects on the planet Earth can be thoroughly examined.

The International Ocean Discovery Program (IODP)—an international scientific consortium of 23 nations—attempts to unravel these complex processes to improve our understanding about our Earth. Deep ocean drilling and coring operations through IODP expeditions provide unprecedented archives of Earth's evolution. These valuable archives are recovered from extremely challenging areas of wide oceans and from ice-covered shallow water to the deepest ocean depths. The two preceding phases of this programme [the Deep-Sea Drilling Project (DSDP) and the Ocean Drilling Programme (ODP)] have witnessed remarkable success in terms of major scientific advancements. Some of these include our in-depth knowledge about plate tectonics, ocean circulation and long- and short-term climate change cause-and-effect relationships etc. During its ongoing phase, IODP aims to unravel complex geodynamic processes through drilling, sampling and experimenting in deep holes in different geological/tectonic regimes. Significant scientific achievements from ocean drilling help us understand and interpret phenomena in various parts of the Earth system.

A broad science plan that drives scientific ocean drilling encompasses 14 major challenging questions under four different themes, e.g. climate change, deep life, planetary dynamics and geohazards. IODP expeditions recover several thousands of metres long cores from the subseafloor that enables us address these challenging questions. Each expedition emanates from specific hypothesis-driven science objectives and typically lasts about two months at sea. Ever since India joined the IODP consortium, Indian scientists started participating in IODP expeditions around the world. Till date, around 50 Indian scientists, each of them have had the '60-day-long exposure at sea'.

The scientific manuscripts presented in this volume are built upon multitude of knowledge and science developed by Indian scientists over the last decade from different IODP expeditions. Each contribution in this seminal collection specifically highlights their role within an overarching scientific objective of the particular IODP expedition. We feel that this collection presents a glimpse of different processes that contribute towards evolution of our dynamic Earth system.

Before we delve into each contribution individually in the forthcoming chapters, we take this opportunity to acknowledge immense support received from the Ministry of Earth Sciences as well as the IODP consortium in providing samples to engage in cutting-edge research concerning topics of global importance.

This special volume would not have been possible without tremendous support received from the different colleagues and organisations. At the outset, we would like to thank each contributing author of this series who took part in their respective expeditions to gather valuable data. Subsequently, they worked tirelessly to come up with excellent findings, which constitute the core of this collection.

The volume editors are highly grateful to Dr. Rajeevan Madhavan Nair, Secretary, Ministry of Earth Sciences, India, for his continuous encouragement and support during production of this monograph. Ever since we came up with the idea of bringing a special collection to commemorate a decade of Indian IODP association, Dr. Satish C. Tripathi from the Society of Earth Scientists (SES) extended immense cooperation and ensured that it came to fruition. We thank him for his continued support towards this endeavour.

Regular motivation from Dr. B. K. Bansal, Adviser, MoES, as well as expedient facilitation from his office towards smooth participation of Indian scientists in IODP expeditions is gratefully acknowledged. Extensive assistance and an amazing coordination received from our colleague Ms. Prerna Ramesh in the manuscript handling, review and final shaping up of this volume are truly acknowledged. Unconditional support from our colleagues in the Geosciences division at NCPOR, Goa, during the course of this monograph needs a special mention.

Last but not least, the production of this monograph has been a long journey and DKP acknowledges the love and support received from his wife Anju and son Abhinav without which it would have been difficult to accomplish this task in time.

Vasco da Gama, India

Dhananjai K. Pandey
M. Ravichandran
Nisha Nair

Contents

About the Editors

Dr. Dhananjai K. Pandey has been working as marine geoscientist at the National Centre for Polar and Ocean Research since 2003. He is presently the Group Director (Geosciences) and looks after various facets of the Indian IODP programme at NCPOR. He holds a doctorate degree in earth sciences from University of Cambridge, UK. His research interests include geodynamics of the continental margins with special emphasis on climate and tectonics linkages. He was the lead proponent and co-chief of the first scientific ocean drilling expedition (IODP 355) in the Laxmi basin of the Arabian Sea. He is also a member of the Science Evaluation Panel of the IODP. He has more than 50 high-impact peer-reviewed scientific publications to his credit. For his distinguished contributions in the Geosciences, he has been awarded 'Certificate of Merit' as well as the National Geoscientist Award.

Dr. M. Ravichandran is the Director, National Centre for Polar and Ocean Research (NCPOR), Goa, since 2 November 2016. He obtained his Ph.D. in physics from University of Pune. He worked as an Atmospheric Scientist in Indian Institute of Tropical Meteorology (ESSO-IITM), Pune (1988–1997); Ocean Technologist at National Institute of Ocean Technology (ESSO-NIOT), Chennai (1997–2001); and a Physical Oceanographer at Indian National Centre for Ocean Information Services (ESSO-INCOIS), Hyderabad (2001–2016), before joining at NCPOR. His research interest includes atmospheric and ocean dynamics, ocean observations and modelling. He is Vice-President of SCAR and Council Member of IASC. He has more than 120 publications in referred journals.

Dr. Nisha Nair completed her Ph.D. from Andhra University and currently working as a marine geophysicist at the National Centre for Polar and Ocean Research (NCPOR), Goa. Her field of interest includes analysis and interpretation of geophysical data and geopotential modelling in order to delineate the sediment-crustal structure of continental margins of India. Through her study, she has been involved in understanding the tectonic evolution of the continental margins of India and adjoining deep-sea basins. She had also been one of the scientific participants in the IODP expedition 362. This expedition was aimed to understand the seismogenesis of the Sumatra subduction zone.

Introduction—Scientific Drilling in the Indian Ocean: An Earth System Process Perspective

Dhananjai K. Pandey, Nisha Nair and M. Ravichandran

Abstract The International Ocean Discovery Program (IODP) addresses the fundamental questions about Earth's climate, deep life, geodynamics, and geohazards, through Earth's climatic, biological, chemical, and geological history buried beneath the ocean floor. Scientific Ocean drilling allows researchers to access these records and explore, analyze, postulate, and test hypotheses that address how our planet evolves on local-to-global spatial scales and on decadal-to-millennial time scales. Based on key scientific challenges, deep sea drilling in the Indian Ocean addressed the broad questions related to palaeoceanographic, tectonic and climatic controls on the Indian Monsoon, evolution of the mid-ocean ridges, formation of large igneous provinces, continental rifting and related deposition, subduction related earthquakes and arc volcanism and how it had influenced the development of deep biosphere.

Keywords Tectonics · Asian monsoon · Indian ocean · IODP expedition

The Indian Ocean hosts exceptional examples of Earth system processes and products that play a vital role in governing Earth's climate. One of the most significant among them is the Asian Monsoon, which acts a major part of the thermohaline conveyor. Prior to 2015, deep sea drilling and coring experiments in the Indian Ocean were scanty. The lack of sufficient sediment coring in the Indian Ocean over several decades resulted in major gaps in geoscientific understanding of this region. In order to address this concern, in 2011 an international ocean drilling workshop with special emphasis on Indian Ocean was organised in Goa, India by IODP-India. The prominent idea that emerged from this workshop was that recovery of buried sedimentary and igneous archives through drilling in the Indian Ocean would substantially augment existing knowledge. In particular about the history of uplift, erosion, deposition and long term monsoonal variations. Such attempts would further improve our understanding about the greenhouse/icehouse and dynamics of ocean gateways. Key questions about precise rift-to-drift transition of the Gondwana super-continent; subduction, tectonic breakup and reorganization and many deep biosphere mysteries

---

D. K. Pandey (✉) · N. Nair · M. Ravichandran
ESSO-National Centre for Polar and Ocean Research (NCPOR), Goa, India
e-mail: pandey@ncpor.res.in

D. K. Pandey et al. (eds.), *Dynamics of the Earth System: Evolution, Processes and Interactions*, Society of Earth Scientists Series,
https://doi.org/10.1007/978-3-030-40659-2_1

Table 1 Summary of recent IODP expeditions in Indian Ocean

IODP	Broad Theme	Dates	Drill Sites	References
353	Indian Monsoon Rainfall	Nov, 2014–Jan, 2015	U1443-U1448	Clemens et al. (2016)
354	Bengal Fan: Himalaya and Climate	Jan, 2015–Mar, 2015	U1449-U1455	France-Lanord et al. (2016)
355	Arabian Sea: Climate-Tectonics-Interactions	Mar, 2015–May, 2015	U1456-U1457	Pandey et al. (2016)
356	Indonesian Throughflow	Jul, 2015–Sept, 2015	U1458-U1464	Gallagher et al. (2017)
359	Maldives Monsoon and Sea Level	Sept, 2015–Nov, 2015	U1465-U1472	Betzler et al. (2017)
360	SW Indian Ridge Lower Crust and Moho	Nov, 2015–Jan, 2016	U1473	MacLeod et al. (2017)
361	South African Climates	Jan, 2016–Mar, 2016	U1474-U1479	Hall et al. (2017)
362	Sumatra Seismogenic Zone	Aug, 2016–Oct, 2016	U1480-U1481	McNeill et al. (2017a, b)

necessitate drilling into the regional sub-seafloor. Based on extensive discussions during 2011 IODP workshop, several promising scientific proposals were submitted to the IODP of which about eight of them culminated into successful drilling during 2015–2016 (see Table 1).

Targeted deep sea drilling expeditions in the Indian Ocean and surrounding margins aimed to provide a regional understanding of the development and evolution of Asian monsoon and its inherent links with the upliftment of the Himalaya and Tibet. Recent IODP Expeditions 353, 354 in the Bay of Bengal and expedition 355 in the Arabian Sea aimed particularly to study long term weathering and erosional patterns in the Himalayan and Tibetan Plateau as a proxy to the tectonic–climatic interactions. Expedition 356 targeted coring for high resolution study of how Indonesian Throughflow acts as a global thermohaline conveyor in transporting heat from the equatorial Pacific to the Indian Ocean and exerts a major control on global climate. The IODP expedition 359 (Maldives Monsoon and Sea Level) aimed at investigating changing ocean currents that affect the Indian monsoon on a large scale. The establishment of the Asian monsoon and its evolution through time and probable linkages between tectonics and climate are conformable with model predictions. The Expedition 360 (Southwest Indian Ridge Lower Crust and Moho) explored the nature of

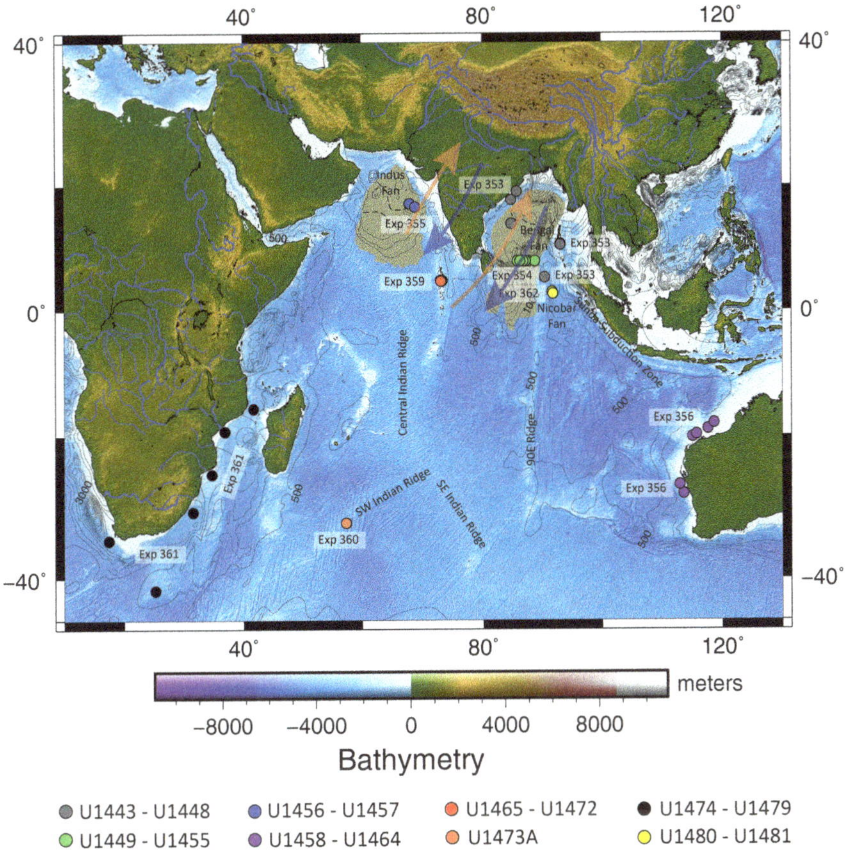

Fig. 1 Bathymetry map of the Indian Ocean (Amante and Eakins 2009). Superposed are the locations of drill holes from IODP expeditions in the Indian Ocean—353, 354, 355, 356, 359, 360, 361 and 362. Arrows indicate the Asian monsoon pattern

lower crust and upper mantle at slow spreading ridges. The South African Climates (Expedition 361) drilled and cores several sites in Mozambique Channel in order to reconstruct the history of the greater Agulhas Current system over the past ~5 My. The Sumatra Seismogenic Zone Expedition 362 aimed to establish the initial and evolving properties of the North Sumatran incoming sediments and their potential effect on seismogenesis and tsunamigenesis. This expedition was inspired by the 2004 Mw 9.2 earthquake and tsunami that affected the coastal communities around the Indian Ocean. This chapter briefly summarizes the early outcomes from recent scientific drilling in the Indian Ocean (Fig. 1). In what follows next, we present an overview of each of the six recently concluded IODP expeditions in the Indian Ocean since 2015 and their preliminary findings to date.

1 Himlayan-Tibetan Orogeny and Long Term Climate Evolution

1.1 Indian Monsoon Rainfall (IODP Expedition 353)

Paleoclimatic/paleooceanographic proxies enable us to document long term evolution of regional monsoon rainfall to changes in the fundamental boundary conditions like insolation, ocean/atmosphere energy exchange, ice volume, and atmospheric greenhouse gas concentrations or seasonal influences. In order to reconstruct variations in the Indian monsoon circulation since the Miocene at tectonic to centennial timescales, Expedition 353 targeted to drill and recover sediments along the Mahanadi Basin, Andaman Islands and the northern Ninety East Ridge (U1443-U1448) (Clemens et al. 2016). The suitable drill sites were picked in the core convective region of the Indian summer monsoon (ISM), which is believed to have been under the direct influence of ISM-induced rainfall and fluvial runoffs receipt from the world's largest river systems.

The geochemical proxy records at submillennial-scale resolution (250–500 years) of ISM-induced rainfall and fluvial runoff sediments obtained from this expedition suggest that the Indian summer monsoon has been a conduit for conveying southern hemispheric latent heat northwards, thereby promoting subsequent northern hemispheric deglaciation (Nilsson-Kerr et al. 2019). The proxy records indicate strong coupling between summer monsoon winds and precipitation across the Indo-Asian monsoon subsystems at the millennial scale. To evaluate changes in the surface ocean salinity response to rainfall and runoff, carbonate oxygen isotope ($\delta^{18}O_c$) and Mg/Ca-derived sea-surface temperatures (SSTs) from the planktic foraminifera *Globigerinoides ruber* were combined to reconstruct seawater $\delta^{18}O$ ($\delta^{18}O_{sw}$). The initial results imply that the ISM underwent two phases of deglacial strengthening (one at ~131 to 130 ka, followed by a strengthening at ~129 ka), with a vigorous interglacial ISM coeval with the development of full deglaciation into the Last Interglacial. This revealed a millennial-scale transient strengthening of the Asian monsoon that punctuates Termination II (i.e. 139–127 thousand years ago (ka) associated with an oscillation of the bipolar seesaw. The progression of deglacial warming across Termination II emerges first in the southern hemisphere, then the tropics in association with Indian summer monsoon strengthening, and finally in the northern hemisphere.

The changes from high calcium carbonate content to silica-rich sediment deposition marks an intense carbonate dissolution episode linking to the global ocean circulation systems (Lyle et al. 1995). Based on an integrated study using benthic foraminiferal stable oxygen and carbon isotope data, X-ray fluorescence (XRF) core scanner-derived elemental ratios and carbonate mass accumulation rates (MARs) from Bengal Fan sediment cores, Lübbers et al. (2019) tried to closely track the development of the carbonate crash in the Indian Ocean and investigated linkages with high-latitude climate evolution (over the interval 13.5 to 8.2 million years ago) with circulation changes in the Pacific and Atlantic Oceans. The declining carbonate percentages along with a marked decrease in carbonate accumulation rates after

~13.2 Ma suggested the onset of a prolonged episode of reduced carbonate deposition lasting till ~8.7 Ma (Fig. 2). After ~8.7 Ma, the carbonate accumulation increased rapidly marking the end of the carbonate crash in the Indian Ocean (Fig. 2) apparently in consonance with an intense phase of the inter-ocean carbonate crash (~11.5 to ~10 Ma). Their findings imply that global changes in the intensity of chemical weathering and riverine input of calcium and carbonate ions into the ocean reservoir

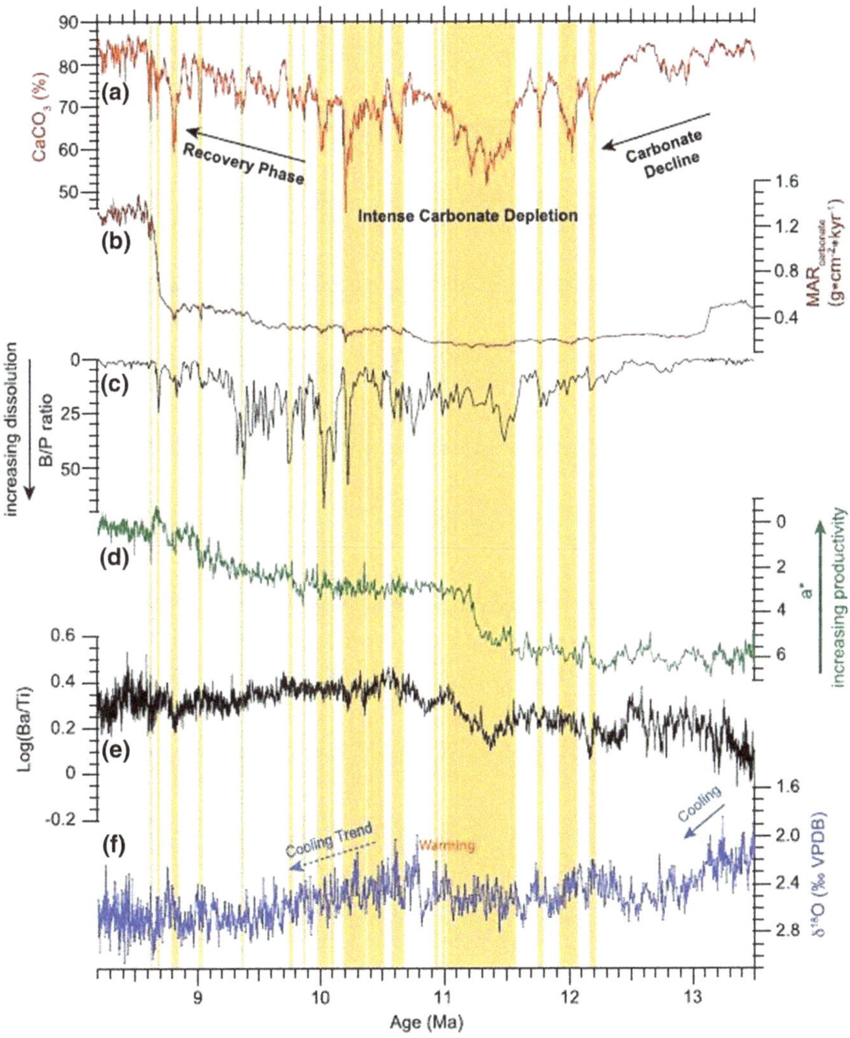

Fig. 2 Paleoceanographic records published by Lübbers et al. (2019) at Site U1443 (Exp 353). **a** XRF scanner-derived $CaCO_3$ (%); **b** Carbonate mass accumulation rate (g cm^{-2} kyr^{-1}); **c** Benthic/Planktonic foraminiferal ratio; **d** Color reflectance a* (Clemens et al. 2015); and **e** XRF scanner Log (Ba/Ti); **f** Benthic foraminiferal $\delta^{18}O$. [Orange shading marks carbonate content under 70%. MAR = mass accumulation rate; VPDB = Vienna PeeDee Belemnite]

might have been the driving force for the carbonate crash. Also an increase in Log (Ba/Ti) together with a change in sediment color from red to green indicated a rise in primary production and organic carbon burial at the sea floor after ~11.2 Ma, prior to the global onset of the biogenic bloom that remained elevated until at least 8.2 Ma. The Ba and sediment color records suggest that the early rise in export flux from biological production might be linked to increased advection of nutrients and intensification of upper ocean mixing, associated with changes in the seasonality and intensity of the Indian Monsoon.

1.2 Bengal Fan

IODP Expedition 354 in the Bay of Bengal drilled a 320 km long transect comprising seven Sites (U1449-U1455). Of these, three deep-penetration and four shallow holes provided a spatial overview of the primarily turbiditic depositional system that comprises the Bengal deep-sea fan. This expedition primarily aimed to reconstruct a time series of erosion from Himalayan region, weathering, and changes in source regions, as well as impacts on the global carbon cycle by coring long sedimentary records (France-Lanord et al. 2016). This expedition also extended the record of an early fan deposition by 10 M.y. into the late Oligocene.

In order to study the Neogene evolution of Bengal Fan and the impact of the monsoonal system and sea-level changes on sediment supply, the fan architecture and development, Weber and Reilly (2018) documented the Pleistocene depositional history (since 1.25 Ma) of the lower Bengal Fan using low-resolution bio- and magnetostratigraphic constraints along with a combination of lithostratigraphy and sediment physical properties. The study identified five distinct depositional regimes (Fig. 3) for the Pleistocene across the lower Bengal Fan comprising of:

(i) Unit 1, consists of Early to Middle Pleistocene turbiditic deposits (1.25 Ma);
(ii) Unit 2, a Middle Pleistocene hemipelagic layer deposited from Marine Isotopic Stages (MIS) 37 to ~17 (~1.24 to 0.68 Ma) during the entire time of the Mid-Pleistocene Transition;
(iii) Unit 3, mostly turbidites with some intercalated hemipelgic sediments deposited from ~MIS 16 to ~13 (~0.68 to 0.48 Ma);
(iv) Unit 4, massive turbiditic sediments dominating deposition with the Mid-Brunhes Transition (~MIS 12) until MIS 7/8 (~0.48 to 0.25 Ma); and
(v) Unit 5, a Late Pleistocene hemipelagic layer at the top that spans MIS 7/8-1 (~0.25 Ma to present).

Their results imply that the regional depositional changes across the lower Bengal Fan occurred at the same time as major changes in the climate system such as the Mid-Pleistocene Transition (Clark et al. 2006) and the Mid-Brunhes Transition (Barth et al. 2018) suggesting that associated external processes such as climate, sea level, and/or monsoon changes might have controlled the sediment accumulation and fan architecture rather than fan-internal, autocyclic mechanisms.

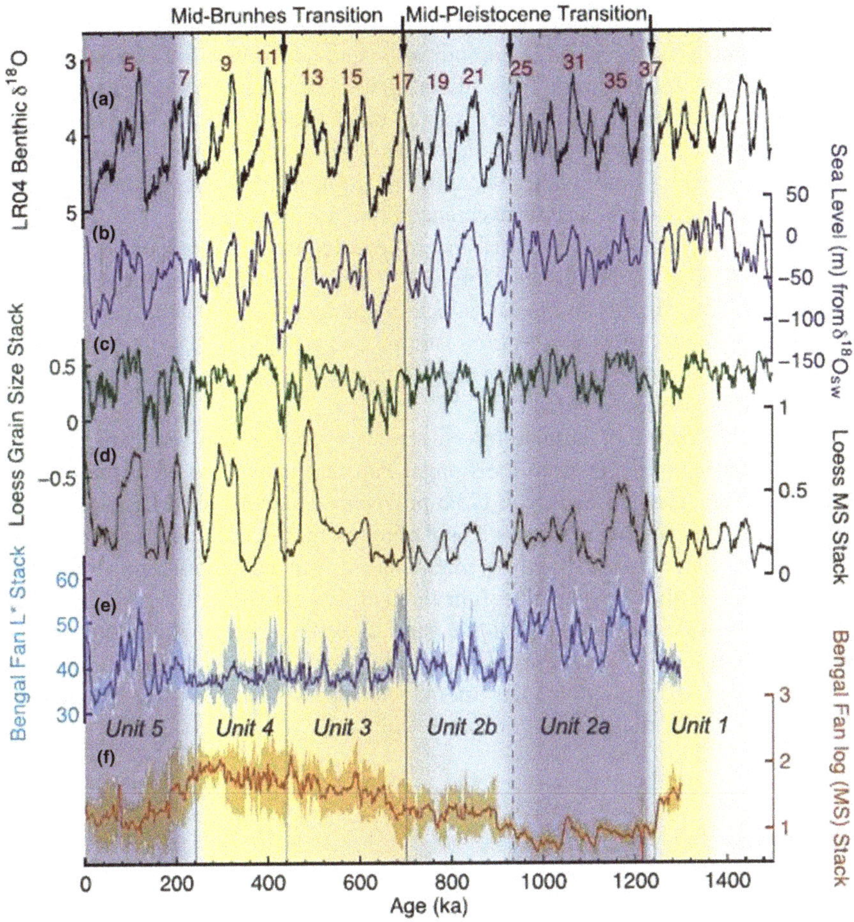

Fig. 3 Comparison of the Expedition 354 Bengal Fan L* and MS stacks with global climate reconstructions spanning the Middle to Late Pleistocene. (a) The LR04 benthic $\delta^{18}O$ stack; Red numbers - Marine Isotope Stages (MIS); (b) Glacio-eustatic sea-level inferred from benthic $\delta^{18}O$; (c) Stacked grain size and (d) MS records from the Chinese Loess Plateau, reflecting changes in the intensity of the East Asian winter and summer monsoons, respectively (Sun et al. 2006); (e) Stacked L*; (f) logarithmically transformed MS values for the IODP Expedition 354 Bengal Fan 8° N transect] (published by Weber and Reilly 2018)

Based on a multi-proxy study using physical and magnetic properties, stable isotopes, organic and inorganic geochemistry, and grain-size analyses, Weber et al. (2018) attempted to decipher the Late Quaternary depositional pattern and its response to monsoonal forcing during glacial-to-interglacial transitions. The variability of total organic carbon, total nitrogen, and the $\delta^{13}C$ composition of organic matter indicated a marine origin. Their results reveal that the marine primary productivity increased during insolation minima, indicating an enhanced NE monsoon during glacials and stadials. The pristine insolation forcing also indicated that darker and coarser-grained material were deposited at higher sedimentation rates during insolation minima. This suggested the mobilization and transport of lithogenic material primarily during lowered sea levels and/or higher influence of the Northern Hemisphere westerlies on the dust transport from the Tibetan Plateau. When compared with the climate record of Antarctic ice cores over multiple glacial cycles, a close relationship between high southern latitude and tropical Asian climate through shifts in position of the Intertropical Convergence Zone was observed.

The entire Bengal Fan monsoonal record revealed two clear responses: an older one from ~200 ka to the Younger Toba Tuff with a very clear and strict response to orbital forcing, mainly to insolation, and a younger part, after the deposition of the Younger Toba Tuff during Marine Isotope Stage (MIS) 7–5, with less distinct response patterns after deposition of the ash during MIS 4–2, consistent with low-amplitude changes in insolation.

The establishment of climate-forced sea-level change and autogenic surface dynamics was carried out from the Bengal Fan turbidites from Miocene to middle Pleistocene. The detrital zircon U-Pb provenance data from Miocene to middle Pleistocene suggested strong tectonic and climatic forcing and long-term drainage integration (Blum et al. 2018) which might have driven sediments from Himalayan source terrains to the continental margin and land-sea boundary. The results implied that the large-scale transfer of turbiditic sand from the continental margin to distal basin-floor fan settings are in episodes as it might have required climate-forced sea-level fall associated with global ice volume increase, so that the Rivers extend their courses to the shelf-margin and connected with canyon-feeder systems.

1.3 Indus Fan

The Indus Fan in the Arabian Sea offers a unique opportunity to study tectonic–climatic interactions and the net impact of the Indian (southwest) summer monsoon (ISM) processes on weathering and erosion of the western Himalaya. Scientific drilling in the eastern Arabian Sea through Expedition 355 (Arabian Sea Monsoon Expedition) aimed to investigate the strengthening/weakening of ISM and it's correlation with the Greater Himalayan exhumation over variable time scales (Pandey et al. 2016). Two sites (U1456-U1457) were drilled into the Laxmi Basin adjoining western continental margin of India (WCMI) in which one of the sites (U1457) also penetrated successfully into the igneous basement. The basement samples recovered from eastern Arabian Sea provided constraints to understand the early rifting history with special emphasis on the continental breakup between India and the Seychelles and its relationship to the hotspot volcanism in the Indian Ocean.

New constraints from Expedition 355 also offer opportunity to determine changing patterns of Himalayan erosion since ~10.8 Ma (Pandey et al. 2015). Clift et al. (2019b) presented new bulk sediment Sr and Nd isotope data, along with other proxies from sediment cored during expedition 355. Their new results allow reconstruction of erosion in the Indus catchment since ~17 Ma. From 17 to 9.5 Ma an increasing εNd values suggest relatively larger erosion from the Karakoram region. Periods between 9.5 and 5.7 Ma show relative stability as evident from long-term decrease in εNd values (Fig. 4). Clift et al. (2019b) inferred that an initial increase in Lesser Himalaya-derived zircon grains after 8.3 Ma is consistent with earlier records from the foreland basin followed by the much larger rise in erosion after 1.9 Ma, which has not previously been reported. This means that the regional erosion increased in the Pleistocene compared to the Pliocene, with shift in source region to the Lesser Himalaya driven by tectonic controls rather than South Asian Monsoon driven erosion.

Expedition 355 successfully sampled through at least two distinct slumped sediment units corresponding to one of the largest mass transport deposits (MTDs) on Earth known as 'Nataraja Slide' (Calvès et al. 2015; Pandey et al. 2016) considered to have occurred during mid-late Miocene period along Indian Continental margin was drilled. Based on biostratigraphy, geochemistry and isotopic signatures Dailey et al. (2019) concluded that these deposits consisted of two dominant phases of mass wasting, each comprising of smaller pulses, with generally fining-upward cycles, and to have been emplaced on Indian margin prior to 10.8 Ma. Some of the most recent scientific contributions about issues like nature of organic matter, turbidite sedimentation and chemical weathering and Himalayan erosion from IODP Expedition 355 could be found as Suzuki et al. (2019), Andò et al. (2019), Kumar et al. (2019), Cai et al. (2019), Chen et al. (2019), Satpathy et al. (2019) and Clift et al. (2019a, b).

The lithostratigraphic variations observed at sites U1456 and U1457 helped in obtaining critical parameters for studying anomalous subsidence along the western continental margin of India and therefore constrains post-rift evolution of western

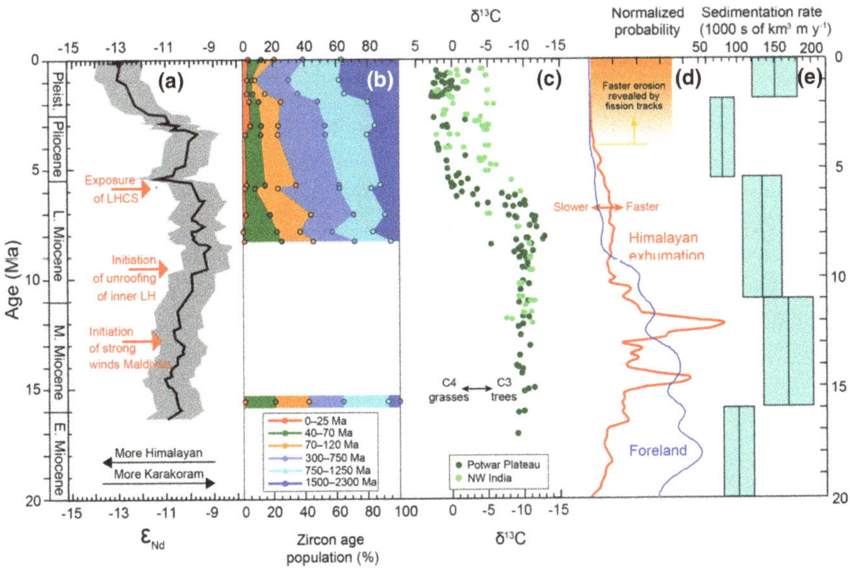

Fig. 4 Changing patterns of various climate, erosion and exhumation proxies since E. Miocene time in the Himalaya (after Clift et al. 2019b)

Indian passive margin. Flexural backstripping of post-rift sediments from interpreted seismic profiles suggest that the underlying crust in the region may have undergone extreme thinning as evident from very large stretching (β) factors (Pandey et al. 2018). A detailed geochemical analysis of basement samples from Site U1457 indicated that these are low-K, high-Mg sub-alkaline tholeiitic basalts that do not represent a standard mid-ocean-ridge basalt (MORB). In-depth geochemical and isotopic examination of samples by Pandey et al. (2019) suggest a supra-subduction zone (SSZ) type affinity of these rocks. Based on new geochemical and isotopic signatures from Site U1457, a new model of crustal evolution has been proposed indicating a relict subduction initiation event in the Laxmi Basin (Pandey et al. 2019). The authors propose that such distinctive geochemical imprints of igneous rocks were made when the Indian plate changed direction and older crust began to sink under an younger oceanic crust about 70 million years ago around the Laxmi basin. This discovery provides first glimpse of a completely new geodynamic setup implying role of convergent plate motion amidst an overall divergent tectonics during the Gondwanaland breakup which separated Madagascar and Seychelles from India (Fig. 5). The findings also suggested that the most spectacular late Cretaceous volcanism that formed well-known Deccan Traps of central India, may not have reached as far as the location of the drill sites in the Laxmi basin.

The South Asian monsoon (SAM)-induced high surface water productivity coupled with poor ventilation of intermediate water resulted in strong denitrification within the oxygen minimum zone (OMZ). The analysis of sediment cores from Laxmi Basin reported the first evidence of denitrification evolution since Late Miocene (~10.2 Ma) and productivity/SAM intensification was reported at ~3.2 to 2.8 Ma (Fig. 6) that coincided with Mid-Pliocene Warm Period (MPWP) (Tripathi et al. 2017). The modern strength of the OMZ was attained at ~1.0 Ma. It was also reported that the SAM was persistently weaker from ~10.2 to 3.1 Ma and did not intensify at Late Miocene (~8 Ma).

The denitrification ($\delta^{15}N$) values during the last 800 kyrs from Site U1456 (Eastern Arabian Sea) when compared with the Site 722B (Western Arabian Sea) showed that $\delta^{15}N$ values were high during interglacial periods, indicating intensified denitrification (Kim et al. 2018). This emphasised that the Eastern Arabian Sea has experienced persistent denitrification without prominent collapse throughout the Mid-Pleistocene. The results implied that the Eastern Arabian Sea has experienced a persistent OMZ to maintain strong denitrification whereas the Western Arabian Sea has undergone OMZ breakdown during some glacial periods. The change in denitrification in the Arabian Sea in response to climatic fluctuations following the 100-ka cyclicity was also represented by global $\delta^{18}O$ values and other multi-proxies.

To constrain the sediment sources and reconstruction of sediment export to the Indus deep-sea fan over the past 600 kyr, Yu et al. (2019) analysed the clay mineralogy and Sr-Nd isotopic compositions of the detrital fraction of sediments. The results revealed that the sediment sources were mainly from the mixing of sediment from the Deccan Traps and the Indus River and excludes a significant contribution of eolian dust to the Laxmi Basin. The wavelet and cross-spectral analyses further revealed that sediment transport to the eastern Arabian Sea since 600 kyr is mainly

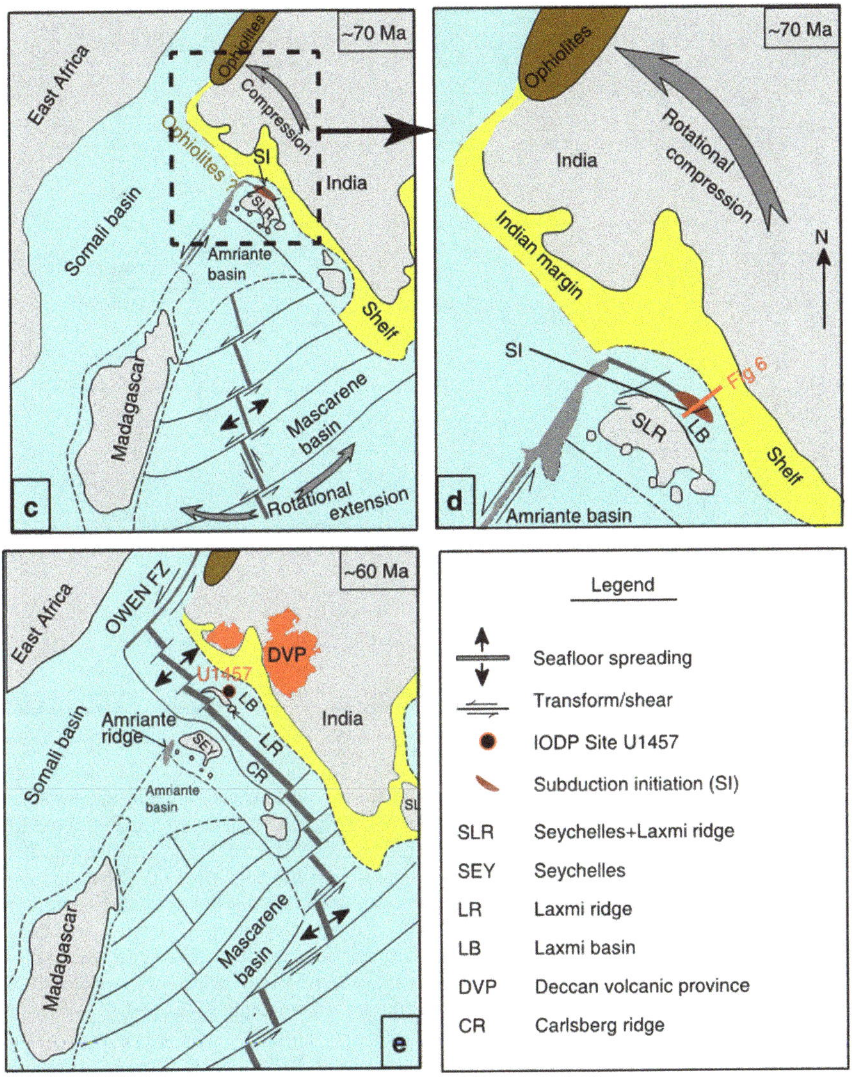

Fig. 5 A generalised plate reconstruction model for the late Cretaceous geodynamic evolution of northwest Indian Ocean (after Pandey et al. 2019) depicting role of convergent tectonics amidst an overall divergent motion between India and Seychelles blocks (not to scale)

dominated by sea-level variations in the eccentricity (100-kyr) band, while the Indian summer monsoon might have contributed moderately to the precession (23-kyr) band. The εNd and turbidite frequency indicated that more turbiditic events are associated with Indus River-sourced sediments at IODP Site U1457 during sea-level lowstands, which might reflect a reconnection of deep-sea channels to the Indus River mouth and/or reworking of previously deposited Indus River sediments located northward.

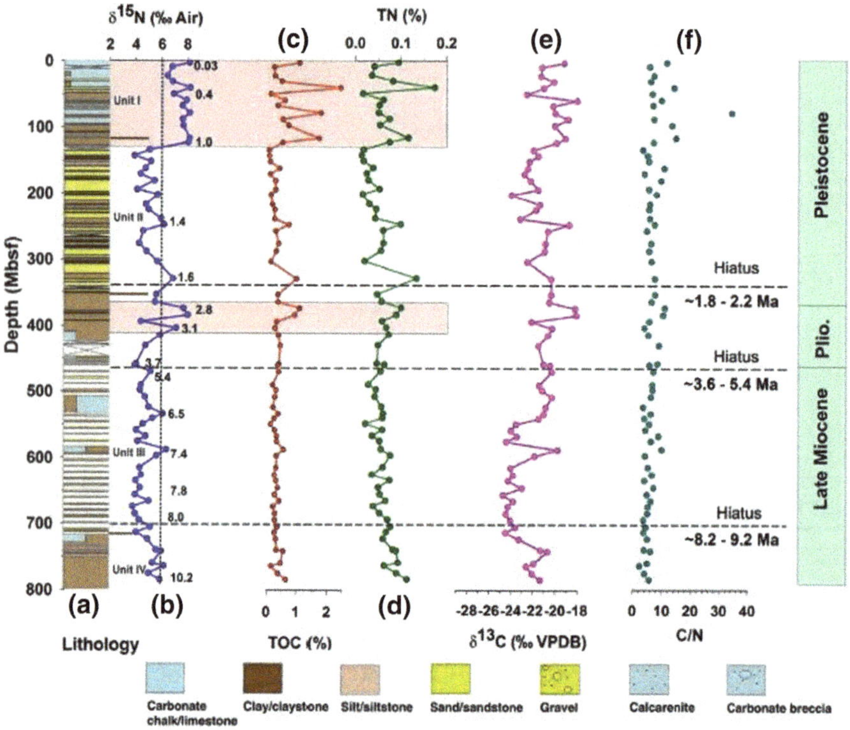

Fig. 6 Record of denitrification, surface water productivity, and provenance of the Sedimentary Organic Matter (SOM) in the Eastern Arabian Sea since Late Miocene. **a** Lithostratigraphy of site U1456, **b** denitrification variability ($\delta^{15}N$ of SOM), **c**, **d** paleoproductivity variability (weight percent total organic carbon [TOC] and total nitrogen [TN] of SOM), **e**, **f** SOM provenance indicators ($\delta^{13}C$ and C/N ratio). The coloured, rectangular boxes show the intensified OMZ coupled with surface water productivity when denitrification occurred in the basin. The horizontal dotted lines indicate the position of the hiatuses. The vertical dashed line over panel 'b' show denitrification threshold and horizontal brown lines separating different lithological units. The age data (in Ma) at Site U1456, shown by the Indo-Arabic numerals in 'panel b', are based on calcareous nannofossil and planktonic foraminifera biostratigraphy, together with magnetostratigraphy. (published by Tripathi et al. 2017)

Recent studies using geochemical proxies of chemical weathering revealed that the Laxmi Basin sediment are generally more weathered than sediment found close to the Indus river mouth during 17 to 8 Ma. Initial assessment of various climatic proxies from IODP-355 are unable to pinpoint significant correlations in time with the C3 to C4 transition. The appearance of dominant C4 biomass in the Himalaya foreland at ~7 Ma reflects dramatic climate change, including enhanced terrestrial aridity (Khim et al. 2018).

2 The Indonesian Throughflow Expedtion

2.1 Indonesian Throughflow (ITF) and the Australian Monsoon

The Indonesian Throughflow (ITF) is a low-latitude oceanic gateway that transports warm, fresh waters from the Pacific to the Indian Ocean as part of the global circulation. Temporal evolution in the ITF has significant impacts on Indo-Pacific oceanography and global climate. The ITF is considered as an important regulator for the global thermohaline circulation through seasonally varying flow patterns in this region. The IODP Expedition 356 (Gallagher et al. 2017) in the year 2015 carried out drilling and coring operations at seven sites off the northwest coast of Australia (Fig. 7). The target was to recover at least 5 My long high resolution records of ITF's past variations and its potential implications on climate evolution in particular the Australian Monsoon. The sediment archives collected during this expedition provide crucial orbital-scale climatic changes in the recent past. Further, the genesis of Australian monsoon is generally linked to the initiation of the East Asian monsoon since the Pliocene time or even earlier. In view of this, the marine climatic records generated during Expedition 356 could be directly compared to the deep-ocean oxygen isotope and ice-core archives, to chart prior variability of the Australian monsoon and the onset of aridity in northwestern Australia. A good regional distribution of the shallow holes would ensure critical data that help in constructing regional subsidence models as well, which provide a geodynamic link between the Australian plate motion and mantle convection over different geological time.

3 Maldives and Equatorial Indian Ocean

3.1 Maldives Monsoon and Sea Level

Deep sea drilling in the Maldives archipelago as part of the IODP expedition 359 (sites U1465-U1472) was aimed to retrieve marine tropical record of the Neogene sea level changes and the onset of the monsoon-related current system in the Indian Ocean. The Maldives archipelago comprises thousands of smaller atolls that are exposed at the sea surface. Typically, shallow lagoons interrupted by deep passages surround these tiny atolls that facilitate strong current circulation and reworking of sediments, which in turn allow growth of carbonate platforms. One of the key objectives of expedition 359 was to determine the precise role of monsoon in migrating sea level-controlled carbonate platforms to a predominantly current-controlled system (Betzler et al. 2017). This objective required precise dating of the onset of current systems. The Maldives archipelago contains unique archives of the evolving Cenozoic icehouse Earth. Therefore, sediment cores from eight sites drilled during expedition

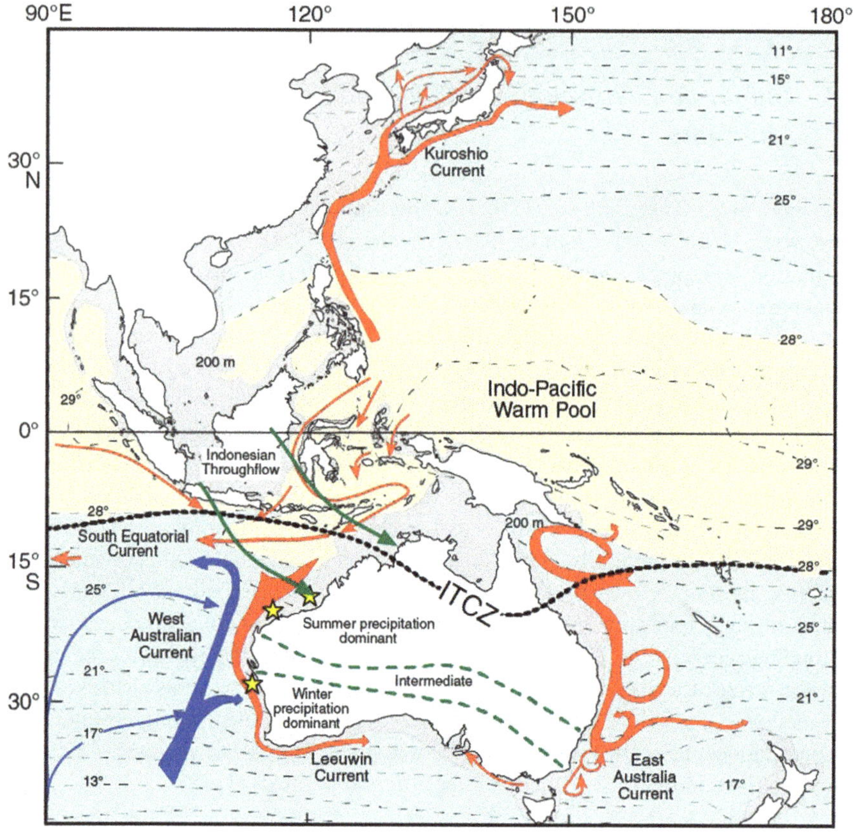

Fig. 7 An overview of the oceanography and climate of the western Pacific (adapted from Gallagher et al. 2017). Red and blue colours represent warm and cold currents respectively. The green arrows represent broad extents of the summer monsoon (see Gallagher et al. 2017 for more details). IODP 356 drill locations are shown by yellow stars

359 around Inner Sea of the Maldives provide important tropical marine records of Neogene global cooling and monsoon evolution. Carbonate rich stratigraphic sequences from Expedition 359 also constrained an extended Neogene $\delta^{13}C$ record of the platforms and marginal sediments and their comparison with corresponding pelagic records (Betzler et al. 2017) as well as records of past oceanic carbonate saturation and dissolution events (Sreevidya et al. 2019). Additional findings from this expedition were discovery of organic-rich layers that imply early Miocene oxygen minima conditions in a relatively shallow basin.

A key outcomes from this expedition was high precision dating of the onset of the current system that is potentially in agreement with the onset or the intensification of the Indian monsoon and further coincides with the onset of the modern current systems within a global framework. Dating confirmed that the onset of the monsoon wind-driven currents in Maldives yields an age to 12.9 Ma (Betzler et al. 2016). It was an abrupt event evolving from a sea-level dominated system. This coincided with the Indian Ocean Oxygen Minimum Zone expansion as revealed by geochemical tracers and the upwelling records (Fig. 8). Proxies of dust deposition fluxes also constrained identification of a weaker 'proto-monsoon' which seemed to have existed between 12.9 and 25 Ma.

The evolution of the South Asian monsoon and sea-level fluctuations during the Neogene was assessed by dating resultant sedimentary alterations that marked the stratigraphic turning points in the Neogene Maldives platform system. The markers during the early and middle Miocene were related to sea-level changes (Betzler et al. 2018). Phases of aggradational platform growth of long-term sea-level high stands during the early Miocene and the early to middle Miocene Climate Optimum was dated between 17 and 15.1 Ma (Fig. 9). The subsequent middle Miocene cooling coincided with the eastern Antarctic ice sheet expansion resulting in a long-term lowering of sea level that is reflected by a progradational platform growth. Thus the changes in platform architecture from aggradation to progradation was marked at 15.1 Ma (Fig. 9). At 12.9–13 Ma, an abrupt change in sedimentation pattern was marked across the entire archipelago switching to a current-controlled mode indicating an abrupt onset of monsoon winds in the Indian Ocean. A major shift in the drift packages is dated with 3.8 Ma (Fig. 9) which coincides with the end of stepwise platform drowning and a reduction of the oxygen minimum zone in the Inner Sea.

The analysis of anisotropy of isothermal remnant magnetization (AIRM) of sediment cores revealed evidences to that the strength of the bottom currents increased suddenly at about 4.2 Myr ago (Lanci et al. 2019). This coincided with the increase of fluvial transport to the Bay of Bengal and appeared compatible with the beginning of the Late Pliocene phase of Himalayan uplift suggesting the Maldives record of the Late Pliocene up lift phase.

Another important outcome was the seismic stratigraphy based hypothesis that the pronounced change in style of the sedimentary carbonate sequence stacking was caused by a combination of relative sea level fluctuations and ocean current system changes.

In addition, a new calcareous sediment drift type with sigmoidal clinoform geometry was identified which deposited under a current-controlled regime in deep to shallow water setting with an aerial extent of 342–384 km^2 and a depositional relief of approximately 500 m (Lüdmann et al. 2018). This drift onset marked the transition from a sea-level controlled to a progressively current dominated depositional regime.

Some other interesting recent scientific contributions from IODP expedition 359 include the Eolian deposition (Kunkelova et al. 2018 and Lindhorst et al. 2019) and the Oligocene-Miocene anoxia and ventilation (Swart et al. 2019 and Reolid et al. 2019).

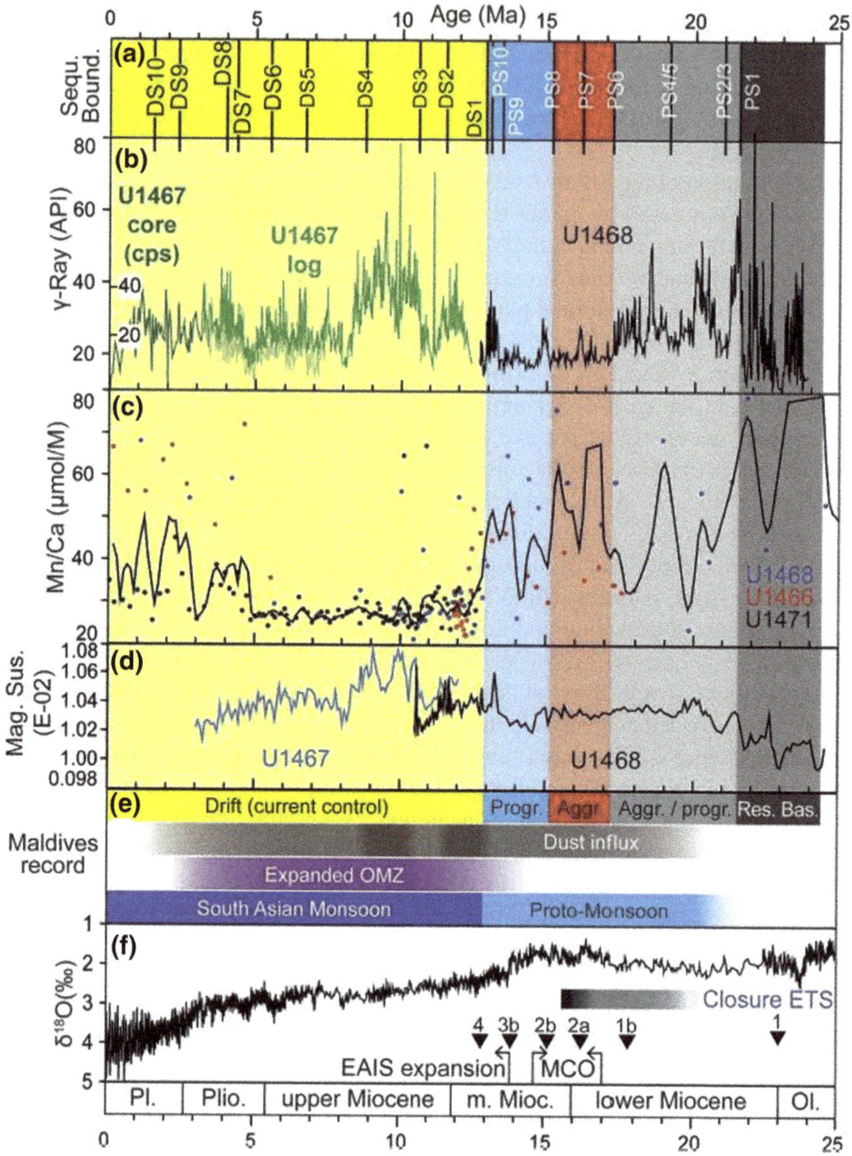

Fig. 8 **a** Age of sequence boundaries; **b** total gamma ray in downhole logs for Sites U1467 and U1468. For Site U1467, core measurements were used for the upper interval of the succession. **c** Mn/Ca ratios at Sites U1466, U1468, and U1471. Black line shows a 3 point running average. **d** Magnetic susceptibility from downhole logs at Sites U1467 and U1468. **e** Stratigraphic breaks and changes, variations of the dust influx and of the OMZ and timing of SAM and Proto-Monsoon. **f** Deep sea oxygen isotope record, timing of the eastern Antarctic ice sheet (EAIS) expansion, timing of the middle Miocene climate optimum (MCO), Mi-events 1–412, and restriction as well as final closure of the eastern Tethys seaway (published by Betzler et al. 2016)

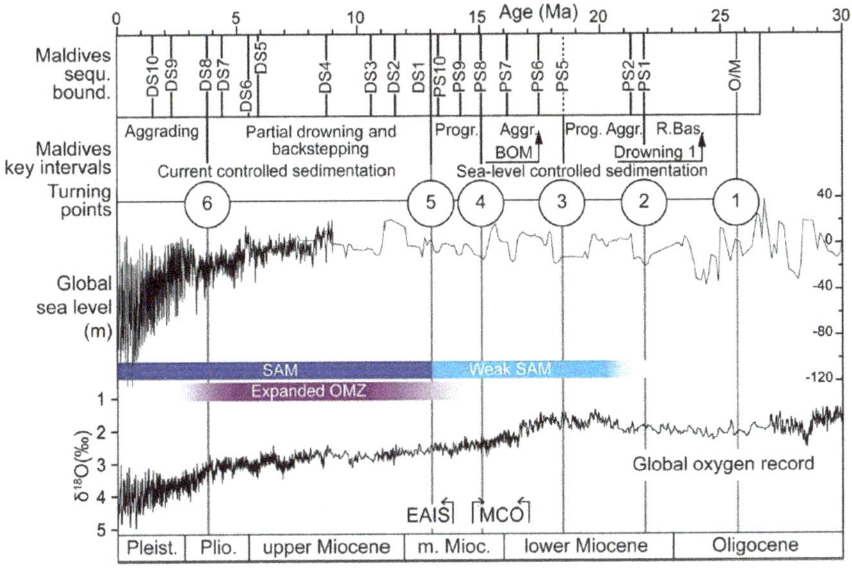

Fig. 9 Oligocene to Miocene evolution of the Maldives carbonate platform with ages of the sequence boundaries and key intervals representing changes of platform configuration. The global sea level curve (Miller et al. 2005). The intensity of the South Asian Monsoon (SAM) and the expansion of the Oxygen Minimum Zone (OMZ) Betzler et al. (2016). The global oxygen record from Zachos et al. (2001). MCO—Miocene Climate Optimum (after Holbourn et al. 2005, 2015), EAIS—East Antarctic Ice Sheet expansion (after Holbourn et al. 2005), BOM—Backstepping of the Outer Margin (published by Betzler et al. 2018)

4 South African Climates

4.1 *The South African Climates and Agulhas Current Density Profile*

The South African Climates or IODP expedition 361 (Hall et al. 2017) carried out using JOIDES Resolution in the early 2016. One of the main objectives of this expedition was to document the level of specific interactions between climate and the Agulhas Current primarily during the Plio/Pleistocene period. The greater Agulhas Current system represents a significant component of the global thermohaline circulation, carrying surface water into the Atlantic Ocean through Indian-Atlantic Ocean gateway. In fact, it plays a vital role in the overall Southern Indian Ocean circulation as it constitutes the strongest western boundary current in the Southern Hemisphere that transports warm, saline tropical surface waters to the tip of Africa (Hall et al. 2017). In total six drill sites in the Southern Indian Ocean and Mozambique Channel (Fig. 10), which are directly affected by the regular sediment discharge from major river catchments. The primary scientific objectives of this expedition included ascertaining the

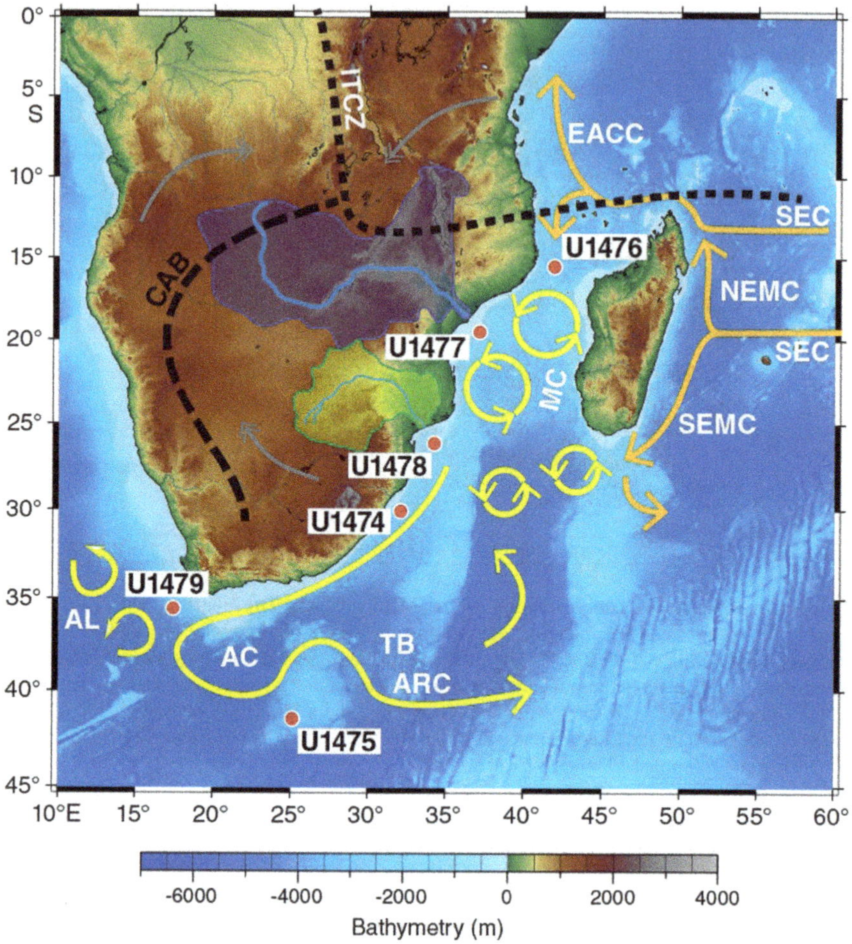

Fig. 10 Locations of IODP Expedition 361 sites (after Hall et al. 2017). A schematic representation of the main surface ocean currents (in yellow) is also marked on the map (see the original reference for more details)

sensitivity of the Agulhas Current, and the dynamics of Indian-Atlantic gateway circulation, during major ocean and climate reorganizations of the past 5 Ma or more. Longer than 5100 m high quality sediment material collected during this expedition could provide decadal-to millennial- scale climatic and paleoceanographic records from the Southern Indian Ocean.

Most of the Expedition 361 sites had good core recovery and therefore helped establishing reasonably well constrained lithological details at least until the late Miocene or ~7 Ma (Hall et al. 2017). Cored material from this expedition would be able to answer some of the fundamental scientific questions such as: (1) What is the effect of the Pliocene–Pleistocene climatic variations, ocean gateways, tropical heat budgets, and the monsoon system to the sensitivity of the Agulhas Current; and (2) What is the dynamics of the Indian-Atlantic Ocean gateway circulation during such climate changes such as regional rainfall and erosional run-off.

5 Plate Deformation Zones

5.1 "The Nature of the Lower Crust and Moho at Slower Spreading Ridges"—SloMo

One of the recent IODP expeditions (IODP-360 or SloMo) in the Indian Ocean targeted to drill and core (U1473A) on the summit of Atlantis Bank through an important globally observed seismic velocity discontinuity- the Mohorovičić (commonly referred as 'Moho'). The primary objective of SloMo expedition was to document lateral stratigraphic variabilities and compare them with legendry Holes 735B and 1105A from earlier ODP Legs (ODP Leg 118, 176 and 179) to demonstrate a continuity of processes and complex interplay of magmatic accretion and steady-state detachment faulting. The preliminary assessment of cored igneous rocks indicated that these lower crustal sections are emplaced through repeated cycles of intrusion by approximately three upwardly differentiated hundreds of meter-thick bodies of olivine gabbro - broadly similar to those encountered in the deeper parts of Hole 735B (MacLeod et al. 2017). Expedition 360 along south west Indian ridge focused on understanding how magmatism and tectonism interacted in accommodating seafloor spreading; how magnetic reversal boundaries are expressed in the lower crust; assessing the role of the lower crust and shallow mantle in the global carbon cycle; and constraining the extent and nature of life at deep levels within the ocean lithosphere.

The retrieved gabbros (from Hole U1473A) revealed a 600 m thick zone of intense crystal-plastic deformation which is cut by a brittle high-angle fault zone (Dick et al. 2017). This represents high-temperature deformation related to the formation of Atlantis Bank oceanic core complex and associated detachment fault zone. The undeformed portions of the Hole U1473A lithology preserved weakly developed magmatic fabric. Microbiological sampling focused on exploring evidence for life in the lower crust and hydrated mantle using culture-based and culture-independent approaches, microscopy, and enzyme assays. Adenosine triphosphate (ATP) was quantified from the samples collected for microbiological analysis. It ranged from below detection level to 5 pg/cm^3 (mostly <1 pg/cm^3), indicating the presence of a subsurface biosphere in Atlantis Bank.

The melts corresponding to the felsic rocks from the ultraslow-spreading Southwest Indian Ridge are inferred to have been derived from a parental mid-ocean ridge basalt (MORB) by a fractional crystallization dominated mechanism (Nguyen et al. 2018). After the formation, these veins acted as channels for hydrothermal fluids infiltration to modify both the host gabbros and felsic rocks, obscuring their magmatic origin. Based on geochemical criteria and texture of the felsic rocks and their relationship with the host gabbros it is suggested that the felsic rock with sharp boundaries is formed when felsic melt intrudes into fractures of host gabbros, resulting in minimal interaction between the melt and the wall minerals. The replacive felsic rock, which is characterized by a pseudomorphic replacement of minerals in the host gabbro. This vein type is caused by the replacement of the host mineralogy by minerals in equilibrium with the felsic melts. Felsic rock with diffused boundaries is formed either by infiltration of felsic melt into the solidifying gabbro body or crystallization of interstitial melts. Infiltration modes of felsic melts are likely controlled by the temperature condition of the cooling host gabbros.

Based on the shear wave imaging results, a fault system ~40 m away from the borehole was imaged corresponding to a steeply dipping fault system, which extends below 570 m depth and strikes dominantly in EW direction (Lee et al. 2019). This method successfully imaged the fault structure as well as helped to determine the orientation of the fault system which will be supportive to apply even to the log data of other expeditions. Some of the major differences in igneous stratigraphy between Atlantis Bank and Atlantis Massif observed during expedition 360 drilling demonstrate that significant variability exists in the mode of accretion and internal differentiation of magmatic bodies in the lower crust between oceanic core complexes. This potentially may indicate variations in magma supply and accompanying tectonic setting, spreading rate, and thermal regime along ocean ridges.

5.2 *To Understand the Role of Input Materials in the Northern Sumatra Subduction*

Another crucial scientific drilling expedition in the Indian Ocean is Expedition 362, the Sumatra Seismogenic Zone drilling. This expedition aimed to establish (1) the initial and evolving properties of the North Sumatran incoming sediments and (2) their potential effect on seismogenesis, tsunamigenesis, and forearc development for comparison with global examples. To address these objectives, the expedition drilled and cored two sites (U1480-U1481) on the Indian oceanic plate ~250 km southwest of the Sumatra subduction zone (McNeill et al. 2017a, b). The two sites were designed to test for local variations in stratigraphy, lithology, physical properties, thermal state, and fluid geochemistry.

5.2.1 Depositional history

Sampling of the full sedimentary section enabled identification of principal lithologies, grain compositions, depositional environments, sediment accumulation rates, and diagenetic and lithification state. The biomagnetostratigraphic age-depth relationships revealed marked changes in sediment accumulation rates between the early pelagic succession (low rates) and overlying submarine fan (very rapid sedimentation

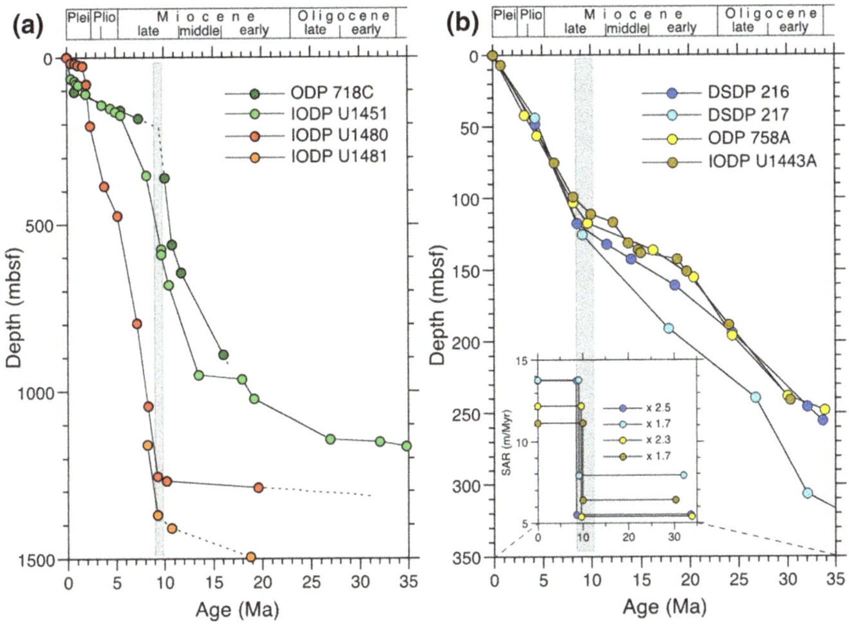

Fig. 11 Biomagnetostratigraphic age-depth relationships (published by McNeill et al. 2017b)

of magnitude), and several potential hiatuses (McNeill et al. 2017a, b). The initial provenance data (zircon U-Pb ages) indicated a Himalayan-West Burman source from the north. An increase in sediment accumulation rate (SAR) was reported between 9 and 10 Ma (Fig. 11). Whereas the Nicobar Fan holds a significant component of the sedimentary record in the Indian Ocean, ~9 to 2 Ma. Sedimentation pattern along the Bengal Fan system from proximal to distal indicated absence of simple fan progradational pattern.

5.2.2 Geochemistry and Fluids of the Input Section

Pore fluid sampling of whole-round and Rhizon sampling helped identify two fluid chemistry anomalies of the sedimentary inputs to the subduction zone (Hüpers et al. 2017). The first is the presence of two phosphate peaks in the upper 150 mbsf, spanning ~20 m intervals with the shallower peak near the seafloor (less magnitude), controlled by organic matter diagenesis and the second peak is sharp and large and may be controlled by a regional flow system. Similar phosphate peaks were also observed in the shallow pore fluids sampled during IODP Expeditions 353 and 354 elsewhere on the Bengal-Nicobar Fan system. The second pore fluid anomaly was observed deeper in the section, where pore fluid freshening was observed. This unit consists of altered volcanic, amorphous silicate-rich (palagonized, tuffaceous mudstone), and increased porosity. The freshening, noted by a very sharp chloride change coincides with the seismic horizon—high amplitude-negative polarity—candidate horizon for décollement development, suggested as a recent phenomenon and has not been significantly impacted by diffusion. Analysis revealed that the diagenesis before subduction may have driven shallow slip which might have resulted in one of the major earthquakes in Sumatra.

6 Concluding Remarks

The brief scientific summary from each of the recent eight IODP expeditions in the Indian Ocean discussed above contribute significantly towards improving our understanding about the dynamic earth system and its response to various tectono-climatic processes. The early outcomes from these expeditions would surely lead to an in-depth investigations having potential to form back-bones for future marine geoscientific research in the Indian Ocean.

Acknowledgements The authors wish to acknowledge co-chiefs and scientific support teams of each of the IODP expeditions discussed above for their significant scientific contributions in form of respective scientific proceedings. The technical summaries presented above are primarily drawn from the Proceedings volumes for respective IODP expeditions (http://publications.iodp.org/proceedings). Accordingly, IODP publications portal and the JOIDES Resolution are acknowledged. Ministry of Earth Sciences, India is thanked for the support towards production of this volume. This is NCPOR contribution no. B-7/2019-20.

References

Amante C, Eakins BW (2009) ETOPO1 1 arc-minute global relief model: procedures, data sources and analysis. In: NOAA technical memorandum NESDIS NGDC-24, 19p. https://doi.org/10.7289/v5c8276m

Andò S, Aharonovich S, Hahn A, George S, Clift P, Garzanti E (2019) Integrating heavy-mineral, geochemical and biomarker analyses of Plio-Pleistocene sandy and silty turbidites: a novel approach for provenance studies (Indus Fan, IODP Expedition 355). Geol Mag 1–10. https://doi.org/10.1017/s0016756819000773

Barth AM, Clark PU, Bill NS, He F, Pisias NG (2018) Climate evolution across the mid-Brunhes transition. Clim Pas 14:2071–2087

Betzler C, Eberli GP, Lüdmann T, Reolid J, Kroon D, Reijmer JJG, Swart PK et al (2018) Refinement of Miocene sea level and monsoon events from the sedimentary archive of the Maldives (Indian Ocean). Prog Earth Planet Sci 5:5. https://doi.org/10.1186/s40645-018-0165-x

Betzler C, Eberli GP, Alvarez Zarikian CA, The Expedition 359 Scientists (2017) Maldives Monsoon and Sea Level. In: Proceedings of the international ocean discovery program, vol 359, (International Ocean Discovery Program), College Station, TX. http://dx.doi.org/10.14379/iodp.proc.359.2017

Betzler C, Eberli GP, Kroon D, Wright JD, Swart PK, Nath BN, Alvarez-Zarikian CA et al (2016) The abrupt onset of the modern South Asian Monsoon winds. Sci Rep 6:29838. https://doi.org/10.1038/srep29838

Blum M, Rogers K, Gleason J, Najman Y, Cruz J, Fox L (2018) Allogenic and autogenic signals in the stratigraphic record of the deep-sea Bengal Fan. Sci Rep 8:7973. https://doi.org/10.1038/s41598-018-25819-5

Cai M, Xu Z, Clift PD, Lim D, Khim B-K, Yu Z, Kulhanek D, Li T, Chen H, Sun R (2019) Depositional history and Indian summer monsoon controls on the silicate weathering of sediment transported to the eastern Arabian Sea: geochemical records from IODP SiteU1456 since 3.8 Ma. Geochem Geophys Geosystems 20. https://doi.org/10.1029/2018GC008157

Cai M, Xu Z, Clift P, Khim B, Lim D, Yu Z, Kulhanek D, Li T (2018) Long-term history of sediment inputs to the eastern Arabian Sea and its implications for the evolution of the Indian summer monsoon since 3.7 Ma. Geol Mag 1–12. https://doi.org/10.1017/s0016756818000857

Calvès G, Huuse M, Clift PD, Brusset S (2015) Giant fossil mass wasting off the coast of West India: the Nataraja submarine slide. Earth Planet Sci Lett 432:265–272. https://doi.org/10.1016/j.epsl.2015.10.022

Chen H, Xu Z, Clift PD, Lim D, Khim B-K, Yu Z (2019) Orbital-scale evolution of the Indian summer monsoon since 1.2 Ma: evidence from clay mineral records at IODP Expedition 355 Site U1456 in the eastern Arabian Sea. J Asian Earth Sci 174:11–22. https://doi.org/10.1016/j.jseaes.2018.10.012

Clark PU, Archer D, Pollard D, Blum JD, Rial JA, Brovkin V, Mix AC, Pisias NG, Roy M (2006) The middle Pleistocene transition: characteristics, mechanisms, and implications for long-term changes in atmospheric pCO_2. Quatern Sci Rev 25:3150–3184

Clemens SC, Kuhnt W, LeVay LJ, The Expedition 353 Scientists (2015) Expedition 353 Preliminary Report: Indian Monsoon Rainfall, International Ocean Discovery Program. http://dx.doi.org/10.14379/iodp.pr.353.2015

Clemens SC, Kuhnt W, LeVay LJ, The Expedition 353 Scientists (2016) Indian Monsoon Rainfall. In: Proceedings of the international ocean discovery program, vol 353. (International Ocean Discovery Program), College Station, TX. http://dx.doi.org/10.14379/iodp.proc.353.2016

Clift P, Kulhanek D, Zhou P, Bowen M, Vincent S, Lyle M, Hahn A (2019a) Chemical weathering and erosion responses to changing monsoon climate in the Late Miocene of Southwest Asia. Geol Mag 1–17. https://doi.org/10.1017/s0016756819000608

Clift PD, Zhou P, Stockli DF, Blusztajn J (2019b) Regional Pliocene exhumation of the Lesser Himalaya in the Indus drainage. Solid Earth 10:647–661

Dailey SK, Clift PD, Kulhanek DK, Blusztajn J, Routledge CM, Calves G, O'Sullivan P et al (2019) Large-scale mass wasting on the Miocene continental margin of western India. Geol Soc Am Bull. https://doi.org/10.1130/B35158.1

Dick HJB, MacLeod CJ, Blum P, Expeditioon 360 Scientists (2017). Expedition 360 summary. In: MacLeod CJ, Dick HJB, Blum P, The Expedition 360 Scientists (eds) Southwest Indian ridge lower crust and moho, proceedings of the international ocean discovery program, vol 360. (International Ocean Discovery Program), College Station, TX. http://dx.doi.org/10.14379/iodp.proc.360.101.2017

France-Lanord C, Spiess V, Klaus A, Schwenk T, The Expedition 354 Scientists (2016) Bengal Fan. In: Proceedings of the international ocean discovery program, vol 354. (International Ocean Discovery Program), College Station, TX. http://dx.doi.org/10.14379/iodp.proc.354.2016

Gallagher SJ, Fulthorpe CS, Bogus K, The Expedition 356 Scientists (2017) Indonesian throughflow. In: Proceedings of the international ocean discovery program, vol 356. (International Ocean Discovery Program), College Station, TX. https://doi.org/10.14379/iodp.proc.356.101.2017

Hall IR, Hemming SR, LeVay LJ, The Expedition 361 Scientists (2017) South African Climates (Agulhas LGM Density Profile). In: Proceedings of the international ocean discovery program, vol 361. (International Ocean Discovery Program), College Station, TX. https://doi.org/10.14379/iodp.proc.361.2017

Holbourn A, Kuhnt W, Schulz M, Erlenkeuser H (2005) Impacts of orbital forcing and atmospheric carbon dioxide on Miocene icesheet expansion. Nature 438:483–487

Holbourn A, Kuhnt W, Kochhann KGD, Andersen N, Meier KJS (2015) Global perturbation of the carbon cycle at the onset of the Miocene Climatic Optimum. Geology 43(2):123–126. https://doi.org/10.1130/G36317.1

Hüpers A, Torres ME, Owari S, McNeill LC, Dugan B, Henstock TJ, Milliken KL et al (2017) Release of mineral-bound water prior to subduction tied to shallow seismogenic slip off Sumatra. Science 356(6340):841–844. https://doi.org/10.1126/science.aal3429

Khim BK, Horikawa K, Asahara Y, Kim JE, Ikehara M (2018) Detrital Sr–Nd isotopes, sediment provenances and depositional processes in the Laxmi Basin of the Arabian Sea during the last 800 ka. Geol Mag 1–13. https://doi.org/10.1017/s0016756818000596

Kim J-E, Khim B-K, Ikehara M, Lee J (2018) Orbital-scale denitrification changes in the eastern Arabian Sea during the last 800 kyrs. Sci Rep 8:7027. https://doi.org/10.1038/s41598-018-25415-7

Kumar A, Dutt S, Saraswat R, Gupta A, Clift P, Pandey D, Yu Z, Kulhanek D (2019) A late Pleistocene sedimentation in the Indus Fan, Arabian Sea, IODP Site U1457. Geol Mag 1–9. https://doi.org/10.1017/s0016756819000396

Kunkelova T, Jung SJA, de Leau ES et al (2018) A two million year record of low-latitude aridity linked to continental weathering from the Maldives. Prog Earth Planet Sci 5:86. https://doi.org/10.1186/s40645-018-0238-x

Lanci L, Zanella E, Jovane L, Galeotti S, Alonso-Garcia M, Alvarez-Zarikian CA, Bejugam NN et al (2019) Magnetic properties of early Pliocene sediments from IODP Site U1467 (Maldives platform) reveal changes in the monsoon system. Palaeogeogr Palaeoclim Palaeoecol 533:109283

Lee S-Q, Tang X-M, Su Y (2019) Shear wave imaging to determine near-borehole faults for ocean drilling exploration. Geophys J Int 217(1):288–293. https://doi.org/10.1093/gji/ggz023

Lindhorst S, Betzler C, Kroon D (2019) Wind variability over the northern Indian Ocean during the past 4 million years–Insights from coarse aeolian dust (IODP exp. 359, site U1467, Maldives). Palaeogeogr Palaeoclim Palaeoecol 536:109371. https://doi.org/10.1016/j.palaeo.2019.109371

Lübbers J, Kuhnt W, Holbourn AE, Bolton CT, Gray E, Usui Y, Kochhann KGD, Beil S, Andersen N (2019) The middle to late Miocene "carbonate crash" in the equatorial Indian Ocean. Paleoceanogr Paleoclimatology 34(5):813–832. https://doi.org/10.1029/2018PA003482

Lüdmann T, Betzler C, Eberli GP, Reolid J, Reijmer JJG, Sloss CR, Bialik OM et al (2018) Carbonate delta drift: a new sediment drift type. Mar Geol 401:98–111. https://doi.org/10.1016/j.margeo.2018.04.011

Lyle M, Dadey KA, Farrell JW (1995) The Late Miocene (11–8 Ma) Eastern Pacific Carbonate Crash: evidence for reorganization of deep–water Circulation by the closure of the Panama Gateway. In Pisias NG, Mayer LA, Janecek TR, Palmer-Julson TH, van Andel (eds) Proceedings of the ocean drilling program, vol 138. Ocean drilling program, College Station, TX, pp 821–838. https://doi.org/10.2973/odp.proc.sr.138.157.1995

MacLeod CJ, Dick HJB, Blum P, The Expedition 360 Scientists (2017) Southwest Indian Ridge Lower Crust and Moho. In: Proceedings of the international ocean discovery program, vol 360. TX (International Ocean Discovery Program), College Station. http://dx.doi.org/10.14379/iodp.proc.360.2017

McNeill LC, Dugan B, Petronotis KE, The Expedition 362 Scientists (2017a). Sumatra Subduction Zone. In: Proceedings of the international ocean discovery program, vol 362. TX (International Ocean Discovery Program), College Station. https://doi.org/10.14379/iodp.proc.362.2017

McNeill LC, Dugan B, Backman J, Pickering KT, Pouderoux HFA, Henstock TJ, Petronotis KE et al (2017b) Understanding Himalayan erosion and the significance of the Nicobar Fan. Earth Planet Sci Lett 475:134–142. https://doi.org/10.1016/j.epsl.2017.07.019

Miller KG, Kominz MA, Browning JV, Wright JD, Mountain GS, Katz ME, Sugarman PJ, Cramer BS, Christie-Blick N, Pekar SF (2005) The Phanerozoic record of global sea-level change. Science 310:1293–1298

Nguyen DK, Morishita T, Soda Y, Tamura A, Ghosh B, Harigane Y, France L et al (2018) Occurrence of Felsic rocks in oceanic gabbros from IODP hole U1473A: implications for evolved melt migration in the lower oceanic crust. Minerals 8(12):583

Nilsson-Kerr K, Anand P, Sexton PF, Leng MJ, Misra S, Clemens SC, Hammond SJ (2019) Role of Asian summer monsoon subsystems in the inter-hemispheric progression of deglaciation. Nat Geosci 12:290–295. https://doi.org/10.1038/s41561-019-0319-5

Pandey DK, Pandey A, Whattam SA (2019) Relict subduction initiation along a passive margin in the northwest Indian Ocean. Nat Commun 10:2248. https://doi.org/10.1038/s41467-019-10227-8

Pandey DK, Pandey A, Clift PD, Nair N, Ramesh P, Kulhanek DK Yadav R (2018) Flexural subsidence analysis of the Laxmi Basin, Arabian Sea and its tectonic implications. Geol Mag. https://doi.org/10.1017/S0016756818000833

Pandey DK, Clift PD, Kulhanek DK, The Expedition 355 Scientists (2016) Arabian Sea Monsoon. In: Proceedings of the international Ocean discovery program, vol 355. TX (International Ocean Discovery Program), College Station. http://dx.doi.org/10.14379/iodp.proc.355.2016

Pandey DK, Clift PD, Kulhanek DK, Andò S, Bendle JAP, Bratenkov S, Griffith EM, Gurumurthy GP, Hahn A, Iwai M, Khim B-K, Kumar A, Kumar AG, Liddy HM, Lu H, Lyle MW, Mishra R, Radhakrishna T, Routledge CM, Saraswat R, Saxena R, Scardia G, Sharma GK, Singh AD, Steinke S, Suzuki K, Tauxe L, Tiwari M, Xu Z, Yu Z (2015) Expedition 355 preliminary report: Arabian Sea Monsoon. Int Ocean Discov Program. http://dx.doi.org/10.14379/iodp.pr.355.2015

Reolid J, Betzler C, Lüdmann T (2019) The record of Oligocene—Middle Miocene paleoenvironmental changes in a carbonate platform (IODP Exp. 359, Maldives, Indian Ocean). Mar Geol 412:199–216. https://doi.org/10.1016/j.margeo.2019.03.011

Satpathy R, Steinke S, Singh A (2019) Monsoon-induced changes in surface hydrography of the eastern Arabian Sea during the early Pleistocene. Geol Mag 1–11. https://doi.org/10.1017/s0016756819000098

Sreevidya E, Sijinkumar AV, Nagender Nath B (2019) Aragonite pteropod abundance and preservation records from the Maldives, equatorial Indian Ocean: inferences on past oceanic carbonate saturation and dissolution events. Palaeogeogr Palaeoclimatol Palaeoecol 534:109313. https://doi.org/10.1016/j.palaeo.2019.109313

Sun Y, Chen J, Clemens SC, Liu Q, Ji J, Tada R (2006) East Asian monsoon variability over the last seven glacial cycles recorded by a loess sequence from the northwestern Chinese Loess Plateau. Geochem Geophy Geosy 7(12):Q12Q02. https://doi.org/10.1029/2006GC001287

Suzuki K, Yamamoto M, Seki O (2019) Late Miocene changes in C3, C4 and aquatic plant vegetation in the Indus River basin: evidence from leaf wax $\delta^{13}C$ from Indus Fan sediments. Geol Mag 1–10. https://doi.org/10.1017/s0016756819001109

Swart PK, Blättler CL, Nakakuni M, Mackenzie GJ, Betzler C, Eberli GP, Reolid J et al (2019) Cyclic anoxia and organic rich carbonate sediments within a drowned carbonate platform linked to Antarctic ice volume changes: late Oligocene-early Miocene Maldives. Earth Planet Sci Lett 521:1–13. https://doi.org/10.1016/j.epsl.2019.05.019

Tripathi S, Tiwari M, Lee J, Khim B-K, IODP Expedition 355 Scientists (2017) First evidence of denitrification vis-à-vis monsoon in the Arabian Sea since Late Miocene. Sci Rep 7:43056. http://dx.doi.org/10.1038/srep43056

Weber ME, Lantzsch H, Dekens P, Das SK, Reilly BT, Martos YM, Meyer-Jacob C et al (2018) 200,000 years of monsoonal history recorded on the lower Bengal Fan-strong response to insolation forcing. Glob Planet Chang 166:107–119

Weber ME, Reilly BT (2018) Hemipelagic and turbiditic deposits constrain lower Bengal Fan depositional history through Pleistocene climate, monsoon, and sea level transitions. Quat Sci Rev 199:159–173. https://doi.org/10.1016/j.quascirev.2018.09.027

Yu Z, Colin C, Wan S, Saraswat R, Song L, Xu Z, Clift P et al (2019) Sea level–controlled sediment transport to the eastern Arabian Sea over the past 600 kyr: clay minerals and Sr–Nd isotopic evidence from IODP Site U1457. Quat Sci Rev 205:22–34. https://doi.org/10.1016/j.quascirev.2018.12.006

Zachos J, Pagani M, Sloan L, Thomas E, Billups K (2001) Trends, rhythms, and aberrations in global climate 65 Ma to present. Science 292:686–693

Morphological and Chemical Properties of Microtektite Grains from Bay of Bengal (IODP Expedition 354)

Masud Kawsar, M. C. Manoj, Kohki Yoshida, Alan T. Baxter and Brendan T. Reilly

Abstract This study reports the presence of Australasian microtektites in a deep-sea core (U1452) retrieved during the International Ocean Discovery Program (IODP) Expedition 354: Bengal Fan. These microtektites are found within a foraminifer-rich calcareous clay layer beneath the Matuyama-Brunhes (M–B) magnetostratigraphic boundary. The majorities are spherical and less than one millimeter in diameter. Typical splash (dumbbell, teardrop, disc etc.) and irregular-shaped forms were recovered. The most abundant microtektites are pale green in color, followed by opaque, pale brown, translucent and transparent varieties. These microtektites are characterized by various surficial attributes including pits, mounds, grooves and fractures. Geochemical analyses suggest that the major oxide compositions are very similar to Australasian tektites and microtektites reported elsewhere and also similar to the average composition of upper crustal rocks. Transparent bottle green microtektites are relatively rich in MgO and low in silica when compared to other microtektites found in U1452. Minor and trace element abundances show a wide range of distribution and individual samples show variations in their concentrations. Differences in minor and trace elements concentration are possibly due to the contamination from the impact ejecta. Other than microtektites, the presence of a possibly polymetallic exsolution structure (Widmanstatten texture), shocked minerals and unmelted and partly melted ejecta within the microtektite-bearing layer in the northern Indian Ocean provides further evidence that the Australasian microtektites might have been formed by the impact of an extraterrestrial projectile at ~0.8 Ma, somewhere in Indochina.

M. Kawsar · M. C. Manoj (✉)
Birbal Sahni Institute of Palaeosciences, Lucknow, India
e-mail: manoj.mcm@gmail.com

K. Yoshida
Shinshu University, Matsumoto, Japan

A. T. Baxter
University of New England, Armidale, Australia

B. T. Reilly
Oregon State University, Corvallis, USA

D. K. Pandey et al. (eds.), *Dynamics of the Earth System: Evolution, Processes and Interactions*, Society of Earth Scientists Series,
https://doi.org/10.1007/978-3-030-40659-2_2

Keywords Microtektite · Bay of Bengal · Matuyama-Brunhes magnetostratigraphic boundary · Geochemical analysis · Australasian tektite

1 Introduction

Tektites are siliceous glass materials up to several cms in size which is produced by the melting and vaporization of the Earth's crust during hypervelocity impacts of extraterrestrial bodies (Glass 1990; Koeberl 1994; Montanari and Koeberl 2000). Earlier studies proposed an extraterrestrial origin for these particles (Barnes 1963, 1964, 1967, 1971; Barnes and Pitakpaivan 1962; O'Keefe (1967–1987). Microtektites are natural glassy materials smaller than 1 mm that were formed by the melting of terrestrial surface deposits/rocks during the impact of an extraterrestrial body (Koeberl et al. 1988). Four known tektite strewn fields such as the North American (35 Ma), Central European (14 Ma), Ivory Coast (1.1 Ma), and Australasian (0.8 Ma) was identified (Zähringer 1963; Gentner et al. 1967; Simonson and Glass 2004; Glass et al. 2004a).

The largest and youngest is the Australasian strewn field, which blankets nearly 10% of the Earth's surface, including parts of Southeast Asia, much of Australia and Tasmania, and the surrounding Indian and Pacific Ocean basins (Barnes 1963; Chapman 1964, 1971; Glass 1967, 1972a, b; Cassidy et al. 1969; O'Keefe 1976; Prasad and Sudhakar 1999). It is believed that the Australasian Comet entered the Earth's gravity field from and N-NW direction at a low angle and impacted the Earth at around 0.8 Ma (Bunopas 1990, 1992; Bunopas et al. Bunopas et al. 1999; Bunopas et al. 2005, 2007). Although most investigators hypothesize that the main source crater was located somewhere in Indochina, the source crater of the Australasian strewn field tektites is still unknown.

Many deep-sea sediment cores from the Indian Ocean and south of Australia have been studied by researchers (Prasad and Sudhakar 1999; Prasad et al. 2007 and references therein, Rudraswami and Prasad 2016). Microtektites preserved in deep-sea sediments provide insight into the geographic extent, frequency, and relative ages of impacts and their associated strewn fields. The present study reports the presence of microtektite grains in a deep-sea core retrieved during the International Ocean Discovery Program (IODP) Expedition 354: Bengal Fan. The Bengal Fan has accumulated erosional material from the Himalaya since the Early Eocene (France-Lanord et al. 2015). The Bengal Fan has developed and extended into the Northern Indian Ocean primarily through turbidity current sedimentation. During times of local fan inactivity, hemipelagic sediment deposition occurs at Bengal Fan. The studied microtektite layer is preserved in one of the hemipelagic sediment units.

The Australasian microtektite layer is closely associated with the Matuyama-Brunhes (M–B) geomagnetic polarity reversal boundary (Glass 1967; Glass and Heezen 1967; Burns 1989; Schneider et al. 1992; Yamei et al. 2000; Trieloff et al. 2007) which occurred around 773 ka, during Marine Isotope Stage (MIS) 19 (Channell et al. 2010). In basins with low sediment accumulation rates (i.e., <1 cm/ka), the microtektite layer is generally on or slightly above the M–B reversal boundary. However, in basins with high sediment accumulation rates, the microtektite layer is below the M–B reversal boundary (Glass and Simonson 2012). The detailed comparison of the microtektite layer and benthic $\delta^{18}O$ from Equatorial Indian Ocean places the layer around the transition from glacial MIS 20 to interglacial MIS 19 (Valet et al. 2014) around 790 ka (Lisiecki and Raymo 2005). At Ocean Drilling Program (ODP) Site 758, the microtektite layer was identified between two ash layers (below the M-B reversal boundary) associated with eruptions of the Toba Caldera, Indonesia (Smit, et al. 1991) and dated using $^{40}Ar/^{39}Ar$ methods to 792.4 ± 0.5 and 785.6 ± 0.7 ka (Mark et al. 2017), consistent with an around 790 ka age. In this study, we have examined the morphology and chemical compositions of microtektites below the M-B boundary from the Bengal Fan. The main objective of the study is to report the morphological and chemical variation of the microtektites, compare them with known occurrences from the Australian strewn field, and assess whether they are derived from a similar source.

2 Materials and Methods

During Expedition 354, seven sites were drilled on an east–west transect at 8° N in the lower Bengal Fan (Fig. 1; France-Lanord et al. 2016). Site U1452 is located in a relatively flat environment with a smooth morphology. Topographic expressions of channels are absent in the vicinity of the site. Hole B at IODP Site U1452 (8° 0.4191′ N and 87° 10.9128′ E, 3670.3 m water depth) was cored to 217.7 m drilling depth below seafloor (DSF) in order to study Upper Pleistocene sedimentation. Site U1452 sediments record the M-B boundary at 184.10 m CSF-A by magnetostratigraphic and biostratigraphic evidence and this is the deepest instances of the polarity transitions identified in this transect (France-Lanord et al. 2016). A white volcanic ash layer is observed at 36F-CC, 29–35 cm at 184.44 m CSF-A, believed to be from the Toba Caldera, Indonesia (France-Lanord et al. 2016). Underneath this ash layer, a 5 cm thick foraminifer-rich calcareous clay layer containing microtektites was identified from the core catcher (36F-CC, 35.5–40.5 cm; 184.52–179.52 m CSF-A). This is found deeper than the M-B boundary and the Toba Ash layer, consistent with previous observations (e.g., Smit et al. 1991).

Approximately 5 g of bulk sediment sample has been used for the investigation. The coarse sediment fraction (>63 μm) from the 36F-CC interval of the site U1452B was separated by sieving, then microtektite, opaque and glass grains handpicked using a stereo-zoom binocular microscope. The distribution of microtektite, opaque and glass grains were studied and the specimens were cleaned using an ultrasonicator to remove the maximum contaminant particles from the surface of the grains. The specimens were mounted on a sample stub and examined under the Field Emission Scanning Electron Microscope (FE-SEM) JEOL 7610 F in association for morphological characteristics. Hand-picked unpolished representative grains were examined

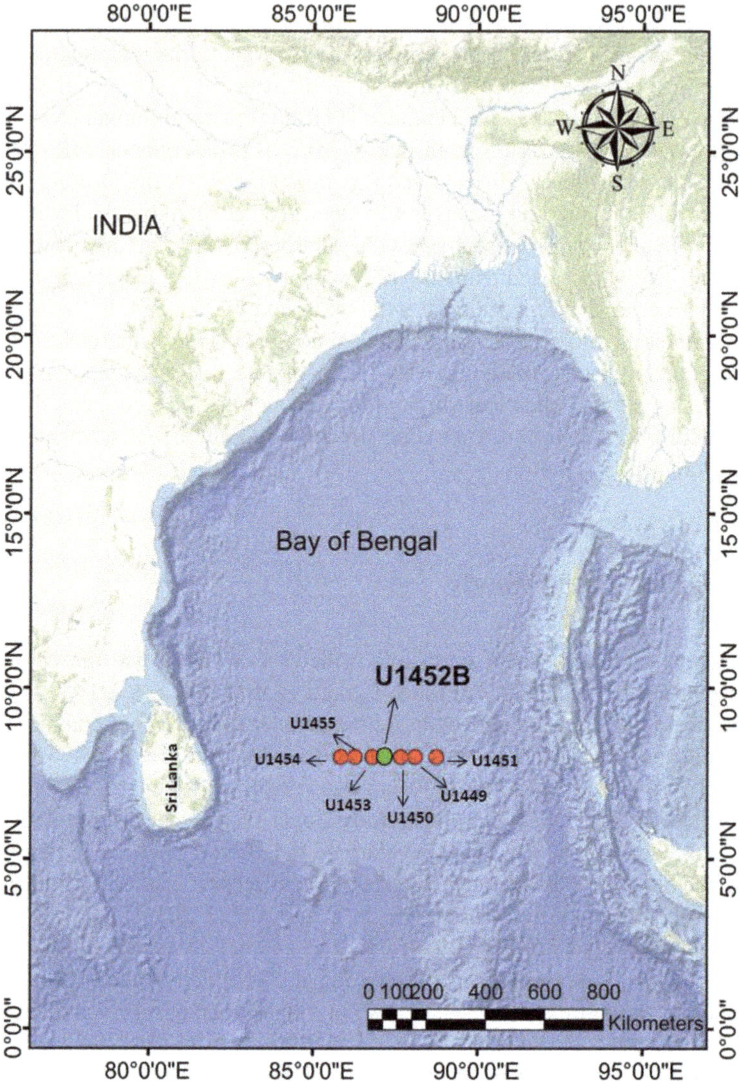

Fig. 1 Map showing the position of Expedition 354 IODP core sites

for elemental distribution, using the Energy Dispersive Spectroscopy (EDS) EDAX Peltier cooled octane plus 30 mm^2 detector attached to the SEM.

3 Results and Discussion

The coarse fraction from the core U1452B-36F-CC is dominated by planktonic foraminifers with microtektites and a minor amount of glass shards. Microtektites and glass shards were recovered in a 20-cm^3 core catcher sample and grains were observed in the 125–250 and >250 μm size fractions. The samples from the top and below core catcher samples are analyzed to check the presence of Australasian microtektites. The Australasian microtektite grains are only observed in the core catcher U1452B-36F-CC sample. Morphological and compositional characteristics of these individual grains are discussed in detail.

4 Morphological Characteristics of the Microtektites

On the basis of morphological features observed by stereoscopic microscope and SEM images, microtektites are categorized with varying shapes, size, color and surface features.

5 Shapes, Size and Color

Australasian microtektites recovered from the core U1452B-35F-CC in the lower Bengal Fan are generally spherical, but other typical splash forms like teardrops, dumbbells and discs are also present (Figs. 2, 3, 4 and 5). Irregular (Fig. 2c) and broken spherical (Fig. 3e) forms with highly fractured surfaces are also present as well. Among the entire range of shapes, spherical shapes are the most abundant, followed by irregular and other splash forms. Some irregular forms have a smooth surface and rounded corners (Fig. 2d) and some appear to be part of broken spherical microtektites. Within the microtektite bearing layer, the glass fragments with very sharp edges and conchoidal fractures (Fig. 7a) were also observed. Large spherical types are mostly broken and some are partly broken or smooth surfaced.

The average size of the recovered microtektite samples was measured from the SEM photomicrographs. The microtektites vary in size, but all are less than a millimeter in diameter. Of all the morphological types, the spherical and the irregular ones are usually larger than the teardrop, dumbbell and disc-shaped forms. The largest of the recovered microtektites is spherical shaped one (Fig. 2a), with diameter of ~700 μm. A large irregular form (Fig. 2c) measuring ~780 μm (long axis) by 600 μm (short axis), was also recovered. Other microtektites are smaller in size and their abundance increase with decreasing size.

The microtektites grains are transparent to translucent and the color varies from colorless to pale-yellow to greenish yellow, to dark brown. Opaque samples and darker colored grains are usually smaller than transparent and translucent variants.

There are also some transparent bottle-green microtektites present along with other microtektites. Pale-yellowish or greenish samples are most abundant among all the recovered microtektites.

6 Surface Sculpturing

These microtektites exhibit different surface features, including pits, mounds, grooves and conchoidal fractures. The surficial features on microtektites are almost the same in shape and (or) size ratios of the tektites reported from the previous studies (Glass 1974). This indicates that the origin of the sculpturing on microtektites is probably the same. Two hypotheses have been proposed for the origin of surfaces

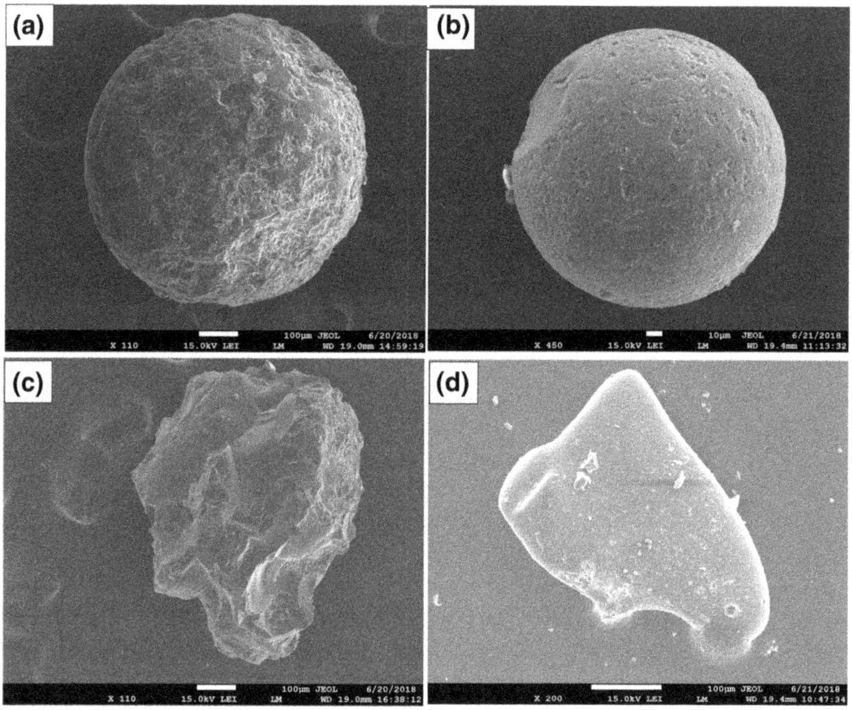

Fig. 2 SEM images of the microtektite grains with surface features in panels (**a**–**f**). **a** pale-greenish microtektite with highly fractured and corroded outer surface, **b** opaque microtektite with relatively smooth surface, **c** pale-greenish, irregular form with conchoidal fracture surface, **d** irregular opaque microtektite with smooth corner and surface, **e** and **f** broken dumbbell shaped microtektite with highly pitted spongy appearance, **f** dumbbell shaped microtektite with flat bottom groove, possibly a fusion boundary between two droplets or tearing point where two teardrops split from (1); etched flow-line on the surface of dumbbell (2); fused line running across the width of the dumbbell diagonally (3); irregular, narrow, branching grooves (4)

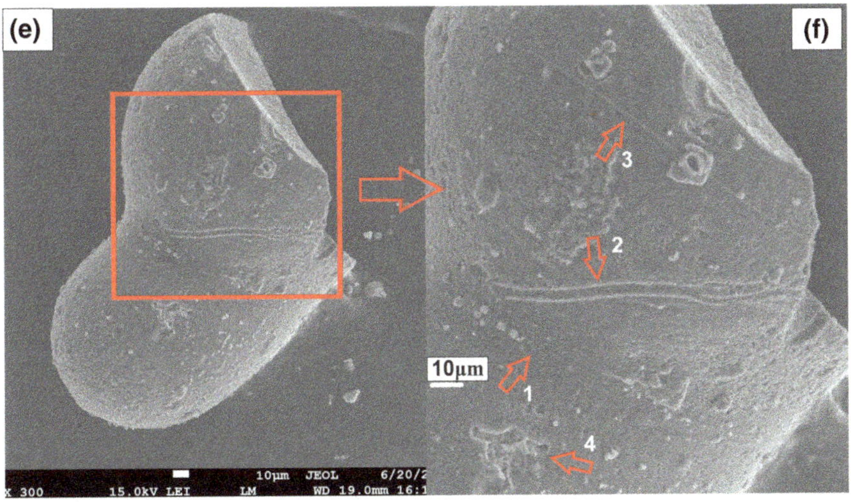

Fig. 2 (continued)

sculpturing on these tektites: (1) aerodynamic ablation or (2) solution by groundwater (Glass 1974). The surface sculpturing on tektites other than the ones shaped by aerodynamic ablation has been the subject of much discussion (Suess 1900; Van der Veen 1923; Baker 1963; O'Keefe 1967).

Pits are the most common surface alterations found on microtektite surfaces. These pits occur in a wide range of sizes, width/depth ratios, shapes (circular to irregular and star-shaped), and abundances (Glass 1974). In some samples the surface is sculpted with numerous fine circular pits, giving it a spongy appearance (Fig. 5a-1 and 5b-1). Irregular pits of much larger size than the fine pits are also present on the surface of this sample (Fig. 5b-2). An eye-catching specimen (Fig. 4a) exhibits "blackberry" like surficial appearance with large and circular overlapping pits (Fig. 4a1) of about 25 μm diameter, giving it a scalloped appearance. There are smaller such pits occurring inside the larger one. On the surface of a broken disk-shaped microtektite (Fig. 4c) circular and conical pits (Fig. 4d-1 and 2) are present with diameters up to ~25 μm, which are more likely the bubble cavities of varying size further modified by the solution. In some microtektite samples (Fig. 4e), the smooth surface is altered with small circular pits. However, some pits are relatively bigger with a micro-sized particle attached to the center (Fig. 4e-1), which making a moat surrounding it. This appears to be the micro-impact crater with the micro-particles striking, still attached to it. It must have taken place during the waning phase of splash ejecta when flight speed was relatively slower and material was in the semi-solid stage. The dimension of the impact crater indicates a slower impact velocity. Two small and two relatively large overlapped impact pits can be identified within the surface of the specimen with conchoidal fractures and radial cracks which seems to be a high velocity solid state projectile impact feature (Fig. 4e-2). This single grain represents ambiguous micro-impact phenomena documenting both low velocity semi-solid impact pitting

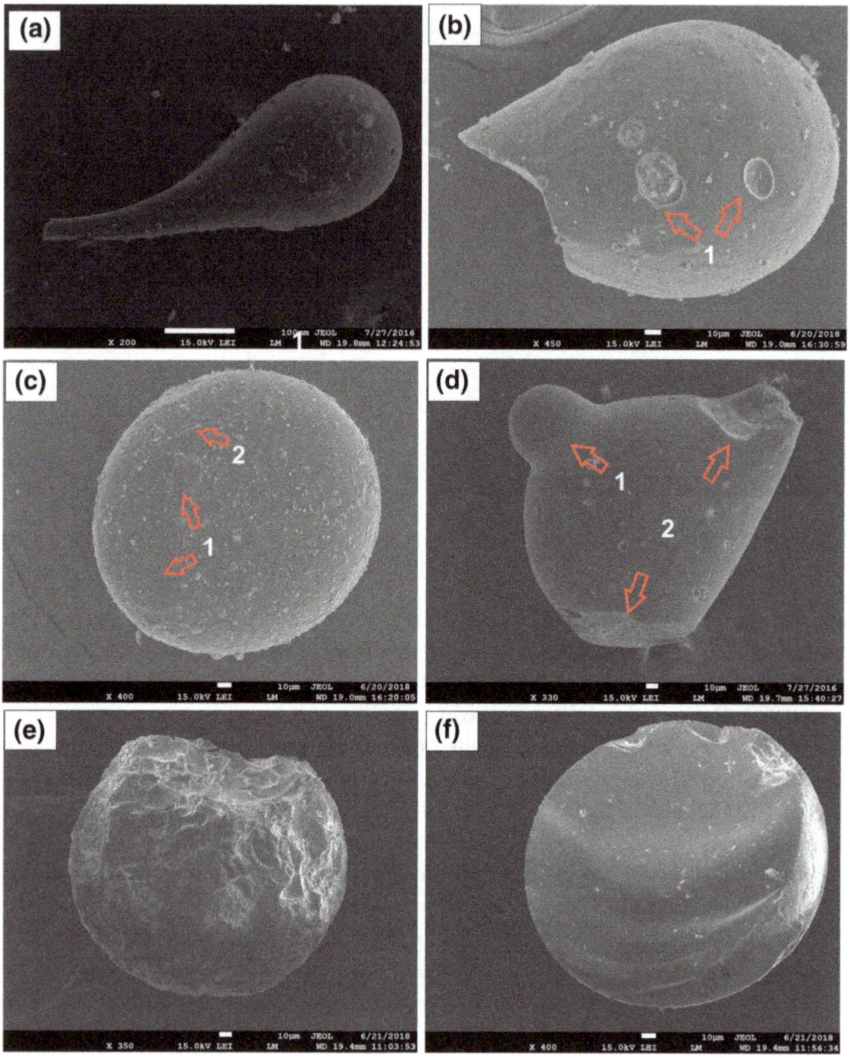

Fig. 3 SEM images of the microtektite grains in panels (**a**–**f**). **a** teardrop form, **b** broken teardrop form with relatively smooth surface and deep etch pits (1), **c** opaque disc shaped microtektite with granular texture, oval or roughly circular patches, slightly elevated from the surrounding rugged surface (1); narrow groove on the surface (2), **d** broken discoid form with a teardrop at its initial phase of development, **e** transparent bottle green microtektite with highly fractured and corroded surface, **f** hemispherical microtektite (possibly a broken half of spherical microtektite)

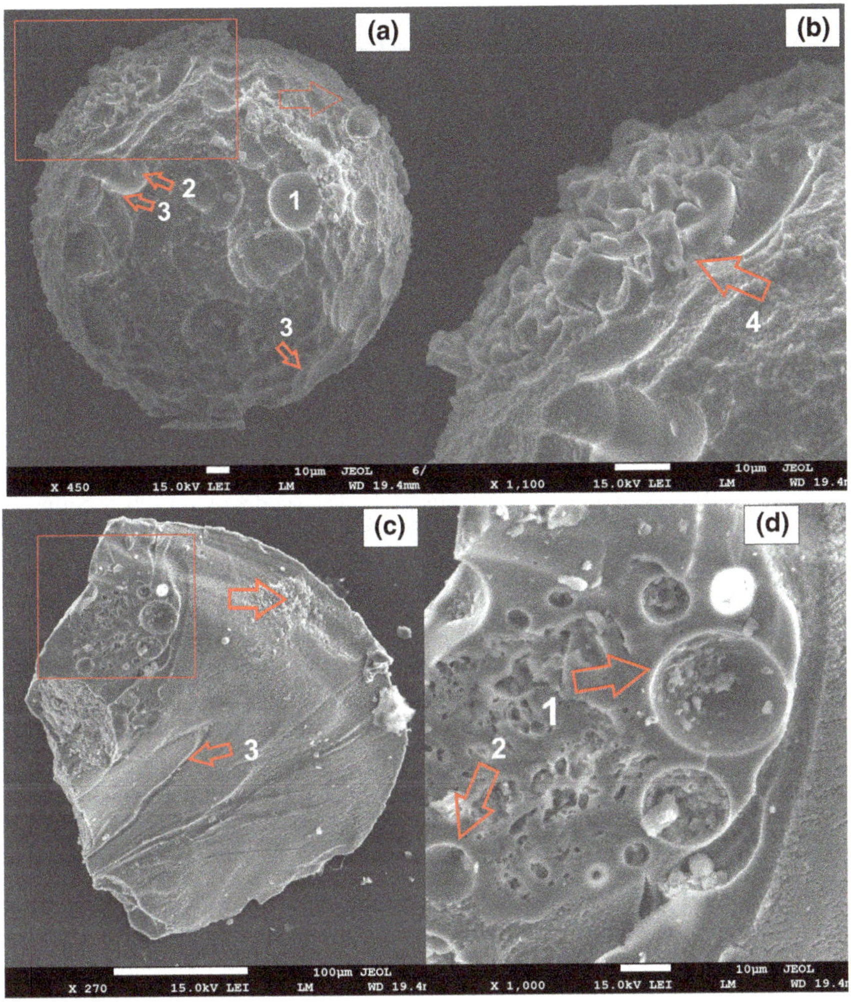

Fig. 4 SEM images of the microtektite grains shows surface sculpturing in panels (**a**–**f**). **a** Opaque grain with "blackberry" like surface texture with shallow-large pits (1); "U" shaped groove (2); ridges between two consecutive parallel U shaped grooves (3), **b** unusual flower like surface texture (4), **c** broken disc, translucent brown microtektite grain with bubble escape pits and flow lines etched out on the surface (3), **d** circular bubble escape pits (1); conical bubble escape pits (2), **e** spherical microtektite with low velocity impact pits with moat structure where the micro-sized particle striking and are still attached (projectile adhering) to the pit (1); Overlapping high velocity impact feature (pitless craters) (2); U shaped groove (3); **f** a pitless crater with concentric fractures (2, 3); and thin, hair like radial micro-fractures with branching (1); conchoidal fracture surface (4)

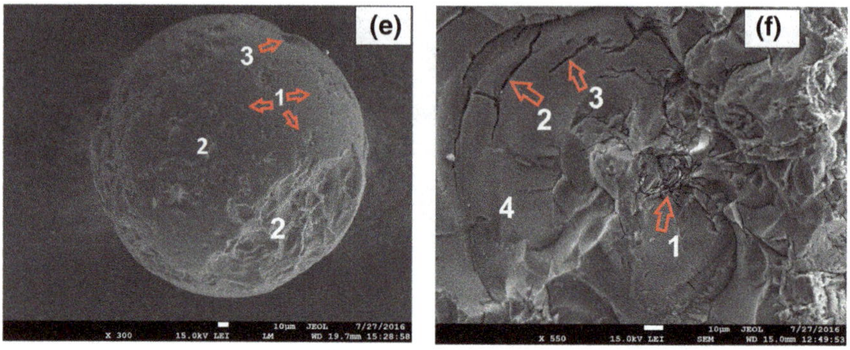

Fig. 4 (continued)

and projectile adhering and high velocity solid state projectile collision followed by pitting and fracturing of microtektite grain. Multiple episodes of impacts or collisions can be postulated for the co-occurrence of both these micro impact pits associated with projectile adhering and high velocity impact fractures and pitless catering which is also reported by Prasad and Khadekar (2003). Exactly when and how these erosive or destructive features developed on the surface is a subject of further investigation.

Circular and mound shaped protrusions have also been observed on the surface of some microtektites from the present study. These protrusions can be of several types. In some samples, it is circular and elevated several micrometers from the surrounding surface. The other types are irregular or roughly circular, which are slightly elevated from the surrounding surface and the third type is conical in shape and appears to be in the form of a mound protruding through the surface of the microtektite. In a particular sample, one or more types are present. Small circular elevated area (Fig. 5b) of around 5 μm diameter, elevated nearly 1–1.5 μm from the surrounding highly pitted surface (Fig. 5b-1) is observed on a spherical microtektite. A relatively larger irregular or roughly circular area (Fig. 5a-3) which is slightly elevated from the surrounding surface also present in the same specimen. A pale-brownish spherical microtektite exhibit both the circular (Fig. 5c-2 and 5d; ~8.5 μm diameter) and mound shaped protrusions (Fig. 5c-1) scattered all over the outer surface. Prasad and Khedekar (2003) described these protrusions as gentle accretionery impacts subjected to projectile adhering. Glass (1974) described the circular elevated protrusions to be of lower refractive indices and confirmed it as silica rich overgrowth by EDAX analyses.

Some minor forms are present on the surface of some microtektites. Some pits occurring in a cluster give rise to ‘U’ shaped groove on the surface of a dark opaque sample (Fig. 4a-2) and ridges are formed between two consecutive parallel grooves running parallel to the circumference of the spherical sample (Fig. 4a-3). Another disc shaped opaque microtektite (Fig. 3c) has a narrow groove along the surface, (Fig. 3c-2) which might have been developed due to etching. On the surface of a broken dumbbell shaped form (Fig. 2e), not very prominent flat bottom groove (Fig. 2f-1) is observed at higher magnification running all across the body of the

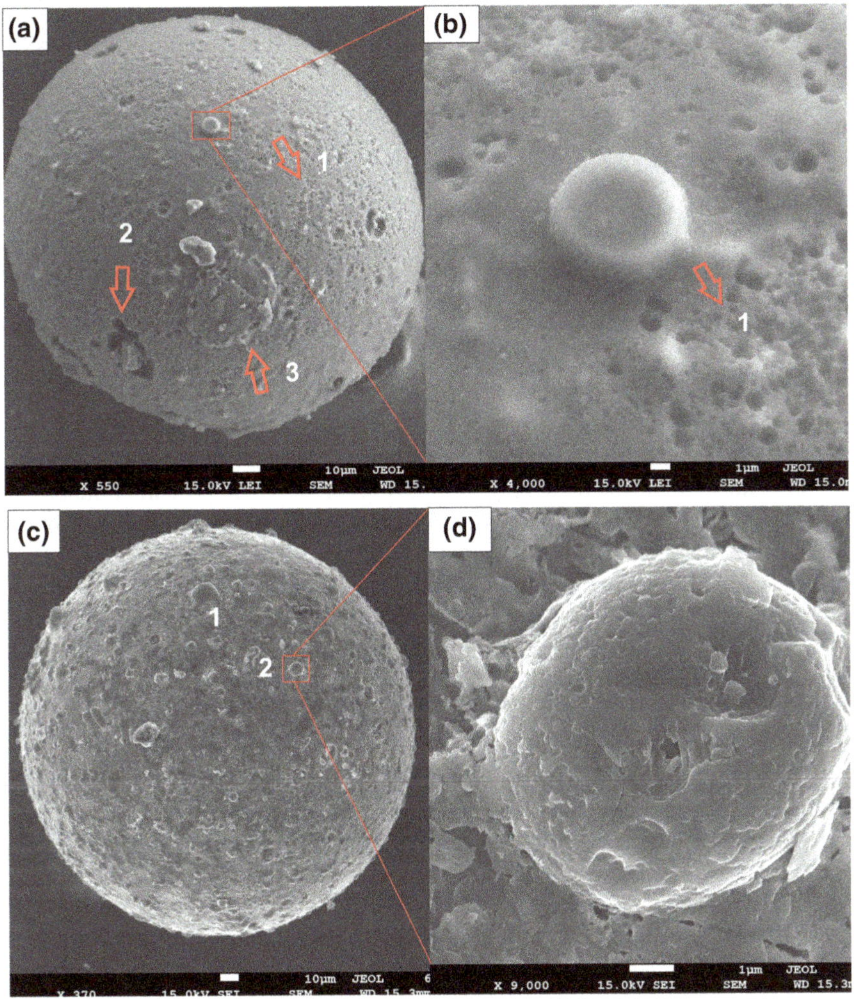

Fig. 5 SEM images of the microtektite grains with different surface sculptures and textures. **a** Oval microtektite grain with highly pitted surface and circular protrusion showing spongy appearance (1); irregular deep pits (2); irregular patch of primordial surface material appearing as elevated from the rest of the surface (3), **b** circular elevated area on the surface of the specimen A showing 1 μm and submicron size circular pits (1); **c** pale- brownish microtektite with circular and mound shaped protrusions (1); circular protrusions (2), **d** Circular protrusion in high magnification. **e** spherical microtektite with rugged up surface, **f** irregular patches of original surface dissected (1); granular texture on the surface of an opaque microtektite sample

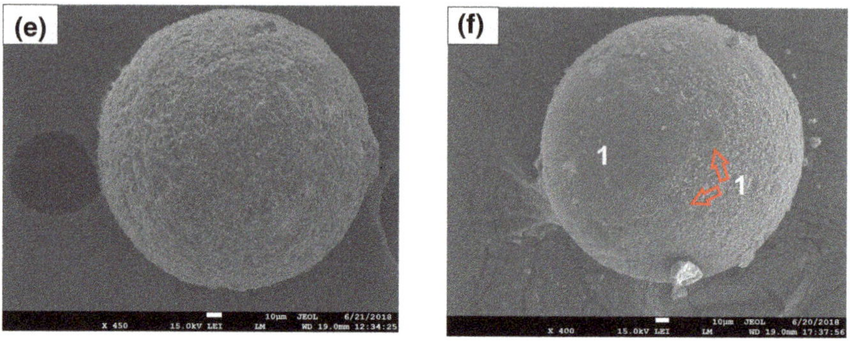

Fig. 5 (continued)

dumbbell microtektite. This appears to be the spot where two pieces of droplets got fused together. A narrow branching groove (Fig. 2f-4) is also present at the bottom of the fused groove. These grooves can be micro fractures further widened by the solution. Another fuse line appears on the surface of the dumbbell (Fig. 2f-3) which runs across entire surface diagonally. Grooves on Australasian microtektites are not as prominent as that of the Ivory Coast and North American microtektites.

Fracturing on the surface is very common in the microtektites. Most of the Australasian microtektites recovered from this core are highly fractured and broken (Figs. 2a–e; 3e and f; 4e and F). Prasad and Khedekar (2003) attributed these fractures and pits to erosive and destructive micro impacts. Numerous small narrow branching micro-fractures can be seen on the conchoidal fracture surface of a microtektite (Fig. 4f-2). Very thin hair like branching fractures (Fig. 4f-1) are initially developed which gradually get wider (Fig. 4f-3) in due course of time due to solution activity. These conchoidal fracture and radial microfractures are features of pitless craters on the surface of the microtektites.

Apart from these distinct surficial features, Australasian microtektites exhibit countless bizarre sculptures on their surfaces that are difficult to explain. Some Australasian microtektites have flowlines that have been etched out on the surface (Smit et al. 1991). The broken dumbbell and the broken disc each of these has one or more etched flowlines of lengths ~110 μm (Fig. 2e-2) and ~140 μm (Fig. 4c-3) respectively on the surfaces. Prasad and Sudhakar (1998) reported that the surface features formed by solution, high-velocity impact pits have been observed on the surface of some of the Australasian microtektites. On the surface of a broken teardrop, several large and deep circular and oval shaped etch pits are developed (Fig. 3b-1). Solution exposed the underneath surface inside the pits which are later filled by other micro-sized particles.

Some Australasian microtektites show unusual surface textures and these textures can directly be attributed to the solution action after the deposition of these microtektites on the ocean floor. An initially smooth surface is dissected into several patches of varying sizes (Fig. 3f). Differential solution left behind the less soluble material and removed the actual surface material leaving behind a rugged granular appearance.

This also proves the heterogeneity of composition of the surface material of these microtektite samples. After solution action, these patches may appear as elevated areas from the surrounding. However, these patches might be more siliceous which is resistant to the solution than the actual material.

7 Compositional Characteristics of Microtektites

The chemical composition of the 33 representative microtektite samples which include transparent, translucent and opaque microtektites and glasses based on EDX measurement was shown in Table 1. Individual microtektite sample differs in the composition regarding the major oxides, minor and trace element concentration. These microtektites have SiO_2 and MgO contents that range between 59.60–75.08% and 4.30–22.95% respectively. The average concentration of the major oxides such as Al_2O_3, Fe_2O_3, CaO, K_2O, Na_2O and TiO_2 are 11.14, 4.05, 2.90, 2.30, 1.50 and 0.41% respectively. On the basis of major oxide content, microtektites are categorized into two types (glasses excluded) (Table 2). The first type is similar to 'normal microtektites' (Glass and Simonson 2012) which have major oxide values comparable to average major oxide composition of other Australasian tektites and microtektites, except for slightly low TiO_2 (average 0.41%). In this study, the normal microtektite group includes almost all the transparent, pale greenish, pale brownish, translucent and some opaque microtektites. A second type is a group of transparent microtektites, which are characterized by exceptionally high MgO (>16%) and low SiO_2 (Fig. 6).

We compared the major oxide content of these microtektites with Australasian, Transantarctic and Australasites from the same strewn field and other fields described from the Ivory Coast and North America (Fig. 2; Table 2; Taylor and Sachs 1964; Taylor and McLennan 1979; Glass et al. 2004b; Folco et al. 2009; Glass and Simonson 2012). Australasian microtektites more depleted in SiO_2, TiO_2 and Fe_2O_3 than both the Ivory Coast and North American microtektites, but richer in MgO, CaO and K_2O. These differences may be due to the difference in the impact object and target rock compositions, which are reflected in the compositions of microtektites of the particular strewn field. The microtektites have a broader range in composition than the tektites and extend down to lower SiO_2 contents (Cassidy et al. 1969; Frey et al. 1970; Glass 1972a, b; Glass et al. 2004b).

The minor and trace element content of these microtektites have a wider range and differ from sample to sample. The microtektites which have been categorized on the basis of morphometric parameters and visual appearances (color, size, surface sculpturing etc.) show dissimilarities in their minor and trace element concentrations. Some of the microtektites show exceptionally high minor and trace element values. Translucent brown, opaque and bottle green samples are richer than the transparent and pale green microtektites in minor and trace elements like Zr, Mo, Sc, La, Cr, Co, Rb and Sr etc. The average major oxide compositions of glasses (Table 1) are quite different from the microtektites. These glasses are characterized by higher SiO_2 (>73%), Na_2O and K_2O and low in MgO and Fe_2O_3 and enriched in minor and trace

Table 1 Major oxides, minor and trace element concentration of microtektites and glasses from the studied core

	Microtektites														
Element	AMT 1	AMT 2	AMT 3	AMT 4	AMT 5	AMT 6	AMT 7	AMT 8	AMT 9	AMT 10	AMT 11	AMT 12	AMT 13	AMT 14	AMT 15
	Wt%	Wt%	Wt%	Wt%	Wt%	Wt%	Wt%	Wt%	Wt%	Wt%	Wt%	Wt%	Wt%	Wt%	Wt%
Major oxides															
Na_2O	0	0.51	0	1.86	0.09	0.11	0.46	0	1.12	0	2.43	1.93	3.26	0	0.99
MgO	0.47	4.12	6.85	4.07	3.14	4	1.9	3.41	1.97	2.42	3.29	4.12	4.78	4.55	16.1
Al_2O_3	11.95	13.73	14.96	10.03	15.82	14.33	13.2	11.53	14.2	12.2	13.18	11.86	9.17	12.32	11.63
SiO_2	68.98	71.58	65.55	70.24	63.58	72.2	69.46	74.27	69.48	74.39	67.07	67.99	68.79	69.93	59.62
P_2O_5	0	0	0	0	0	0	0	0	0	0	0	0	0	0	0
K_2O	2.42	1.56	1.35	1.94	3.21	2.06	3.33	0.85	3.39	0.94	3.33	2.37	2.93	2.27	1.23
CaO	4.25	2.83	3.09	2.88	2.48	1.9	2.6	3.14	2.18	3.63	2.62	5.34	2.97	3.85	2.57
TiO_2	0.91	0.27	0.58	0.36	0.66	0.16	0.6	0.13	0.73	0.14	0.57	0.14	0.43	0.62	0.55
Fe_2O_3	10.97	3.21	3.7	2.64	3.82	3.8	3.51	4.32	4.11	3.76	3.23	2.97	3.47	3.38	3.27
Minor/Trace elements															
Zr	0	1.07	0.73	1.31	1.45	0.81	0.86	0.41	0.7	0.63	1.22	1.05	1.15	0.83	1.46
Mo	0	0.79	0.9	1.26	1.22	0.15	0.92	0.23	1.22	0.3	1.31	1.12	1.7	1.18	1.42
Sc	0	0	0.04	0	0.14	0	0.09	0	0.04	0	0.07	0	0.1	0.13	0.15
La	0	0	0.11	0.06	0.3	0	0.22	0.04	0.2	0	0.14	0	0.25	0.22	0.22
V	0	0	0.01	0	0.07	0	0.03	0	0.05	0	0.01	0	0	0.04	0.03
Cr	0	0.02	0.05	0.04	0.16	0	0.04	0.02	0.09	0.01	0.07	0	0.13	0.05	0.15
Mn	0.02	0.01	0.08	0.04	0.09	0	0.05	0.03	0.07	0.01	0.13	0.04	0.05	0.16	0.08
Co	0	0	0.07	0.04	0.13	0.02	0.12	0.19	0.06	0.11	0.07	0.07	0.07	0.07	0.06
Ni	0	0	0.07	0.08	0.1	0.03	0.05	0.09	0.03	0.06	0.05	0.05	0.02	0.03	0.03

(continued)

Table 1 (continued)

	Microtektites														
Element	AMT 1	AMT 2	AMT 3	AMT 4	AMT 5	AMT 6	AMT 7	AMT 8	AMT 9	AMT 10	AMT 11	AMT 12	AMT 13	AMT 14	AMT 15
	Wt%	Wt%	Wt%	Wt%	Wt%	Wt%	Wt%	Wt%	Wt%	Wt%	Wt%	Wt%	Wt%	Wt%	Wt%
Cu	0	0	0.06	0.11	0.09	0.05	0.06	0.13	0.02	0.07	0.06	0.04	0.02	0.03	0.02
Zn	0	0	0.13	0.11	0.18	0.04	0.14	0.15	0.02	0.08	0.07	0.06	0.03	0.03	0.01
Ir	0	0	0.09	0.21	0.25	0	0.17	0.06	0	0.1	0.07	0.03	0	0	0
Ga	0	0	0.06	0.12	0.1	0.02	0.06	0.14	0.01	0.05	0.03	0.04	0	0.01	0.01
Ge	0	0.02	0.29	0.31	0.41	0.15	0.33	0.31	0.11	0.3	0.24	0.15	0.05	0.05	0.05
As	0	0	0.09	0.17	0.29	0.05	0.18	0.12	0	0.12	0.09	0.03	0	0	0
Se	0	0	0.18	0.39	0.48	0.06	0.33	0.07	0	0.21	0.17	0.05	0	0.01	0
Br	0	0	0.15	0.26	0.44	0	0.36	0.03	0	0.15	0.05	0.03	0	0	0
Rb	0.01	0	0.12	0.28	0.38	0	0.24	0.02	0	0	0	0	0	0	0
Sr	0.01	0	0.19	0.29	0.52	0	0.26	0	0	0.01	0	0	0	0	0

	Microtektites														Glass			
Element	AMT 16	AMT 17	AMT 18	AMT 21	AMT 22	AMT 25	AMT 26	AMT 27	AMT 28	AMT 29	AMT 31	AMT 33	AMT 34	AMT 35	AMT 19	AMT 20	AMT 23	AMT 24
	Wt%	Wt%	Wt%	Wt%	Wt%	Wt%	Wt%	Wt%	Wt%	Wt%	Wt%	Wt%	Wt%	Wt%	Wt%	Wt%	Wt%	Wt%
Major oxides																		
Na_2O	0.01	0.00	0.00	0.94	2.28	1.45	0.83	0.00	0.00	0.00	2.00	2.13	0.00	1.19	0.05	17.92	0.04	8.64
MgO	0.07	0.47	22.95	5.29	3.37	5.81	16.38	0.86	1.37	4.79	3.94	3.15	2.87	1.60	0.48	4.30	0.00	0.62
Al_2O_3	4.68	11.95	0.00	13.89	12.35	16.18	12.39	11.60	2.41	12.98	12.83	8.50	10.59	12.23	9.84	7.46	11.32	11.22

(continued)

Table 1 (continued)

	Microtektites														Glass			
Element	AMT 16	AMT 17	AMT 18	AMT 21	AMT 22	AMT 25	AMT 26	AMT 27	AMT 28	AMT 29	AMT 31	AMT 33	AMT 34	AMT 35	AMT 19	AMT 20	AMT 23	AMT 24
	Wt%	Wt%	Wt%	Wt%	Wt%	Wt%	Wt%	Wt%	Wt%	Wt%	Wt%	Wt%	Wt%	Wt%	Wt%	Wt%	Wt%	Wt%
SiO_2	35.24	68.98	65.78	69.33	68.03	64.23	59.68	71.02	28.90	67.14	68.57	75.04	74.37	74.20	74.13	60.42	75.09	70.86
P_2O_5	0.00	0.00	0.00	0.00	0.00	0.00	0.00	0.00	0.00	0.00	0.00	0.00	0.00	0.00	0.00	0.00	0.00	0.00
K_2O	4.38	2.42	0.22	1.76	3.09	2.47	1.01	2.65	3.12	1.83	2.88	2.34	2.99	3.02	1.93	0.64	4.80	3.50
CaO	1.40	4.25	8.18	2.46	2.49	2.52	2.69	3.30	1.11	6.83	2.12	1.86	1.36	2.47	1.69	4.12	0.61	1.12
TiO_2	0.25	0.91	0.10	0.51	0.61	0.52	0.61	0.63	0.17	0.44	0.53	0.41	0.61	0.51	0.51	0.04	0.08	0.00
Fe_2O_3	15.52	10.97	0.60	3.04	2.70	3.17	2.62	4.47	18.67	3.15	2.77	1.75	2.05	2.58	4.75	0.35	1.81	0.71
Minor/Trace elements																		
Zr	0.14	0.00	0.47	1.04	0.94	1.14	1.19	0.64	0.25	0.62	1.01	0.80	1.42	0.64	0.47	0.60	0.05	0.94
Mo	27.86	0.00	0.48	0.84	1.56	1.28	1.14	1.42	27.04	0.96	1.13	1.23	1.29	0.80	0.44	1.13	0.22	1.23
Sc	0.06	0.00	0.00	0.07	0.18	0.15	0.18	0.14	0.05	0.09	0.12	0.11	0.24	0.05	0.00	0.06	0.05	0.00
La	0.17	0.00	0.00	0.10	0.32	0.21	0.30	0.32	0.18	0.33	0.21	0.20	0.39	0.14	0.22	0.13	0.36	0.01
V	0.04	0.00	0.00	0.01	0.08	0.02	0.03	0.09	0.04	0.01	0.03	0.07	0.11	0.04	0.03	0.04	0.05	0.00
Cr	0.07	0.00	0.00	0.05	0.10	0.09	0.13	0.25	0.10	0.08	0.10	0.12	0.12	0.06	0.19	0.09	0.35	0.01
Mn	0.06	0.02	0.00	0.09	0.07	0.05	0.09	0.25	0.07	0.11	0.04	0.13	0.13	0.04	0.33	0.10	0.47	0.01
Co	0.08	0.00	0.00	0.03	0.09	0.07	0.09	0.37	0.14	0.10	0.08	0.13	0.10	0.04	0.52	0.14	0.67	0.02
Ni	0.21	0.00	0.02	0.01	0.13	0.03	0.04	0.19	0.43	0.01	0.05	0.11	0.05	0.02	0.43	0.13	0.43	0.02
Cu	0.00	0.00	0.00	0.02	0.08	0.02	0.05	0.29	0.02	0.04	0.06	0.13	0.05	0.01	0.40	0.17	0.50	0.04
Zn	0.00	0.00	0.00	0.02	0.06	0.03	0.02	0.29	0.01	0.04	0.09	0.09	0.02	0.00	0.55	0.16	0.74	0.04

(continued)

Table 1 (continued)

Element	Microtektites														Glass			
	AMT 16	AMT 17	AMT 18	AMT 21	AMT 22	AMT 25	AMT 26	AMT 27	AMT 28	AMT 29	AMT 31	AMT 33	AMT 34	AMT 35	AMT 19	AMT 20	AMT 23	AMT 24
	Wt%	Wt%	Wt%	Wt%	Wt%	Wt%	Wt%	Wt%	Wt%	Wt%	Wt%	Wt%	Wt%	Wt%	Wt%	Wt%	Wt%	Wt%
Ir	0.01	0.00	0.03	0.00	0.07	0.00	0.00	0.11	0.00	0.00	0.04	0.11	0.02	0.00	0.33	0.13	0.60	0.00
Ga	0.01	0.00	0.02	0.01	0.02	0.00	0.01	0.05	0.01	0.04	0.07	0.05	0.01	0.00	0.22	0.08	0.37	0.01
Ge	0.04	0.00	0.21	0.02	0.14	0.00	0.00	0.27	0.05	0.05	0.12	0.21	0.01	0.02	0.80	0.29	1.00	0.09
As	0.00	0.00	0.22	0.01	0.06	0.00	0.00	0.21	0.00	0.00	0.08	0.09	0.00	0.00	0.24	0.06	0.36	0.00
Se	0.00	0.00	0.19	0.00	0.06	0.00	0.01	0.19	0.00	0.00	0.16	0.15	0.00	0.00	0.33	0.23	0.47	0.04
Br	0.00	0.00	0.06	0.00	0.15	0.00	0.00	0.13	0.00	0.00	0.13	0.18	0.00	0.00	0.18	0.14	0.23	0.00
Rb	0.00	0.01	0.03	0.00	0.04	0.00	0.00	0.01	0.00	0.00	0.06	0.02	0.00	0.00	0.02	0.34	0.02	0.01
Sr	0.00	0.01	0.02	0.00	0.02	0.00	0.00	0.00	0.00	0.00	0.02	0.02	0.00	0.00	0.00	0.21	0.07	0.00

Table 2 Comparison chart of the compositional data from the present study, Australasian microtektites—normal, high-Ni, and high-Mg (Glass et al. 2004b), Ivory Coast (Glass et al. 2004b), Transantarctic Mountain microtektites (Folco et al. 2009), Australites (Taylor and Sachs 1964; Taylor and McLennan 1979)

	Present study					Australasian			Transantarctic	Ivory Coast	North American	Australasite
	Normal				Bottle green							
Major oxides	Translucent brown	Opaque	Transparent	Plae green	Bottle green	Normal	High Ni	High Mg				
	Avg. wt%	Avg. wt%	Avg. wt%	Avg. wt%	Avg. wt%	wt%	wt%	wt%	wt%	Avg. wt%	Avg. wt%	wt%
SiO_2	67.20	69.52	70.32	70.25	59.65	60.48–78.05	66.69–72.01	51.83–56.35	59.8–77.6	66.4	73.4	70.4–78.8
TiO_2	0.55	0.49	0.42	0.41	0.58	0.64–1.18	0.69–0.81	0.72–1.17	0.71–1.29	0.55	0.73	0.55–0.81
Al_2O_3	12.79	13.01	10.22	12.50	12.01	10.25–21.46	12.48–14.94	14.66–20.66	12.2–23.4	16.9	14.1	9.36–13.70
FeO/Fe_2O_3	3.23	3.10	2.79	4.22	2.94	2.90–7.23	5.11–6.77	4.50–7.78	2.60–5.49	6.65	4.37	3.67–5.51
MgO	3.05	3.27	4.24	3.92	16.24	1.67–5.07	2.90–4.17	11.34–21.82	1.89–5.57	4.27	1.73	1.50–2.48
CaO	2.44	3.00	3.89	3.50	2.63	1.63–6.56	2.37–4.30	4.37–5.72	2.17–5.3	1.73	1.64	2.13–5.09
Na_2O	0.73	1.21	1.80	0.58	0.91	0.22–1.70	0.98–1.59	0.43–0.79	0.15–0.42	1.92	1.06	1.12–1.52
K_2O	2.89	2.79	2.37	1.72	1.12	0.67–3.21	2.22–2.86	0.04–0.17	0.38–1.49	1.6	2.92	2.08–2.55

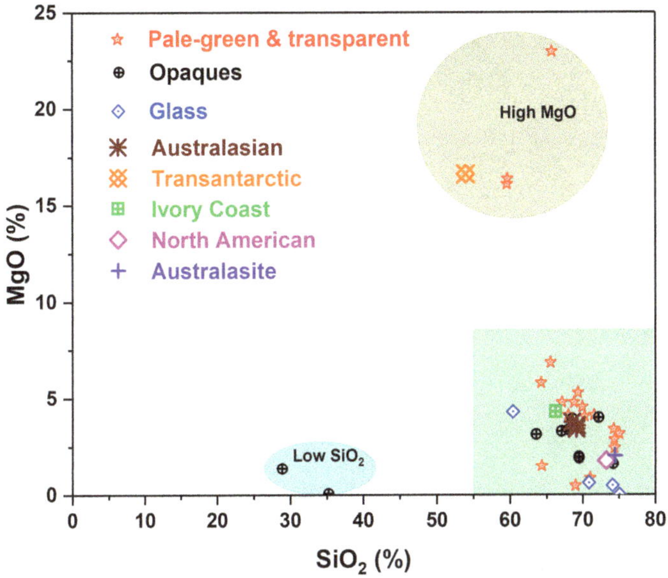

Fig. 6 A plot between SiO_2 and MgO of the Microtektite, opaques and glass grains (studied here) and records from Australasian microtektites-normal, high-Ni, and high-Mg (Glass et al. 2004b), Ivory Coast (Glass et al. 2004b), Transantarctic Mountain microtektites (Folco et al. 2009), Australites (Taylor and Sachs 1964; Taylor and McLennan 1979)

elements (> 5%). There is a glass-like substance (Fig. 7b) which has a low SiO_2 (<65%) and Na_2O as high as 17.91% (Table 1, AMT-20).

Three microtektite samples AMT-16, AMT-17 and AMT-28 show different values of major, minor and trace element distribution than other microtektites found from this site (Table 1). AMT-16 and AMT-28 are opaque samples with very low SiO_2 (Fig. 6), high in Fe_2O_3 and are enriched in Mo, Co and Ni. Another pale brownish translucent grain (AMT-17) has a comparatively high SiO_2 content (>64%) than the previous two and average values for the other major oxides but is enriched in almost every element listed in the table. The normal Australasian microtektites have trace element contents similar to the average trace element content of the upper continental crust. Glass et al. (2004b) identified a high-Ni group of normal Australasian microtektites with higher Ni and Co and generally higher Cr contents than the other normal Australasian microtektites. The Cr, Co, and Ni content of the high-Ni Australasian microtektites are positively correlated with each other and with the FeO and MgO content, and are negatively correlated with the other major oxides. This is suggestive of meteoritic contamination; but the ratios between Cr, Co, and Ni contents are closer to mantle ratios than meteoritic.

Correlation analysis provides an effective way to find relationships among the major oxides, minor and trace elements. The results of Pearson's correlation coefficients and their significance levels ($n = 33$; $p < 0.005$ and $p < 0.01$) are shown in Table 3. The significant strong positive correlations observed between Al_2O_3/SiO_2,

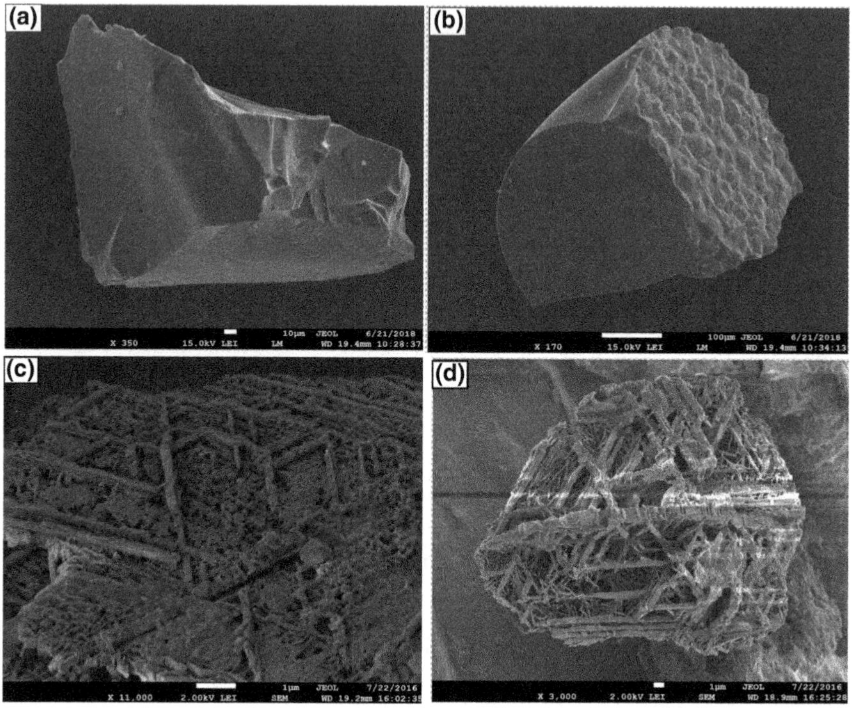

Fig. 7 SEM images of the **a** broken transparent glass with sharp corners and conchoidal fracture, **b** glass like transparent substance with rugged surface texture. **c** and **d** Possibly remnant of "Widmanstatten structure" which is found on meteorite due to exsolution of Ni and Fe or a part of metal micrometeorite

Al_2O_3/TiO_2 and MgO/CaO. The higher correlation between Al_2O_3, SiO_2 and TiO_2 is probably due to having the same source along with their chemical affinity. K_2O is highly negatively correlated with MgO/CaO. Fe_2O_3 is not showing any significant relationship with the terrestrial elements and showing highly negative correlation with SiO_2. The elemental concentrations of Cr, Mn, Co, Ni, Cu, Zn, Ir, Ga, Ge, As, Se, Br, Rb and Sr shows highly positively correlated with each other. Sc is correlated well with Zr, V, Cr and La and La is correlated with V, Cr, Mn and Co. There is also a good correlation observed between Ni and Mo and Cr and V. The inter-relationship between the major oxides and the trace elemental concentration suggests that these are originated from the same sources and processes.

Table 3 Correlation analyses of major oxides (a), minor and trace elements (b) of the microtektites from the core U1452B-36F-CC

(a)

	Na_2O	MgO	Al_2O_3	SiO_2	K_2O	CaO	TiO_2	Fe_2O_3
Na_2O	1.00							
MgO	−0.05	1.00						
Al_2O_3	−0.14	−0.23	1.00					
SiO_2	−0.03	−0.08	**0.51**	1.00				
K_2O	−0.14	**−0.63**	0.09	−0.16	1.00			
CaO	0.02	**0.51**	−0.22	0.11	**−0.59**	1.00		
TiO_2	−0.36	−0.02	**0.47**	0.13	0.15	−0.07	1.00	
Fe_2O_3	−0.30	−0.30	−0.36	**−0.78**	0.31	−0.24	0.06	1.00

(b)

	Zr	Mo	Sc	La	V	Cr	Mn	Co	Ni	Cu	Zn	Ir	Ga	Ge	As	Se	Br	Rb	Sr
Zr	1.00																		
Mo	−0.37	1.00																	
Sc	**0.46**	−0.02	1.00																
La	0.15	0.05	**0.81**	1.00															
V	0.12	0.11	**0.70**	**0.74**	1.00														
Cr	−0.02	0.00	**0.42**	**0.75**	**0.61**	1.00													

(continued)

Table 3 (continued)

(b)

	Zr	Mo	Sc	La	V	Cr	Mn	Co	Ni	Cu	Zn	Ir	Ga	Ge	As	Se	Br	Rb	Sr
Mn	−0.20	−0.08	0.16	**0.56**	**0.46**	**0.86**	1.00												
Co	−0.28	−0.05	−0.02	**0.40**	0.37	**0.81**	**0.91**	1.00											
Ni	−0.36	**0.41**	−0.12	0.26	0.34	**0.64**	**0.71**	**0.83**	1.00										
Cu	−0.11	−0.18	−0.10	0.23	0.31	**0.69**	**0.80**	**0.92**	**0.77**	1.00									
Zn	−0.16	−0.18	−0.14	0.24	0.25	**0.70**	**0.83**	**0.94**	**0.78**	**0.98**	1.00								
Ir	0.00	−0.16	−0.14	0.15	0.24	**0.57**	**0.66**	**0.79**	**0.71**	**0.92**	**0.93**	1.00							
Ga	−0.12	−0.17	−0.22	0.13	0.15	**0.57**	**0.70**	**0.85**	**0.73**	**0.93**	**0.95**	**0.96**	1.00						
Ge	−0.15	−0.20	−0.27	0.11	0.15	**0.57**	**0.72**	**0.86**	**0.73**	**0.92**	**0.96**	**0.94**	**0.95**	1.00					
As	0.07	−0.18	−0.15	0.03	0.20	**0.43**	**0.48**	**0.66**	**0.60**	**0.84**	**0.82**	**0.94**	**0.87**	**0.86**	1.00				
Se	0.17	−0.16	−0.15	0.00	0.17	0.35	**0.40**	**0.57**	**0.55**	**0.80**	**0.77**	**0.93**	**0.83**	**0.81**	**0.97**	1.00			
Br	0.23	−0.13	−0.06	0.01	0.21	0.28	0.29	**0.46**	**0.48**	**0.71**	**0.66**	**0.85**	**0.74**	**0.68**	**0.92**	**0.97**	1.00		
Rb	0.26	−0.09	−0.08	−0.07	0.15	0.17	0.18	0.34	**0.40**	**0.62**	**0.55**	**0.76**	**0.65**	**0.56**	**0.85**	**0.92**	**0.97**	1.00	
Sr	0.27	−0.09	−0.07	−0.05	0.15	0.19	0.18	0.34	**0.40**	**0.62**	**0.56**	**0.78**	**0.66**	**0.57**	**0.86**	**0.93**	**0.98**	**1.00**	1.00

8 Evidence of Extraterrestrial Materials Within the Microtektite Layer

Tektites form by melting of terrestrial continental rocks during the hypervelocity impact of an extraterrestrial object (Koeberl 1986, 1993). The melt produced during the impact of a cometary projectile with the continental upper crustal rocks led to the formation of Australasian tektites and microtektites (Taylor 1962; Barnes 1964, 1989; Taylor and Kaye 1969; Koeberl 1986; Wasson 1991; Koeberl 1994). An iridium anomaly (Schmidt et al. 1993; Koeberl 1993), discovery of shocked minerals (Glass and Wu 1993), and finding of shock lithified rock fragments associated with the Australasian microtektite layer (Glass and Koeberl 2006) have demonstrated the impact origin of this strewn field is beyond doubt. Geographic distribution of Australasian tektites and microtektites shows radial and concentric patterns and suggests that source is located somewhere in Indochina. Though, many studies carried out to study the source, process and origin of these impacts, evidence for the extraterrestrial components from the deep sea cores are not well established.

Strange cross-hatched structures on the surface of some substances were found within the microtektite bearing layer from this core while scanning the acid (HCl and HF) treated macerated sediment sample. These structures have a strong resemblance with the crisscrossing physical crystalline structure of octahedrites (A type of Iron meteorite). The distinctive structures otherwise termed as "Widmanstatten texture" appear from exsolution of two Fe-Ni mineral phases namely kamacite and taenite during very slow cooling (Owen and Burns 1939). Octahedrites have a double pyramid crystal structure (octahedron) and the hatched structure is best observed in sections parallel to the octahedral planes, with three set of bands at 60^0 angles. The structures observed on the surface of one specimen could possibly be the remnant of metallic exsolution (Fig. 7c–d). Prasad et al. (2017), reported a variety of cosmic metal particles such as kamacite, plessite, taenite, and Fe-Ni beads from deep-sea sediments of the Indian Ocean. The chemistry of the specimen and individual crystal bands are subjects to further investigation with precise analytical techniques. Their exact chemistry and origin, whether they are individual micrometeorites or parts of a metallic parent meteorite or parts of the ejecta itself are yet to be confirmed.

Glass et al. (2004b) identified a high-Ni group of Australasian microtektites with higher Co and Cr contents than the other normal Australasian microtektites. An anomalous Fe content of some opaque microtektites (AMT-16 and AMT-28) is recorded in the present study. Folco et al. (2018), based on Cr/Ni versus Co/Ni ratios in the Australasian microtektites and the previous observation of Glass et al. (2004b) suggested an impactor of primitive composition, either a chondritic or primitive achondritic responsible for Australasian tektite/microtektite strewn field and considered the possible contamination of a LL ordinary chondrite giving rise to such minor and trace distribution pattern in Australasian microtektites.

Though the exact impact crater associated with Australasian strewn filed has yet not been confirmed most authors suggest that it should be located somewhere in Indochina to explain geometric pattern in occurrences of tektite/microtektite, their geochemical signatures and that it's diameter should exceed ~30 km (e.g., Glass and Pizzuto 1994; Ma et al. 2004; Glass and Koeberl 2006; Prasad et al. 2007). Reversed polarity in mud lens closely associated with in situ tektites bearing flood deposits found in Ban Ta Chang and Chum Phuang in northeast Thailand are penecontemporaneous with the impact event at ~0.8 ma and suggestive of a crater location in vicinity (Hanies et al. 2004). Yet more detail investigations based on geomorphic features, geochemical signatures, geographic and geometric distribution patterns of microtektite occurrences and abundance are needed for exact crater identification and preparing impact models which may explain the enigma of formation of these artistic artifacts of nature and their distribution throughout the Australasian strewn field.

9 Conclusions

Microtektites, recovered from lower Bengal Fan sediments, occur beneath the M-B boundary, signifying the last geomagnetic reversal at around ~0.8 Ma. SEM images show the presence of mostly spherical forms and also other typical splash forms like teardrops, dumbbells and discs. These are also characterized with different colors—pale greenish, opaque, pale brown translucent and transparent and are characterized by various surficial attributes in the combination of pits, mounds, grooves and fracturing. Individual microtektites show dissimilarities in the chemical composition and are similar to the average composition of upper crustal rocks. The interrelationship between the major oxides and the trace elemental concentration also point out that these microtektites may have originated from the same sources and processes. Geochemical records from the present study also posses the similar geochemical signature of the previously studied Australasian microtektites. Presence of polymetallic exsolution texture within the microtektite bearing layer again re-establish the impact origin of these grains. This supports the formation of Australasian tektites and microtektites by the impact of an extraterrestrial projectile (possibly one or more large chondrite) with the upper crustal rock somewhere in northeast Thailand.

Acknowledgements The authors are grateful to the Director, Birbal Sahni Institute of Palaeosciences (BSIP), India, for providing the support and facilities extended. We acknowledge the support from IODP-India (NCAOR/IODP/20.15/5(IV)). We are extremely grateful to all IODP Expedition 354 Scientists and the samples and data were provided by IODP. We are thankful to Dr. Subodh Kumar for the timely help in carrying out the SEM-EDS analysis work at BSIP. This is BSIP contribution No. BSIP/RDCC/26/2018-19.

References

Baker (1963) Form and sculpture of tektites. In: O'Keefe JA (ed) Tektites: Chicago, Univ. Chicago Press, pp 1–24

Barnes VE (1963) Tektite strewn-fields. In: O'Keefe JA (ed) Tektites. University of Chicago Press, Chicago, pp 25–50

Barnes VE (1964) Variations in petrographic and chemical characteristics of indochinite tektites within their strewn-field. Geochim Cosmochim Acta 28:893–913

Barnes VE (1967) Tektites. In: Runcorn K (ed) International dictionary of geophysics, vol 2. Pergamon Press, Oxford, pp 1507–1518

Barnes VE (1971) Description and origin of a large tektite from Thailand. Chem Erde 30:13–19

Barnes VE (1989) Origin of tektites. Tex J Sci 41:5–33

Barnes VE, Pitakpaivan I (1962) Origin of indochinite tektites. Proc Nat Acad Sci 1 JSA 48:947–955

Bunopas S (1990) Tektites-Their origin and the continental catastrophic destruction in NE Thailand and Indochina. In: Proceedings of the 16th conference on sciences and technology of Thailand, 25–27 Oct 1990, Bangkok:512–513

Bunopas S (1992) Tektites-Their origin and the Early Quaternary catastrophic destruction in North-eastern Thailand and south Indochina. In: Proceedings of 29th Intern Geol Congress 24 Aug–3 Sept 1992. Kyoto, Japan Abstr Vol. 3(3), II-12-4 0-8 2114, p 650

Bunopas S, Wasson JT, Vella P, Fontaine H, Hada S, Burret C, Suphajunya T, Khositanont S (1999) Catastrophic loess, mass mortality and forest fires suggest that a Pleistocene cometary impact in Thailand caused the Australasian tektite field. J Geol Soc Thailand 1:1–17

Bunopas S, Vella P, Fontaine H, Hada S, Burrett C, Haines P, Howard Kieren T, Khositanont S (2005) Latest Pliocene to early M. Pleistocene Inthanon Epeirogenesis, 2.0–0.65 Ma, in north Thailand connecting to late Himalayan extrusion continuum: An application of extraterrestrial/terrestrial real-time layer of catastroloess from Buntharik Cometary Impact Event (new). In: Proceedings of the international conference on geology, geotecnology and mineral resources of Indochina. GEOINDO 2005, Kosa Hotel, Khon Kaen, 28–30 Nov 2005

Bunopas S, Vella P, Burrett C, Fontaine H, Hada S, Haines P, Khositanont' S, Prathuang Chintasakul, Chaodamrong P, Charusiri P, Howard KT (2007) Ausiralasian Cometary Impact, 0.8ma Catastroloess Buried Alive Miocene, Pliocene 2 Faunas In Thailand And Central Australia and Tektite-Bearing Flood Deposits In NE. Thailand. New world's discovery extinction age in Thailand (Indochina) and Australia impact fields. In: GEOTHAI'07 international conference on geology of Thailand: towards sustainable development and sufficiency economy. Bangkok, Thailand. 1–9 Nov 2007

Burns CA (1989) Timing between a large impact and a geomagnetic reversal and the depth of NRM acquisition in deep-sea sediments. In: Lowes FJ et al (eds) Geomagnetism and paleomagnetism. Kluwer Academic Publishers, Dordrecht, pp 253–261

Cassidy WA, Glass BP, Heezen BC (1969) Physical and chemical properties of Australasian microtektites. J Geophys Res 74:1008–1025

Channell JET, Hodell DA, Singer BS, Xuan C (2010) Reconciling astrochronological and 40Ar/39Ar ages for the Matuyama-Brunhes boundary and late Matuyama Chron. Geochem Geophys Geosyst 11, Q0AA12. https://doi.org/10.1029/2010GC003203

Chapman DR (1964) On the unity and origin of the Australasian tektites. Geochim Cosmochim Acta 28(6):841–880

Chapman DR (1971) Australasian tektite geographic pattern, crater and ray of origin, and theory of tektite events. J Geophys Res 76(26):6309–6338

Folco L, D'Orazio M, Tiepolo M, Tonarini S, Ottolini L, Perchiazzi N, Rochette P, Glass BP (2009) Transantarctic mountain microtektites: geochemical affinity with Australasian microtektites. Geochim Cosmochim Acta 73:3694–3722

Folco L, Glass BP, D'Orazio M, Rochette P (2018) Australasian microtektites: Impactor identification using Cr, Co and Ni ratios. Geochim Cosmochim Acta 222(1):550–568. https://doi.org/10.1016/j.gca.2017.11.017

France-Lanord C, Spiess V, Klaus A, the Expedition 354 Scientists (2015) Expedition 354 Preliminary Report: Bengal Fan. International Ocean Discovery Program. http://dx.doi.org/10.14379/iodp.pr.354.2015

France-Lanord C, Spiess V, Klaus A, Schwenk T, the Expedition 354 Scientists (2016) Bengal Fan. In: Proceedings of the international ocean discovery program 354: College Station, TX (International Ocean Discovery Program). http://dx.doi.org/10.14379/iodp.proc.354.2016

Frey FA, Spooner CM, Baedecker PA (1970) Microtektites and tektites: a chemical comparison. Science 170:845–847

Gentner W, Kleinmann B, Wagner GA (1967) New K-Ar and fission track ages of impact glasses and tektites. Earth Planet Sci Lett 2:83–86

Glass BP (1967) Microtektites in deep-sea sediments. Nature 214:372–374

Glass BP (1972a) Australasian microtektites in deep-sea sediments. In: Hayes DE (ed) Antarctic oceanology II: the Australian–New Zealand sector. American Geophysical Union, Antarctic Res Ser 19:335–348

Glass BP (1972b) Bottle-green microtektites. J Geophys Res 77:7057–7064

Glass BP (1974) Microtektite surface sculpturing. Geol Soc Am Bull 85:1305–1314

Glass BP (1990) Tektites and microtektites: key facts and inferences. Tectonophysics 171:393–404

Glass BP, Heezen BC (1967) Tektites and geomagnetic reversals. Sci Am, 33–38

Glass BP, Wu J (1993) Coesite and shocked quartz discovered in the Australasian and North American microtektite layers. Geology 21:432–438

Glass BP, Pizzuto JE (1994) Geographic variation in Australasian microtektite concentrations: implications concerning the location and size of the source crater. J Geophys Res 99:19075–19081

Glass BP, Koeberl C (2006) Australasian microtektites and associated impact ejecta in the South China Sea and the Middle Pleistocene supereruption of Toba. Meteorit Planet Sci 41:305–326

Glass BP, Simonson BM (2012) Distal impact ejecta layers: spherules and more. Elements 8:43–48. https://doi.org/10.2113/gselements.8.1.43

Glass BP, Liu S, Montanari A (2004a) Impact ejecta in upper Eocene deposits at Massignano, Italy. Meteorit Planet Sci 39:589–597

Glass BP, Huber H, Koeberl C (2004b) Geochemistry of Cenozoic microtektites and clinopyroxene-bearing spherules. Geochim Cosmochim Acta 68:3971–4006

Haines PW, Howard KT, Ali JR, Burrett CF, Bunopas S (2004) Flood deposits penecontemporaneous with ~0.8 Ma tektite fall in NE Thailand: impact-induced environmental effects? Earth Planet Sci Lett 225:19–28

Koeberl C (1986) Geochemistry of tektites and impact glasses. Ann Rev Earth Planet Sci 14:323–350

Koeberl C (1993) Extraterrestrial component associated with Australasian microtektites in a core from ODP Site 758B. Earth Planet Sci Lett 119:453–458

Koeberl C (1994) Tektite origin by hypervelocity asteroidal or cometary impact: target rocks, source craters, and mechanisms. In: Dressler BO, Grieve RAF, Sharpton VL (eds) Large meteorite impacts and planetary evolution. Spec Pap Geol Soc Am 293:133–151

Koeberl C, Brandstaetter F, Niedermayr G, Kurat G (1988) Moldavites from Austria. Meteoritics 23:325–332

Lisiecki LE, Raymo ME (2005) A Pliocene-Pleistocene stack of 57 globally distributed benthic $\delta^{18}O$ records. Paleoceanography 20, PA1003. https://doi.org/10.1029/2004PA001071

Ma P, Aggrey K, Tonzola C, Schnabel C, De Nicola P, Herzog GF, Wasson JT, Glass BP, Brown L, Tera F, Middleton R, Klein J (2004) Beryllium-10 in Australasian tektites: constraints on the location of the source cra.ter. Geochim Cosmochim Acta 68:3883–3896

Mark DF, Renne PR, Dymock R, Smith VC, Simon JI, Morgan LE, Staff RA, Ellis BS (2017) High-precision $^{40}Ar/^{39}Ar$ dating of pleistocene tuffs and temporal anchoring of the Matuyama-Brunhes boundary. Quater Geochronol 39:1–23. https://doi.org/10.1016/j.quageo.2017.01.002

Montanari A, Koeberl C (2000) Impact stratigraphy: the Italian record. Springer, Heidelberg, p 364

O'Keefe JA (1967) Tektite sculpturing. Geochimica et Cosmochimica Acta 31(10):1931–1933

O'Keefe JA (1976) Tektites and their origin. Elsevier, New York, p 254

O'Keefe JA (1987) Zhamanshin and Aouelloul: craters produced by impact of tektite-like glasses? Meteorites 22:219–228
Owen EA, Burns BD (1939) X-ray study of some meteoric irons. Philos Mag 28:497–512
Prasad MS, Rudraswami NG, de Araujo AA, Khedekar VD (2017) Unmelted cosmic metal particles in the Indian Ocean. Meteorit Planet Sci 52(6):1060–1081. https://doi.org/10.1111/maps.12858
Prasad MS, Sudhakar M (1998) Microimpact phenomena on Australasian microtektites: implications for ejecta plume characteristics and lunar surface processes. Meteorit Planet Sci 33:1271–1279
Prasad MS, Sudhakar M (1999) Australasian minitektites discovered in the Indian Ocean. Meteorit Planet Sci 34:179–184
Prasad MS, Mahale VP, Kodagali VN (2007) New sites of Australasian microtektites in the central Indian Ocean: implications for the location and size of source crater. J Geophys Res 112:E06007. https://doi.org/10.1029/2006JE002857
Rudraswami NG, Prasad MS (2016) Extraterrestrial components from deep sea sediments of Indian Ocean. Proc Indian Natn Sci Acad 82(3):653–661. https://doi.org/10.16943/ptinsa/2016/48474
Schmidt G, Zhou L, Wasson JT (1993) Iridium anomaly associated with the Australasian tektite-producing impact: masses of the impactor and of the Australasian tektites. Geochim Cosmochim Acta 57:4851–4859
Schneider DA, Kent DV, Mellow GA (1992) A detailed chronology of the Australasian impact event, the BrunesMatuyama geomagnetic polarity reversal, and global climate change. Earth Planet Sci Lett 111:395–405
Simonson BM, Glass BP (2004) Spherule layers—records of ancient impacts. Annu Rev Earth Planet Sci 32:329–361
Smit J, van Eijden AJM, Troelstra SR (1991) Analysis of the Australasian microtektite event, the Toba Lake event, and the Cretaceous/Paleogene boundary, Eastern Indian Ocean. Proc ODP Sci Results 121:489–503
Suess FE (1900) Die Herkunft der Moldavite. Jahrb Reichanst Vienna 50:193–382
Taylor SR, McLennan SM (1979) Chemical relationships among irghizites, zhamanshinites, Australasian tektites and Henbury impact glasses. Geochim Cosmochim Acta 43:1551–1565
Taylor SR, Sachs M (1964) Geochemical evidence for the origin of australites. Geochim Cosmochim Acta 28:235–264
Taylor SR (1962) Fusion of soil during meteorite impact and the chemical composition of tektites. Nature 195:32–33
Taylor SR, Kaye M (1969) Genetic significance of the chemical composition of tektites: a review. Geochim Cosmochim Acta 33:1083–1100
Trieloff M, Bollinger K, Kunz J, Jessberger EK (2007) ^{40}Ar-^{39}Ar ages of Australasian tektites [abs]. Meteorit Planet Sci 42(Suppl):A150
Valet J-P, Bassinot F, Bouilloux A, Bourlès D, Nomade S, Guillou V, Lopes F, Thouveny N, Dewilde F (2014) Geomagnetic, cosmogenic and climatic changes across the last geomagnetic reversal from Equatorial Indian Ocean sediments. Earth Planet Sci Lett 397:67–79. https://doi.org/10.1016/j.epsl.2014.03.053
Van der Veen RW (1923) Origin of tektite sculpture and some consequences. Geol Mijnbouwkd. Genootsch Ned Kolonien. Geol Ser 7:15–42
Wasson JT (1991) Layered tektites: a multiple impact origin for the Australasian tektites. Earth Planet Sci Lett 102:95–109
Yamei H, Potts R, Baoylin Y, Zhengtang G, Deino A, Wei W, Clark J, Guangmao X, Weiwan H (2000) Mid-Pleistocene Acheulean-like stone technology of the Bose Basin, South China. Science 287:1622–1626
Zähringer J (1963) K-Ar measurements of tektites. In: Radioactive dating. Proceedings IAEA symposium, international atomic energy agency Athens, Nov 19–23, 1962, Vienna, pp 289–305

Assessing Mid-pleistocene to Holocene Sea-Ice Extent and Carbonate Compensation Depth Fluctuations in the Japan Sea: A Multiproxy Approach

Manisha Das, Nishant Vats, Raj K. Singh, Sibasish Mishra, S. Sova Barik, R. V. Divya, S. Sengupta, A. Ranjan and D. K. Pandey

Abstract The extent and magnitude of sea-ice varied extensively since middle Pleistocene in northern Japan Sea affecting the climate, ecologic conditions and carbonate compensation depth (CCD) of the region. Sediment core samples of Integrated Ocean Drilling Program site U1423 from north-eastern Japan Sea, have been quantitatively analyzed for angular and faceted surface Ice Rafted Debris (IRD), foraminifera and detrital fragments to reconstruct permanent sea-ice history, paleoecological conditions and CCD fluctuations in the Japan Sea. The enhanced cooling at the beginning of middle Pleistocene caused the southward movement of sea-ice melting zone suggested by higher abundance of angular and faceted surface IRD between 880 and 450 ka. The site was above the CCD and melting of ice significantly cooled the surface water resulting in abundance of the low salinity and extremely low temperature tolerant species, *Neogloboquadrina pachyderma* (sinistral). The next phase of extreme cooling in the Japan Sea started after 450 ka that shoaled the CCD; caused complete absence or decrease in calcareous planktic foraminifera species till 150 ka. The abundance of detrital material during this period suggests the presence of seasonal sea-ice linked with East Asian Winter Monsoon wind strength and Tsushima Warm Current inflow into the Japan Sea. A significant decrease in the permanent and seasonal ice sheet and deepening of CCD was observed during the last 150 ka. The IRD and foraminifera respond to 100 kyr variability, attributed to ice sheet dynamics and global ice volume changes.

Keywords Ice Rafted Debris · Carbonate compensation depth · Foraminifera · Tsushima Warm Current · East Asian Winter Monsoon

M. Das · N. Vats · R. K. Singh (✉) · S. Mishra · S. S. Barik · R. V. Divya · S. Sengupta · A. Ranjan
School of Earth, Ocean and Climate Sciences,
Indian Institute of Technology Bhubaneswar, Argul, Jatni 752050, India
e-mail: rksingh@iitbbs.ac.in

D. K. Pandey
ESSO-National Centre for Polar and Ocean Research, Goa 403804, India

D. K. Pandey et al. (eds.), *Dynamics of the Earth System: Evolution, Processes and Interactions*, Society of Earth Scientists Series,
https://doi.org/10.1007/978-3-030-40659-2_3

1 Introduction

The expansion of sea-ice in the Japan Sea is not only controlled by global climate, but also affected by regional climate (Martinson et al. 1987; Sakamoto et al. 2005). The extent and magnitude variation of sea-ice is related to glacial-interglacial sea-level fluctuation and associated climatic phenomenon like intensification of East Asian Winter Monsoon (EAWM). The cold, strong EAWM winds lead to sea-ice formation in marginal Okhotsk Sea, which readily occurs along its north-western margin (Talley et al. 2003). The oceanographic and metrological factors also influence the formation, shift, and retreat of sea-ice. The advection of sea-ice can be controlled by wind and the maximum sea-ice coverage variation is affected mainly by wind and air temperature conditions over the sea (Kimura and Wakatsuchi 1999). The relationship between sea-ice and climate change in longer time scales (orbital cycles) provides an opportunity to understand the sea-ice formation and its expansion in the higher latitude (e.g., Takahashi 1998). Ice Rafted Debris (IRD) is one such useful proxy that help to reconstruct the sea-ice history from marine sediments (Sakamoto et al. 2005). IRD consists of the various material deposited on ice that may eventually be embedded in the ice. These embedded materials drift up to certain distance into the sea along with large ice sheets, get deposited on the sea floor after the ice melts, and are preserved in the geological sequence (Bischof 2000). Ice rafting was the prime mechanism of sediment transport during the glacial cycles of the Pleistocene when sea level was low and much of the areas, especially at higher latitudes, were covered with ice (Jansen et al. 2000). The IRD is more rounded and discoid in the region of seasonal sea-ice and is much more angular with faceted surfaces if IRD is delivered by more extensive ice sheets that extended past the paleo-shoreline via calving and icebergs (Goldschmidt et al. 1992; Dadd and Foley 2016). Thus, IRD can be useful proxies to reconstruct the sea-ice history from marine sediments; the angular and faceted surface IRD indicate more extensive ice sheet formation and drifting of large ice sheets as iceberg into the Japan Sea (Sakamoto et al. 2005; Dadd and Foley 2016). Further, during the glacial intervals, lowering of sea-level cause shoaling of Carbonate Compensation Depth (CCD), which varied from <1000 to >3000 m during the early middle Pleistocene to Holocene, and causes periodic dissolution of calcareous foraminifera, whenever CCD shoaled in the Japan Sea (Oba et al. 1991; Sagawa et al. 2018; Das et al. 2018). Hence, this study attempts to reconstruct middle Pleistocene to Holocene sea-ice history of northern Japan Sea and its effect on CCD by assessing angular and faceted surface IRD and foraminifera of IODP site U1423 of Japan Sea.

2 Location, Lithology, and Hydrography of Japan Sea

Integrated Ocean Drilling Program (IODP) Site U1423 is situated on north-eastern Japan Sea at 41° 41.95′ N, 139° 4.98′ E and 1785 m below sea level (Fig. 1). The site is ~100 km north-west of the entrance of the Tsugaru Strait and close to the Japanese

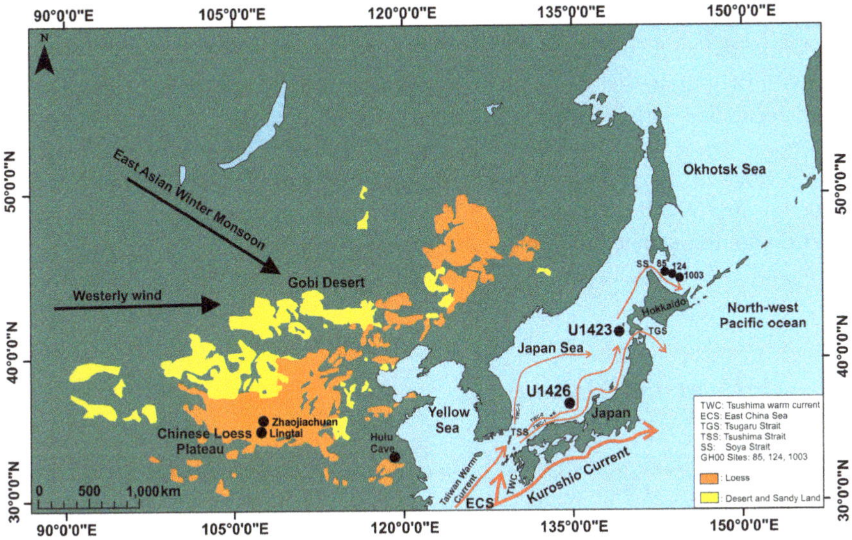

Fig. 1 Physiographic map showing locations of IODP Sites U1423, U1426 (Tada et al. 2015) in Japan Sea, GH00 Sites: 85, 124, 1003 (Sakamoto et al. 2005) in the Okhotsk sea, and Zhaojiachuan, Lingtai sections on the south central Chinese Loess Plateau (Clemens et al. 2008). Red arrows indicate surface current directions and black arrow indicate the East Asian Winter Monsoon and Westerly wind direction

Archipelago, the Korean Peninsula, Eurasian continent and a small volcanic island is present 30 km south-east of the studied site (Tada et al. 2015; Igarashi et al. 2018). A branch of Tsushima Warm Current (TWC), that flows further north beyond the Tsugaru strait and Soya strait through the Tsushima strait is the only warm current affecting this site (Fig. 1). This surface warm current mixes with the nutrient-rich coastal water and supplies nutrients, heat and organic matter to the Japan Sea during the interglacial periods (Sagawa et al. 2018). The deep water below 200–300 m water depths in the Japan Sea is called the Japan Sea Proper Water (JSPW), which is characterized by high dissolved oxygen concentrations >210 μmol/kg (Gamo 1999; Das et al. 2018). The JSPW is well ventilated, cold oxygen-rich water that promotes the hyper-oxic deep water condition and shallow CCD (Gamo et al. 2014) and its reduction causes anoxic bottom water and deepening of CCD in the Japan Sea (Tada et al. 2015; Sagawa et al. 2018). The bottom water oxygenation and CCD fluctuation were observed over the past 775 kyr at various glacial-interglacial periods and affected the sediment composition of Japan Sea (Oba et al. 1991; Sagawa et al. 2018; Tada et al. 2018; Das et al. 2018). The collected sediment core is divided into two major lithologic units, distinguished by sediment composition (Tada et al. 2015). The Unit I (0–2.2 Ma) is mostly characterized by fine grain terrigenous materials and is further subdivided into Subunit IA (0–1.8 Ma) and Subunit IB (1.8–2.2 Ma). The studied samples belong to Subunit IA, and are composed of terrigenous, volcanic and biogenic sediments. The terrigenous sediments are dominated by angular and faceted surface IRD and various other detrital sediments, while biogenic sediments are dominated by diatom and have calcareous foraminifera during the selected time interval (Tada et al. 2015). The terrigenous fine-grained sediments were supplied to the Japan Sea during the period of intense EAWM via the atmospheric pathway, icebergs or come from the East China Sea (ECS) along with TWC. The sediment supply from ECS was limited during the glacial period (Um et al. 2017) due to reduction in TWC flow (Das et al. 2018). The main lithology at this site is interbedded with frequent tephra (volcanic ash) layers originating from the Japanese Arc (Irino and Tada 2000; Tada et al. 2015) and constitutes volcanic sediments.

2.1 *Material and Methods*

We analyzed 130 samples (10 cc) from revised sediment splice between Holes U1423A and U1423B at ~1 m interval from the top of the sediments succession down to 28 m core composite depth below sea floor (CCSF) and at ~0.5 m interval below that down to 72 m CCSF. The dry weight of the sample was noted and soaked overnight in water to which a few drops of Hydrogen Peroxide (2%) was added for better disintegration of clay matrix. The soaked samples were washed over a 63 μm size sieve using a hand shower and dried in an oven at ~45 °C temperature. The dried samples were then transferred into glass vials and labeled properly with sample numbers. Adequate measures were taken to avoid contamination. Sieves were washed in the ultrasonicator for 15 min and methylene blue was applied on the sieve net after washing each sample to identify contaminated samples. Each sample was dry-sieved over 125 μm size sieve and analyzed using stereozoom microscope. A quantitative aliquot, obtained with Otto splitter, containing >300 specimens of angular and faceted surface IRD (henceforth, termed as IRD), Detrital fragments (include sub-angular to sub-rounded IRD from seasonal sea-ice and other terrigenous material), Foraminifera (planktic and benthic) and other microfossils (includes large size Diatom, Radiolaria, Spores, Pollens and glass materials) were manually counted and converted to final numbers for the whole sample. The IRD particles identified and counted under the stereozoom microscope have typical sharp angular and faceted surfaces as described by Watkins et al. (1974) and Ledbetter and Watkins (1978) (Fig. 2). Among the foraminifera (Fig. 2), individual Planktic foraminifera species (Fig. 3) were identified using taxonomic reference (Kennett and Srinivasan 1983; Hemleben et al. 1989) and benthic foraminifera were identified to the species level by following published taxonomic descriptions (van der Zwaan et al. 1986; Scott et al. 2000; Holbourn et al. 2013). Detrital grains are characterized by their sphericity and roundness, pertaining to the degree of the transportation from the source areas; thus sub-angular to sub-rounded terrestrial grains are identified as detrital, which may be deposited by seasonal sea-ice and/or by wind and/or river. Most of the diatom and radiolarian species are very small in size and are studied using smear slides. However, some of the bigger diatom and radiolarian species are present along with

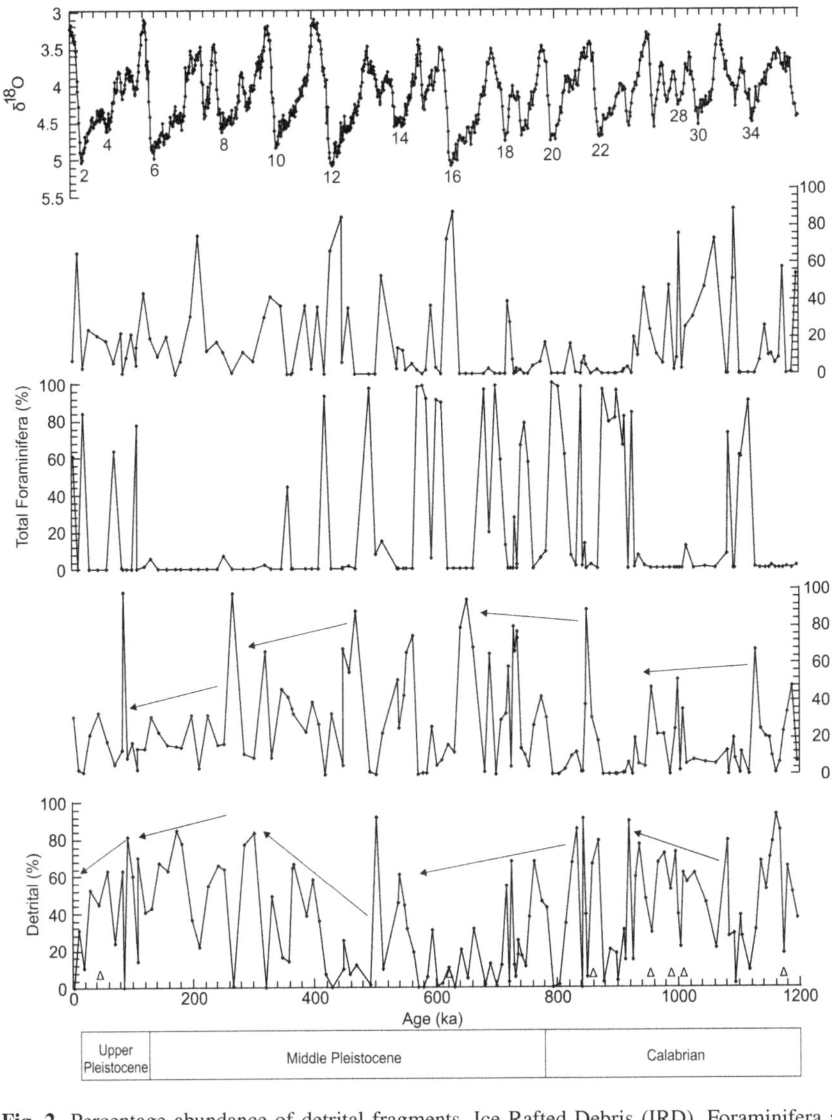

Fig. 2 Percentage abundance of detrital fragments, Ice Rafted Debris (IRD), Foraminifera and others (includes large size diatom and Radiolaria species, Spores, Pollens, and glass materials) at site U1423 plotted against interpolated numerical ages. The global benthic stack $\delta^{18}O$ (Lisiecki and Raymo 2005) plotted against the numerical ages and number within the plots are marine isotope stage (MIS). The triangle sign (Δ) shows the prominent tephra events encountered at this site

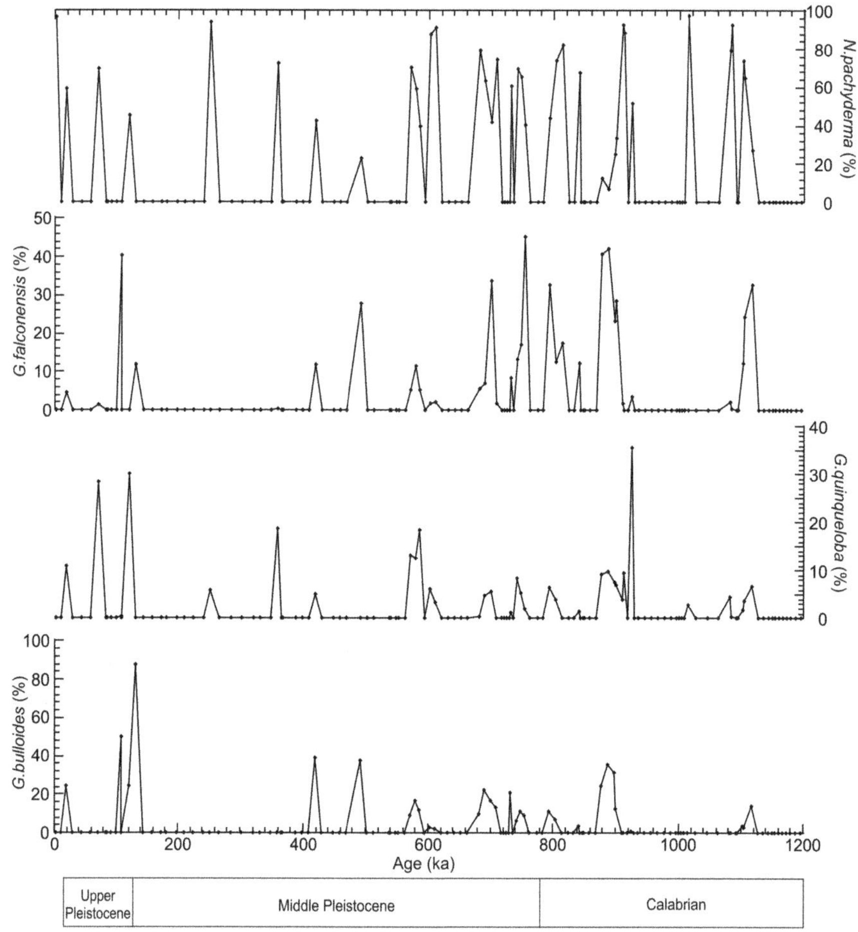

Fig. 3 Percentage abundance of dominant planktic foraminifera species *N. pachyderma* (Sinistral), *G. falconenesis*, *G. quinqueloba*, and *G. bulloides* within the total planktic foraminifera abundance at site U1423 plotted against interpolated numerical ages

glasses, spores, and pollens in an aliquot of >125 μm fraction and all these clubbed as others, as we are not focussing our result and interpretation on these data.

The ages are linearly interpolated for each sample using the initial age model of Tada et al. (2015) and ages of tephra layers given in Irino et al. (2018). The basal age of the succession studied is 1195 ka based on interpolated age model and average time resolution per sample is ~10 kyr.

Spectral analysis was carried out for IRD data to know the periodicity of extensive ice sheet formation and for planktic foraminifera data to assess the periodicity in CCD fluctuation over past 1195 ka using PAST 3.22 software (Hammer et al. 2001) and red noise was calculated using REDFIT (Figs. 4a and 5a; Schulz and Mudelsee 2002).

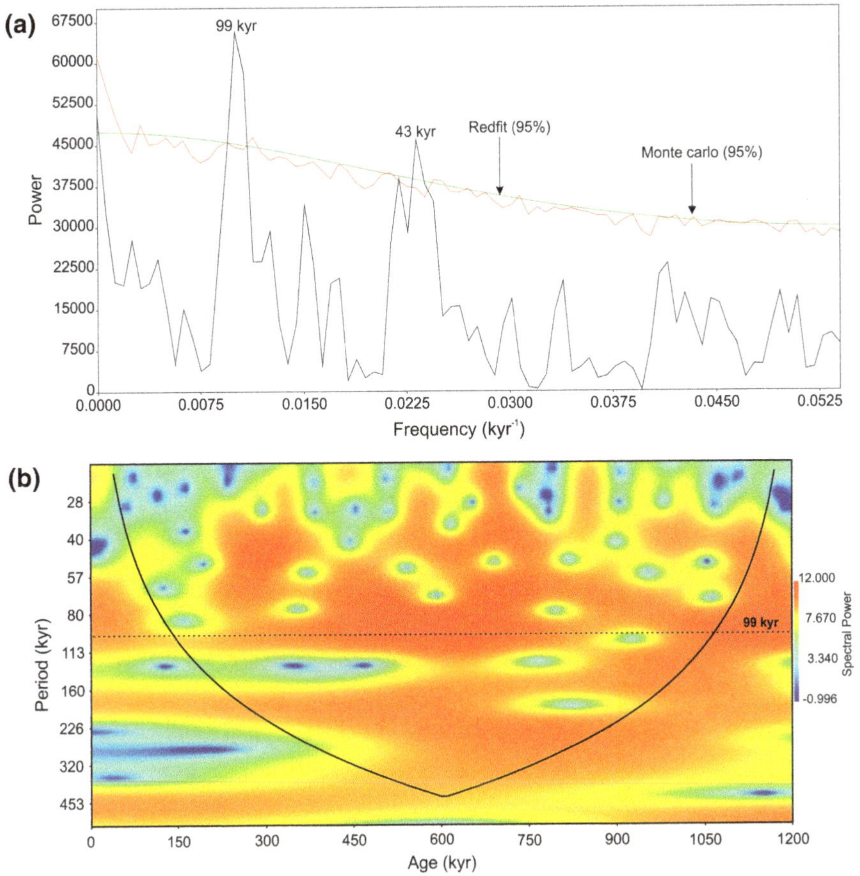

Fig. 4 **a** Spectral analysis of IRD for the period 0–1195 ka showing most significant periodicities centered at 99 and 43 kyr. **b** Continuous wavelet transform (CWT) analysis of IRD for the period 0–1195 ka on linearly interpolated data showing periodicity centered at 99 kyr. Spectral power is indicated by the color bar on the right

For the spectral analysis, two segments with two frequency overlapping were selected to obtain the spectra. The window parameter was chosen to 'Rectangle' as the variance is different. The Monte Carlo simulation option allows the spectrum to be bias-corrected (Hammer et al. 2001; Schulz and Mudelsee 2002). The IRD and planktic foraminifera data were made evenly distributed at 10 kyr using linear interpolation technique and were used for continuous wavelet transform (CWT) analysis using "Morlet" window parameter in the PAST 3.22 software (Figs. 4b and 5b; Hammer et al. 2001).

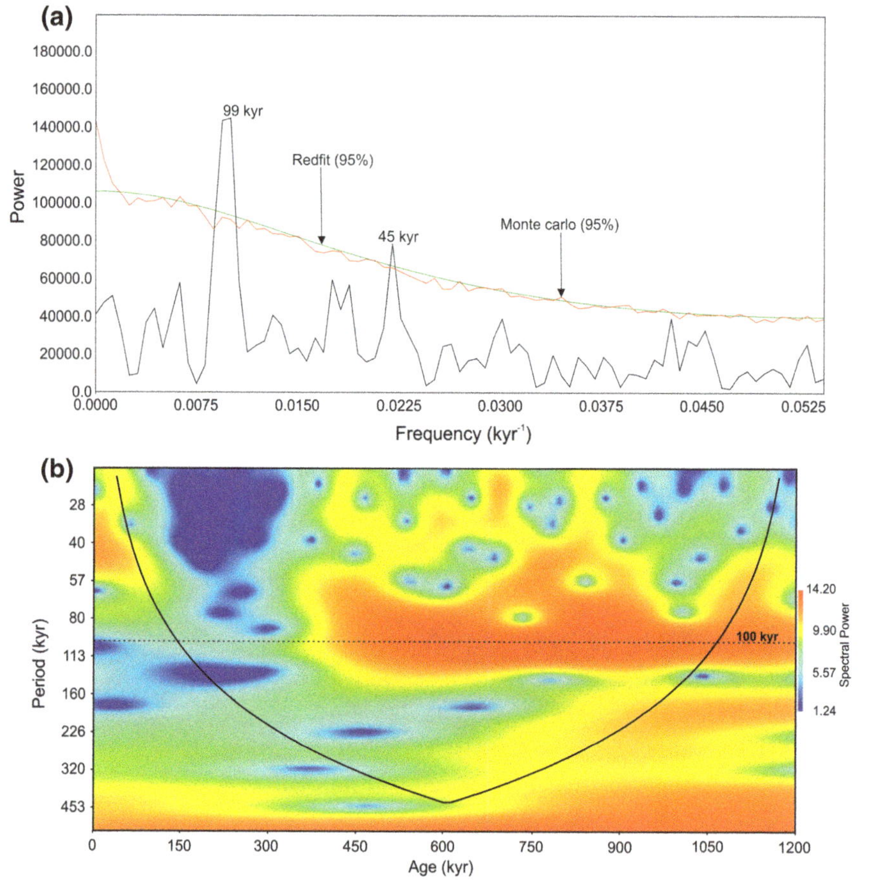

Fig. 5 **a** Spectral analysis of planktic foraminifera for the period 0–1195 ka showing most significant periodicities centered at 99 and 45 kyr. **b** Continuous wavelet transform (CWT) analysis of planktic foraminifera for the period 0–1195 ka on linearly interpolated data showing periodicity centered at 100 kyr. Spectral power is indicated by the color bar on the right

3 Results

3.1 Distribution of IRD and Detrital

The origin of IRD can be constrained from mineral assemblages of the surface sediments. The IRD particles characterized by their angular size and faceted surfaces are released just after the extensive sea-ice melt and deposited in the sea floor. Significant variations are observed in the IRD abundance over past 1195 ka. The IRD abundance is low between 1195 and 880 ka, increase significantly between 880–450 ka and decreased thereafter, except for two high abundant peaks observed at 300 and 71 ka (Fig. 2). The high IRD abundance is observed near the glacial-interglacial boundary, except few exceptions and in the samples nearer to the tephra layers (Figs. 2 and 6; Das et al. 2018; Irino et al. 2018). The spectral analysis shows peaks above 95% confidence level at 99, and 43 kyr for IRD (Fig. 4a) and CWT analysis for IRD shows 99 kyr time embedded periodic component and time varying amplitude (Fig. 4b).

The detrital fragments are a combination of sub-angular to sub-rounded grains carried by EAWM winds, rivers and streams from nearby landmass and islands and/or melting of seasonal sea-ice (Fig. 2). These detrital fragments have an inverse relationship with IRD at all the Marine Isotope Stage (MIS) boundaries (Fig. 2; Lisiecki and Raymo 2005). The detrital fragments are high between 1195 and 840 ka, low between 840 and 450 ka and fluctuating trend thereafter up to 100 ka (Fig. 2).

3.2 Distribution of Foraminifera

Out of 130 analyzed samples, foraminifera are present only in 78 samples and constitute >1% of sediment fraction only in 58 samples. The foraminifera occurrences are most consistent between 900 and 450 ka, and only in a few samples between 1195 and 900 ka (Fig. 2). Seven samples from the tephra layers are entirely devoid of any foraminifera or contain very few specimens of foraminifera. We have identified a total of 46 species of foraminifera, of which 22 species are planktic and remaining 24 species are benthic. Planktic foraminifera have a much higher population abundance in comparison to benthic foraminifera. Out of 22 species of planktic foraminifera, only 4 species (viz., *Globigerina bulloides*, *Neogloboquadrina pachyderma* (sinistral), *Globigerina falconensis* and *Globigerina quinqueloba*) are present in at least two or more samples with 10% or more abundance (Fig. 3), while only two agglutinated benthic foraminifera species (*Rhabdamina* sp. and *Milliammina echigoensis*) are present in at least one or more samples with 10% or more abundance. Hence we have considered the total planktic foraminifera population abundance and the abundance of the 4 dominant planktic species for CCD and paleo-ecological reconstruction over north-eastern Japan Sea. The ecological conditions and abundance of these dominant planktic foraminifers are mainly influenced by glacial-interglacial sea level changes and fluctuation in CCD depth. The spectral analysis of planktic foraminifera shows peaks above 95% confidence level at 99 and 45 kyr (Fig. 5a) and CWT analysis shows 100 kyr time embedded periodic component up to 450 ka and their time-varying amplitude as limited foraminifera found between 450 and 150 ka.

3.3 Environmental Preferences of Dominant Planktic Species

The left-coiling (sinistral) variant of *Neogloboquadrina pachyderma* is dominant in extremely cold water masses, and it prefers low temperature and low surface salinity (Darling and Wade 2008). *N. pachyderma* (S) is a polar species that occur in both

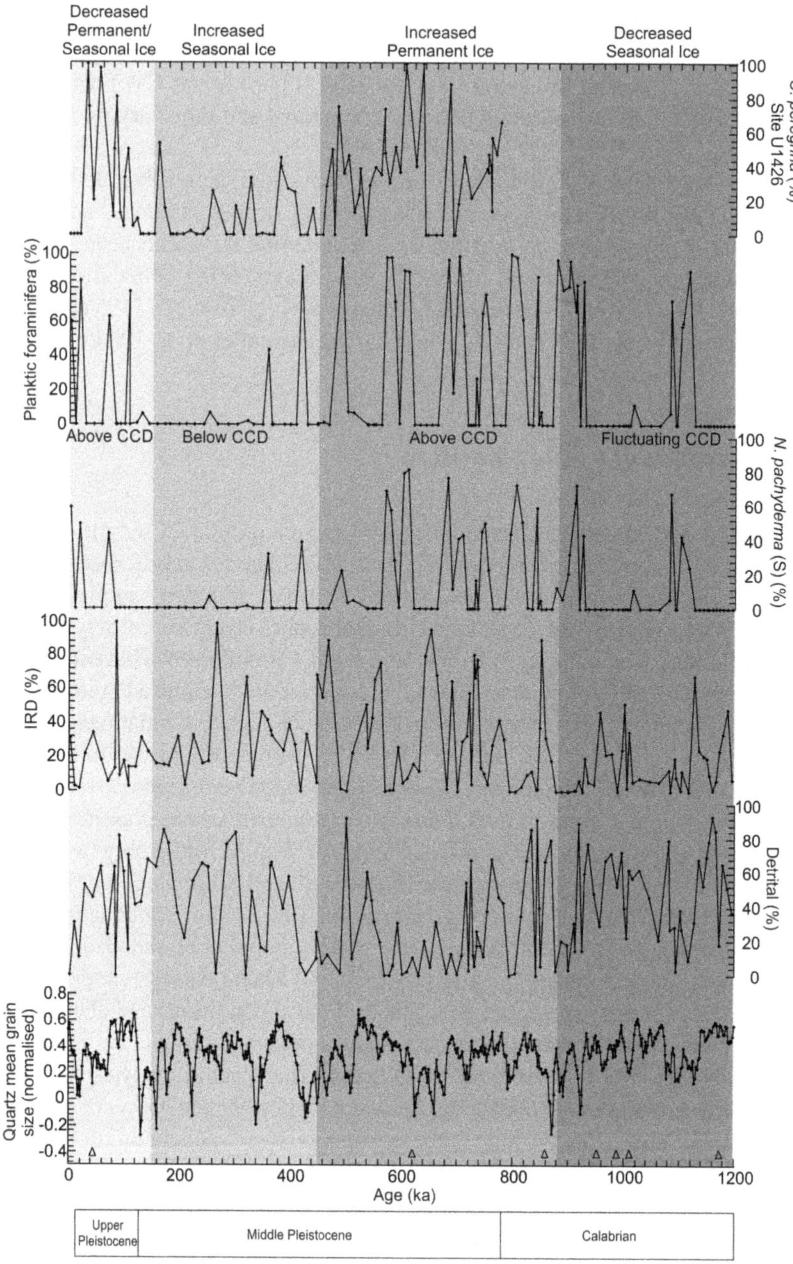

Fig. 6 Normalized quartz mean grain size (Clemens et al. 2008) of Zhaojiachuan, Lingtai sections on the south central Chinese Loess Plateau; detrital fragments (%), IRD (%), *N. pachyderma* and Planktic foraminifera percent abundance in whole samples at site U1423 and *U. peregrina* percent abundance at site U1426 are plotted against interpolated numerical ages. The triangle sign (Δ) shows the tephra events encountered at this site. The different grey bar indicates the CCD level and status of seasonal and permanent ice

northern and southern hemisphere (Be 1977). The standing stocks for *N. pachyderma* (S) are highest than any modern species from melted sea-ice samples (~190 individuals/liter; Spindler and Dieckmann 1986). *N. pachyderma* (S) is widely distributed and most dominant species of the studied site between 1195 and 0 ka, (Fig. 3) when the studied site was above the CCD. The abundance of this species indicates that the region was influenced by cold polar water and/or melting of extensive sea-ice.

G. bulloides is associated with temperate to sub-polar water. It is a non-symbiotic species that dwell within the upper 150 m of water column (Schulz et al. 2002; Iwasaki et al. 2017). *G. bulloides* was dominant when water was relatively warm at the beginning of the interglacial time period. This species is a well-known representative of upwelling in middle and low latitudes (Prell and Curry 1981; Gupta et al. 2003; Iwasaki et al. 2017) and is also present in coastal waters of East China Sea (Xu and Oda 1999). At this site, *G. bulloides* has sporadic occurrences at 1120–450, 420, 150–110 and 30–20 ka, reflecting relatively warm water conditions at this site that can be attributed to the beginning of an interglacial event. The higher abundance of this species in the studied holes indicates the transition to the interglacial period (Fig. 3).

G. falconensis is morphologically similar to *G. bulloides* and generally follow the abundance pattern of *G. bulloides* (Schulz et al. 2002; Munz et al. 2015). This species has higher abundance at 1150–1100, 910–450, and 131–110 ka (Fig. 3) and suggesting environmental preferences similar to that of *G. bulloides*.

G. quinqueloba occurs in subtropical to subarctic waters of Pacific Ocean (André et al. 2014; Darling and Wade 2008; Iwasaki et al. 2017) and dwells over a wide temperature range (2–20 °C) in the western North Pacific. The higher abundance of this species is related to low salinity and low temperature of surface water of ECS (Xu and Oda 1999). The presence of *G. quinqueloba* at 920–870, 580–570, 450 and 120–70 ka (Fig. 3) suggest higher inflow of TWC.

4 Discussion

The East Asian Winter Monsoon is characterized by cold, dry wind, which cools the surface water over Japan Sea bordering along the far eastern coast of Russia. This cold, dry north westerly wind blows from the Siberian high towards the low lying Japan Sea (Ikehara and Itaki 2007; Tada et al. 2015). The warm surface water of Japan Sea provides humidity and moisture to this cold, dry winter wind, which nourishes almost all glaciers and causes heavy snowfall in the Japan Sea side of the islands (ONO 1984; Ikehara and Itaki 2007). In the geological past, during the ice ages, the IRD were locked within the glaciers and were transported as sea-ice/icebergs to the sampling location, where they melted and IRD were deposited at the sea floor in the beginning of interglacial or during the inter-stadial periods. There are no direct evidences of sea-ice in the Japan Sea in last few decades, but its influence in ocean circulation and material transport has been noticed over past 150 ka (Ikehara 2003; Ikehara and Itaki 2007) affecting the paleoclimate and oceanographic circulation in the northern Japan Sea and may have continued through the middle Pleistocene. The occurrence of sea-ice at north-eastern Japan Sea is evident from greater than sand-sized IRD's that were transported along with icebergs (Ikehara 2003; Ikehara and Itaki 2007).

The glaciations started during late Calabrian at 1200 kyr, but significant increase in global ice volume was observed at 940–890 ka (Head and Gibbard 2005), which increased the sea-ice/icebergs at higher latitudes and caused, a significant increase in the angular IRD at north-eastern Japan Sea (Figs. 2 and 6). Recently, the International Commission of Stratigraphy moved the base of middle Pleistocene to 781 from 1200 ka (ICS 2018) on the basis occurrences of permanent ice sheets in the various parts of the world. The higher abundance of IRD between 880 and 450 ka reflects, that the melting zone of sea-ice might have moved southward in response to enhanced global cooling in two phases, at 940–890 ka and 725–650 ka (Ikehara 2003; Head and Gibbard 2005). The dominance of cold air mass over the central and northern Japan causes the southern shift of the polar front which varies with the reduction of summer temperature (ONO 1984). Most sea-ice originates from the northern part of the Okhotsk Sea and southerly wind transports sea-ice southwards (Sakamoto et al. 2005), and during the period of enhanced glaciations of middle Pleistocene, the sea-ice was carried further south into the northern Japan Sea, the melting of which caused colder SST over Japan Sea. The colder SST over Japan Sea was not able to provide sufficient moisture to the cold, dry EAWM wind, which caused the reduction in seasonal ice, suggested by decrease detrital fragments input into the north-eastern Japan Sea (Figs. 2 and 6). The increase in source water (Kuroshio water) temperature, changes in the volume of transport of the TWC and/or the north-south oscillation of the sub-polar front position within the Japan Sea, leads to an increase in SST variation (Fujine et al. 2009). The increased flow of TWC melted the sea-ice/iceberg and caused deposition of IRD and spread of extremely cold water. The study site was above the CCD for most of the time between 900 and 450 ka suggested by the presence of planktic foraminifera (Fig. 6). The melting of sea-ice cooled the surface water and significantly reduced the salinity and SST of surface water and caused the dominance of *N. pachyderma* (S) (Figs. 3 and 6). It was the period of enhanced organic flux to the sea floor with variations in oxygenation due to intensification of the TWC flow during the interglacial high stands (Kitamura and Kimoto 2007; Das et al. 2018). The presence of other planktic foraminifera like *G. quinqueloba*, *G. bulloides*, and *G. falconensis*, (Fig. 3) also suggest strong TWC inflow into the Japan Sea during the interglacials (Gallagher et al. 2015; Das et al. 2018) and local temporal variations in the paleoecological condition over the north-eastern Japan Sea. Our data from the studied site is not sufficient to capture these local temporal variations in the paleoecological condition over north-eastern Japan Sea. The abundance of the detrital fragment is observed during the interglacial periods, which is associated to increased sea-level (Sagawa et al. 2018) and is concordant with winter monsoon loess data (Fig. 6; Clemens et al. 2008).

Prior to a significant increase in global ice volume at 900 ka (Head and Gibbard 2005), the large ice sheets carrying angular IRD had not reached up to the south of Okhotsk Sea, and had limited influence over the northern Japan Sea as suggested by low IRD content (Figs. 2 and 6). The stronger EAWM wind and relatively warmer SST over Japan Sea between 1200 and 900 ka supplied enough moisture to nourish all glaciers, cause heavy snowfall in the Japan Sea side of the islands and its melting causes deposition of higher number of sub-angular to sub-rounded detrital fragments in the Japan Sea (Figs. 2 and 6).

The onset of large-amplitude glacial-interglacial cycles at 490–470 ka can be linked to more extreme glaciations in the Japan Sea (e.g., Clark et al. 2006; Das et al. 2018) characterized by prominent dark and light cycles that reflect climatic and/or oceanographic perturbations and can be correlated across the entire Japan basin (Tada et al. 2015, 2018; Das et al. 2018). The transition period to more extreme glacial conditions was marked by the presence of all the dominant planktic foraminifera species and non-dominance of any single species (Fig. 3), which suggests, mixed conditions at the northern Japan Sea. After the transition phase, the detrital particles shows periodic fluctuation correlated with the EAWM wind strength marked by variability in quartz mean grain size of Chinese loess deposits (Fig. 6; Clemens et al. 2008). The extreme glaciations of late middle Pleistocene began at 450 ka induced shoaling of the CCD, leading to complete absence or decrease in calcareous foraminifera species at this site (Figs. 2 and 6). The shoaling of CCD is also observed in various other shallow sites over the Japan Sea (Oba et al. 1991; Sagawa et al. 2018; Das et al. 2018). When the CCD deepened during strong interglacials, *N. pachyderma* (sinistral) (Figs. 3 and 6) was found, which indicates the presence of relatively cooler water along with *G. quinqueloba* associated with the higher inflow of TWC. *G. bulloides* and *G. falconensis* have minimal presence between 450 and 150 ka. The significant decrease in IRD is also observed (Figs. 2 and 6) during the period of extreme glaciations except at 300 and 71 ka, suggesting a substantial reduction of the permanent ice sheet presence in the northern Japan Sea started since 450 ka. This period was dominated by seasonal sea-ice, whose formation was controlled by the strength of EAWM and TWC. The variability in the detrital fragments is well correlated with the normalized quartz mean grain size of south central Chinese loess Plateau, which varies with EAWM wind strength (Fig. 6; Clemens et al. 2008) and abundance of *Uvigerina peregrina* at IODP site U1426, which suggest enhanced inflow of TWC during the interglacials (Fig. 6; Das et al. 2018). During the last 150 ka, the CCD deepened and the study site is positioned above the CCD; there are significant decrease in IRD and detrital fragments at north-eastern Japan Sea and increase in calcareous foraminifera is observed, which suggest that the Japan Sea is retaining its modern ecological conditions since 150 ka (Ikehara 2003; Ikehara and Itaki 2007). All the dominant planktic foraminifera show sporadic occurrences at the site since late Pleistocene, suggesting periodic glacial-interglacial fluctuation in the CCD.

The early to middle Pleistocene transition represents major episode in earth history during which low amplitude 41 kyr obliquity forced cycles of the earlier Pleistocene replaced by high amplitude 100 kyr cycles (Head and Gibbard 2005; Clark et al. 2006; Das et al. 2018), but persistent 41 kyr cyclicity continued in its initial phase (Head and Gibbard 2005) and 100 kyr cyclicity dominated in the subsequent phase. This second phase of enhanced glaciations in the Japan Sea began at 450 ka (Das et al. 2018).

The paleoceanographic and palaeoclimatic conditions of the Japan Sea have been affected by global sea-level fluctuations during the Quaternary, linked with orbital to sub-orbital changes in the earth positions and it responded in terms of glacio-eustatic sea level changes (Oba et al. 1991; Irino and Tada 2000; Kido et al. 2007; Das et al. 2018). The IRD abundance shows response towards the orbital cycle of paleoclimatic changes, which affected the ice volume in the Japan Sea. The calcareous foraminifera constitutes most of the Calcium Carbonate ($CaCO_3$) content, preservation of which indicates the CCD level in the deep-sea water (Khim et al. 2012). The spectral analyses of IRD and planktic foraminifera shows cyclicity of ~100 kyr in response to changes in the dominant glacial-interglacial cyclicity from 41 to 100 kyr (Figs. 4a, b and 5a, b) attributed to ice sheet dynamics and internal oscillation associated with tropical process, which controlled the TWC inflow into the Japan Sea or carbon cycle affecting sea-ice production (Kitamura and Kimoto 2007; Ford et al. 2016; Li et al. 2017; Nie 2017; Das et al. 2018). This 100 kyr cyclicity in foraminifera is limited to1200–450 ka, due to shoaling of CCD after 450 ka, which was unconducive for the preservation of calcareous foraminifera (Fig. 5b). Both these proxies also responds to precession cycle suggesting dominance of 41 kyr cyclicity after the beginning of the first phase of glaciations (Head and Gibbard 2005); the 100 kyr high amplitude cyclicity marked the subsequent enhanced phase of glaciations, which began after 450 ka in the Japan Sea (Das et al. 2018).

5 Conclusion

In the northern Japan Sea, the middle Pleistocene transition was characterized by the severity of glaciations during which terrigenous sediments was locked within the glaciers are deposited to the sea floor. The prominent cooling began at around 940–890 ka caused the formation and southward movement of extensive ice sheets into the Japan Sea, which is evident by the higher abundance of IRD between 880 and 450 ka. The melting of sea-ice cooled SST over the Japan Sea that reduced the seasonal ice over northern Japanese Island suggested by decreased detrital material and dominance of N. *pachyderma* (S). The detrital material presence in the north-eastern Japan Sea is linked with the EAWM strength and TWC inflow towards the studied site, which caused the formation of seasonal sea-ice. The onset of more extreme glacial-interglacial climate cycles in the Japan Sea began at 490–470 ka. The absence or decrease of foraminifera was observed during 450–150 ka suggesting shoaling of the CCD induced by extreme glaciations. The northern Japan Sea is retaining its modern ecological condition since 150 ka suggested by a significant decrease in IRD and detrital. The studied site is above the CCD since 150 ka marked by an abundance of calcareous foraminifera. Both IRD and planktic foraminifera strongly respond to the global ice volume change and paleoclimatic shift to 100 kyr cycle after the middle Pleistocene transition.

Acknowledgements We would like to thank the Integrated Ocean Drilling Program (IODP) for providing core samples to RKS (Request No. #4201 and 13522). We acknowledge NCPOR Goa (IODP-India) and IIT Bhubaneswar for their support and providing Assistantship to SSB, SM, RVD, SS, AR. MD acknowledge DST for INSPIRE fellowship and NV acknowledge CSIR for providing CSIR-JRF. We are grateful to the Editors and anonymous reviewer for thoughtful reviews.

References

André A, Quillévéré F, Morard R, Ujiié Y, Escarguel G, de Vargas C, de Garidel-Thoron T, Douady CJ (2014) SSU rDNA divergence in planktonic foraminifera: molecular taxonomy and biogeographic implications. PLoS ONE 9(8):e104641. https://doi.org/10.1371/journal.pone.0104641

Be AW (1977). An ecological, zoogeographic and taxonomic review of recent planktonic foraminifera. Oceanic Micropaleontol:1–100

Bischof J (2000) Ice drift, ocean circulation, and climate change. Springer-Verlag, London, p 215

Clark PU, Archer D, Pollard D, Blum JD, Rial JA, Brovkin V, Mix AC, Pisias NG, Roy M (2006) The middle Pleistocene transition: characteristics, mechanisms, and implications for long-term changes in atmospheric pCO_2. Quatern Sci Rev 25(23–24):3150–3184. https://doi.org/10.1016/j.quascirev.2006.07.008

Clemens SC, Prell WL, Sun Y, Liu Z, Chen G (2008) Southern hemisphere forcing of Pliocene $\delta^{18}O$ and the evolution of Indo-Asian monsoons. Paleoceanography Paleoclimatol 23(4). https://doi.org/10.1029/2008PA001638

Dadd K, Foley K (2016) A shape and compositional analysis of ice-rafted debris in cores from IODP Expedition 323 in the Bering Sea. Deep-Sea Res Part II 125–126:191–201. https://doi.org/10.1016/j.dsr2.2016.02.007

Darling KF, Wade CM (2008) The genetic diversity of planktic foraminifera and the global distribution of ribosomal RNA genotypes. Mar Micropaleontol 67(3–4):216–238. https://doi.org/10.1016/j.marmicro.2008.01.009

Das M, Singh RK, Vats N, Holbourn A, Mishra S, Farooq SH, Pandey DK (2018) Changes in the distribution of Uvigerinidae species over the past 775 kyr: implications for the paleoceanographic evolution of the Japan Sea. Palaeogeogr Palaeoclimatol Palaeoecol 507:201–213. https://doi.org/10.1016/J.PALAEO.2018.07.019

Ford HL, Sosdian SM, Rosenthal Y, Raymo ME (2016) Gradual and abrupt changes during the Mid-Pleistocene Transition. Quatern Sci Rev 148:222–233. https://doi.org/10.1016/j.quascirev.2016.07.005

Fujine K, Tada R, Yamamoto M (2009) Paleotemperature response to monsoon activity in the Japan Sea during the last 160 kyr. Palaeogeogr Palaeoclimatol Palaeoecol 280(3–4):350–360. https://doi.org/10.1016/J.PALAEO.2009.06.022

Gallagher SJ, Kitamura A, Iryu Y, Itaki T, Koizumi I, Hoiles PW (2015) The Pliocene to recent history of the Kuroshio and Tsushima Currents: a multi-proxy approach. Prog Earth Planet Sci 2(1):17. https://doi.org/10.1186/s40645-015-0045-6

Gamo T (1999) Global warming may have slowed down the deep conveyor belt of a marginal sea of the northwestern Pacific: Japan Sea. Geophys Res Lett 26(20):3137–3140. https://doi.org/10.1029/1999GL002341

Gamo T, Nakayama N, Takahata N, Sano Y, Zhang J, Yamazaki E, Taniyasu S, Yamashita N (2014) The sea of Japan and its unique chemistry revealed by time-series observations over the last 30 years. Monogr Environ Earth Planets 2(1):1–22. https://doi.org/10.5047/meep.2014.00201.0001

Goldschmidt PM, Pfirman SL, Wollenburg I, Henrich R (1992) Origin of sediment pellets from the Arctic seafloor: sea ice or icebergs? Deep Sea Res Part A 39(2 PART 1):539–565. https://doi.org/10.1016/S0198-0149(06)80020-8

Gupta AK, Anderson DM, Overpeck JT (2003) Abrupt changes in the Asian southwest monsoon during the holocene and their links to the North Alantic Ocean. Nature. https://doi.org/10.1038/nature01340

Hammer Ø, Harper DAT, Ryan PD (2001) PAST: paleontological statistics software package for education and data analysis. Palaeontol Electron 4(1)(1):1–9. https://doi.org/10.1016/j.bcp.2008.05.025

Head MJ, Gibbard PL (2005) Early-Middle Pleistocene transitions: an overview and recommendation for the defining boundary. Geol Soc London Spec Publ 247(1):1–18. https://doi.org/10.1144/GSL.SP.2005.247.01.01

Hemleben C, Spindler M, Anderson OR (1989) Collecting and culture methods. In: Modern planktonic foraminifera. Springer, New York, NY, pp 33–55. https://doi.org/10.1007/978-1-4612-3544-6_3

Holbourn A, Henderson AS, MacLeod N (2013) Atlas of benthic foraminifera. Wiley-Blackwell, Oxford, UK. https://doi.org/10.1002/9781118452493

Igarashi Y, Irino T, Sawada K, Song L, Furota S (2018) Fluctuations in the East Asian monsoon recorded by pollen assemblages in sediments from the Japan Sea off the southwestern coast of Hokkaido, Japan, from 4.3 Ma to the present. Glob Planet Change 163(July 2017):1–9. https://doi.org/10.1016/j.gloplacha.2018.02.001

Ikehara K (2003) Late Quaternary seasonal sea-ice history of the northeastern Japan Sea. J Oceanogr 59(5):585–593. https://doi.org/10.1023/B:JOCE.0000009588.49944.3d

Ikehara K, Itaki T (2007) Millennial-scale fluctuations in seasonal sea-ice and deep-water formation in the Japan Sea during the late Quaternary. Palaeogeogr Palaeoclimatol Palaeoecol 247(1–2):131–143. https://doi.org/10.1016/j.palaeo.2006.11.026

Irino T, Tada R (2000) Quantification of aeolian dust (Kosa) contribution to the Japan Sea sediments and its variation during the last 200 ky. Geochem J 34(1):59–93. https://doi.org/10.2343/geochemj.34.59

Irino T, Tada R, Ikehara K, Sagawa T, Karasuda A, Kurokawa S, Seki A, Lu S (2018) Construction of perfectly continuous records of physical properties for dark-light sediment sequences collected from the Japan Sea during Integrated Ocean Drilling Program Expedition 346 and their potential utilities as paleoceanographic studies. Prog Earth Planet Sci 5(1):23. https://doi.org/10.1186/s40645-018-0176-7

Iwasaki S, Kimoto K, Kuroyanagi A, Kawahata H (2017) Horizontal and vertical distributions of planktic foraminifera in the subarctic Pacific. Mar Micropaleontol 130:1–14. https://doi.org/10.1016/J.MARMICRO.2016.12.001

Jansen E, Fronval T, Rack F, Channell JET (2000) Pliocene-Pleistocene ice rafting history and cyclicity in the Nordic Seas during the last 3.5 Myr. Paleoceanography 15(6):709–721. https://doi.org/10.1029/1999PA000435

Kennett JP, Srinivasan MS (1983) Neogene planktonic foraminifera: a phylogenetic atlas. Hutchinson Ross

Khim BK, Ikehara K, Irino T (2012) Orbital- and millennial-scale paleoceanographic changes in the north-eastern Japan Basin, East Sea/Japan Sea during the late Quaternary. J Quat Sci 27(3):328–335. https://doi.org/10.1002/jqs.1554

Kido Y, Minami I, Fujine K, Irino T, Ikehara K, Chun J-H (2007) Orbital-scale stratigraphy and high-resolution analysis of biogenic components and deep-water oxygenation conditions in the Japan Sea during the last 640 kyr. Palaeogeogr Palaeoclimatol Palaeoecol 247(1–2):32–49. https://doi.org/10.1016/J.PALAEO.2006.11.020

Kimura N, Wakatsuchi M (1999) Processes controlling the advance and retreat of sea ice in the Sea of Okhotsk. J Geophys Res Oceans 104(C5):11137–11150. https://doi.org/10.1029/1999JC900004

Kitamura A, Kimoto K (2007) Eccentricity cycles shown by early Pleistocene planktonic foraminifers of the Omma Formation, Sea of Japan. Glob Planet Change 55(4):273–283. https://doi.org/10.1016/J.GLOPLACHA.2006.09.001

Ledbetter MT, Watkins ND (1978) Separation of primary ice-rafted debris from lag deposits, utilizing manganese micronodule accumulation rates in abyssal sediments of the Southern Ocean.

Bull Geol Soc Am 89(11):1619–1629. https://doi.org/10.1130/0016-7606(1978)89%3c1619:SOPIDF%3e2.0.CO;2

Li T, Liu F, Abels HA, You C-F, Zhang Z, Chen J, Ji J, Li L, Li L, Liu H-C, Ren C, Xia R, Zhao L, Zhang W, Li G (2017) Continued obliquity pacing of East Asian summer precipitation after the mid-Pleistocene transition. Earth Planet Sci Lett 457:181–190. https://doi.org/10.1016/J.EPSL.2016.09.045

Lisiecki LE, Raymo ME (2005) A Pliocene-Pleistocene stack of 57 globally distributed benthic $\delta^{18}O$ records. Paleoceanography 20(1):1–17. https://doi.org/10.1029/2004PA001071

Martinson DG, Pisias NG, Hays JD, Imbrie J, Moore TC, Shackleton NJ (1987) Age dating and the orbital theory of the ice ages: development of a high-resolution 0 to 300,000-year chronostratigraphy. Quatern Res 27(01):1–29. https://doi.org/10.1016/0033-5894(87)90046-9

Munz PM, Siccha M, Lückge A, Böll A, Kucera M, Schulz H (2015) Decadal-resolution record of winter monsoon intensity over the last two millennia from planktic foraminiferal assemblages in the northeastern Arabian Sea. Holocene 25(11):1756–1771. https://doi.org/10.1177/0959683615591357

Nie J (2017) The Plio-Pleistocene 405-kyr climate cycles. Palaeogeogr Palaeoclimatol Palaeoecol. https://doi.org/10.1016/J.PALAEO.2017.07.022

Oba T, Kato M, Kitazato H, Koizumi I, Omura A, Sakai T, Takayama T (1991) Paleoenvironmental changes in the Japan Sea during the last 85,000 years. Paleoceanography 6(4):499–518. https://doi.org/10.1029/91PA00560

Ono Y (1984) Last glacial paleoclimate reconstructed from glacial and periglacial landforns in Japan. Geogr Rev Japan 57(1):87–100

Prell WL, Curry WB (1981) Faunal and isotopic indices of monsoonal upwelling-western Arabian Sea. Oceanol Acta 4(1):91–98

Sagawa T, Nagahashi Y, Satoguchi Y, Holbourn A, Itaki T, Gallagher SJ, Saavedra-Pellitero M, Ikehara K, Irino T, Tada R (2018) Integrated tephrostratigraphy and stable isotope stratigraphy in the Japan Sea and East China Sea using IODP Sites U1426, U1427, and U1429, Expedition 346 Asian Monsoon. Prog Earth Planet Sci 5(1). https://doi.org/10.1186/s40645-018-0168-7

Sakamoto T, Ikehara M, Aoki K, Iijima K, Kimura N, Nakatsuka T, Wakatsuchi M (2005) Ice-rafted debris (IRD)-based sea-ice expansion events during the past 100 kyrs in the Okhotsk Sea. Deep-Sea Res Part II 52(16–18):2275–2301. https://doi.org/10.1016/j.dsr2.2005.08.007

Schulz M, Mudelsee M (2002) REDFIT: estimating red-noise spectra directly from unevenly spaced paleoclimatic time series. Comput Geosci 28(3):421–426. https://doi.org/10.1016/S0098-3004(01)00044-9

Schulz H, Emeis K-C, Erlenkeuser H, von Rad U, Rolf C (2002) The Toba volcanic event and interstadial/stadial climates at the marine isotopic stage 5 to 4 transition in the Northern Indian Ocean. Quatern Res 57(01):22–31. https://doi.org/10.1006/qres.2001.2291

Scott DB, Takayanagi Y, Hasegawa S, Saito T (2000) Illustration and taxonomic reevaluation of Neogene foraminifera described from Japan. Palaeontol Electron 3(2):1–41

Spindler M, Dieckmann GS (1986) Distribution and abundance of the planktic foraminifer Neogloboquadrina pachyderma in sea ice of the Weddell Sea (Antarctica). Polar Biol 5(3):185–191. https://doi.org/10.1007/BF00441699

Tada R, Murray RW, Alvarez Zarikian CA, Anderson WTJ, Bassetti M-A, Brace BJ, Clemens SC, da Costa Gurgel MH, Dickens GR, Dunlea AG, Gallagher SJ, Giosan L, Henderson ACG, Holbourn AE, Ikehara K, Irino T, Itaki T, Karasuda A, Kinsley CW, Kubota Y, Lee GS, Lee KE, Lofi J, Lopes CICD, Peterson LC, Saavedra-Pellitero M, Sagawa T, Singh RK, Sugisaki S, Toucanne S, Wan S, Xuan C, Zheng H, Ziegler M (2015) Expedition 346 summary, 346. https://doi.org/10.2204/iodp.proc.346.101.2015

Tada R, Irino T, Ikehara K, Karasuda A, Sugisaki S, Xuan C, Sagawa T, Itaki T, Kubota Y, Lu S, Seki A, Murray RW, Alvarez-Zarikian C, Anderson WT, Bassetti MA, Brace BJ, Clemens SC, da Costa Gurgel MH, Dickens GR, Dunlea AG, Gallagher SJ, Giosan L, Henderson ACG, Holbourn AE, Kinsley CW, Lee GS, Lee KE, Lofi J, Lopes CICD, Saavedra-Pellitero M, Peterson LC, Singh RK, Toucanne S, Wan S, Zheng H, Ziegler M (2018) High-resolution and high-precision correlation of dark and light layers in the Quaternary hemipelagic sediments of the Japan Sea recovered during IODP Expedition 346. Prog Earth Planet Sci 5(1):19. https://doi.org/10.1186/s40645-018-0167-8

Takahashi K (1998) The Bering and Okhotsk Seas: modern and past paleoceanographic changes and gateway impact. J Asian Earth Sci 16(1):49–58. https://doi.org/10.1016/S0743-9547(97)00048-2

Talley LD, Lobanov V, Ponomarev V, Salyuk A, Tishchenko P, Zhabin I, Riser S (2003) Deep convection and brine rejection in the Japan Sea. Geophys Res Lett 30(4). https://doi.org/10.1029/2002GL016451

Um IK, Choi MS, Bahk JJ, Chun JH (2017) Provenance of late Quaternary sediments on the southwestern slope of the Ulleung Basin, East/Japan Sea. Quatern Int 459:153–164. https://doi.org/10.1016/j.quaint.2017.04.013

van der Zwaan GJ, Jorissen FJ, Verhallen PJJM, von Daniels CH (1986) Atlantic-European oligocene to recent uvigerina: taxonomy, paleoecology and paleobiogeography. Utrecht micropaleontological bulletins, vol 35. Schotanus & Jens

Watkins ND, Keany J, Ledbetter MT, Huang TC (1974) Antarctic glacial history from analyses of ice-rafted deposits in marine sediments: new model and initial tests. Science 186(4163):533–536. https://doi.org/10.1126/science.186.4163.533

Xu X, Oda M (1999) Surface-water evolution of the eastern East China Sea during the last 36, 000 years. Mar Geol 156:285–304

Modern Radiolarians Recovered from the Mudline Samples at IODP-341 Sites in the South Alaska Basin, North East Pacific Ocean

Shyam Murti Gupta

Abstract The Integrated Ocean Drilling Program (IODP) Expedition 341 drilled the 5 sites (U1417-U1421) across the Surveyor channel near the Aleutian trench in the Gulf of Alaska to study the sedimentation during middle Miocene to Recent periods for the climatic changes in the region. Out of the 5 sites drilled, the radiolarians microfossils were found in the deep water sites (namely U1417-19) and provided the biostratigraphy of the sediment cores ranging from the middle Miocene to Pleistocene periods. It motivated and encouraged the author to find out the radiolarians species present over the drill sites in the modern time. Radiolarians recovered from the watery sediment slurry (termed as mudline) present at the sediment-water interface on the top of the core barrels are studied at the species level, and are documented herein.

Keywords IODP expedition 341 · Mudline radiolarians · Gulf of Alaska

1 Introduction

During the year 2013, the Integrated Ocean Drilling Program (IODP) Expedition 341 drilled a transect of 5 sites (U1417-U1421) across the Surveyor Fan in the Gulf of Alaska and Bering-Malaspina slope and shelf offshore the St. Elias Mountains (Fig. 1) to study the sedimentation during middle Miocene to Recent periods encompassing climatic changes in the region. The details on the physical, oceanic and tectonic settings of the area can be referred in Jaeger et al. (2014) and Gulick et al. (2015). Out of 5 sites drilled, the sediments recovered from three sites located in deeper waters, i.e. U1417-1419, yielded highly rich radiolarian and diatom microfossils from late Miocene to Recent times. Besides the sediment cores, the watery sediment slurry, which is termed as mudline samples, from the sediment-water interface was also collected at these sites. Researcher recovered water rich slurry sediment-water inter-phase material as mudline samples from Hole U1417D,

S. M. Gupta (✉)
CSIR-National Institute of Oceanography, Dona Paula, Goa, India
e-mail: smurti_gupta@yahoo.com

D. K. Pandey et al. (eds.), *Dynamics of the Earth System: Evolution, Processes and Interactions*, Society of Earth Scientists Series,
https://doi.org/10.1007/978-3-030-40659-2_4

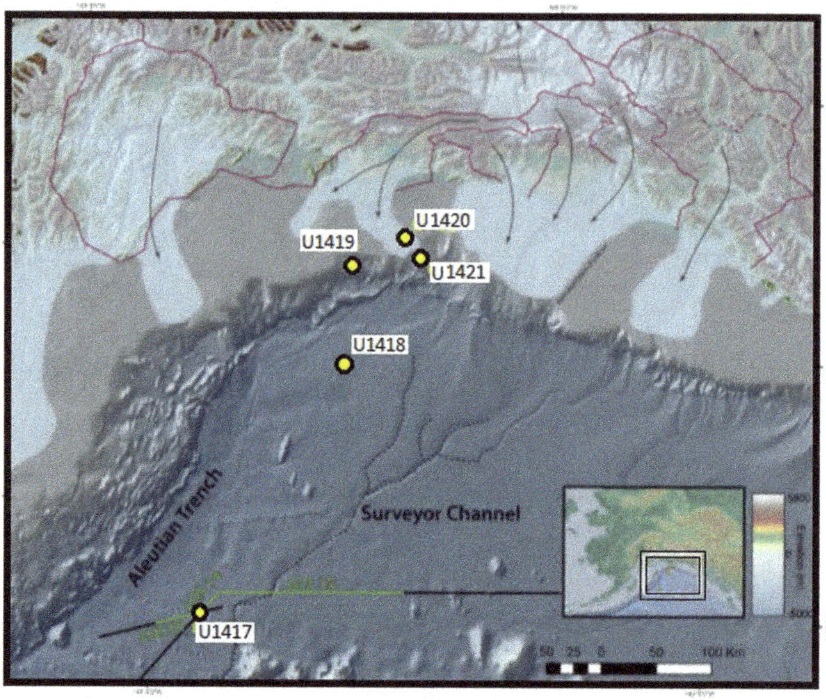

Fig. 1 Location of the IODP sites U1417–1421 (aft. Gulick et al. 2015) situated on the continental shelf, slope and to the abyssal plain between the Aleutian Trench and the Surveyor channel in the south Alaska Basin, North East Pacific Ocean. Inset is the regional area depicting north stern part of Alaska

during the ocean drilling in the south Alaska basin. Objective of the present study was to record the radiolarian species present in the mudline samples and make their relative abundance data (%) available. Radiolarian taxomony is after Gupta (1988, 1996, 2000, 2002), Gupta and Fernandes (1997), Johnson and Nigrini (1980), Nigrini and Moore (1979). The material was analyzed for recovery of modern plankton fauna believed to be living in and around the surface water over the drill sites. This data reports, illustrates and documents the occurrences of modern radiolarian species encountered in the mudline samples recovered during the IODP-341 expedition.

2 Radiolarians from Mudline Samples

Mudline sediment sample was mixed with a small amount of rose Bengal dye to stain the living or dead specimens (if protoplasm is intact) that have traces of protoplasm. There after mudline fluid was sieved on 63 μm mesh and the coarse fraction was used to make the smear slide for identifying and counting the radiolarians. Total radiolarian abundances were determined based on light-microscopic observations at 40 × magnification using a ZEISS AXIOSKOP microscope by counting the individual species in mudline slides (Tables 1, 2, 3 and 4). Radiolarian specimens encountered from mudline are illustrated in Figs. 2, 3, 4, 5 and 6.

3 Results

3.1 *Radiolarians in the Mudline, Hole U1417D*

A total of 45 species were identified and counted (Table 1) from the mudline sample in Hole U1417D, and they are illustrated in Figs. 2, 3, 4. The abundance data are shown in Table 1. We noticed that some of the radiolarian tests were stained by rose Bengal, suggesting that they had been living in the overlying seawater at the site in immediate past. Most of the species found in the mudline sample were not present in samples older than 1 Ma.

3.2 *Radiolarians in the Mudline, Hole U1418C*

A total of 65 species were identified and counted from the mudline sample of Hole U1418C. Most of the species found (Table 2) are illustrated in Figs. 2, 3 and 4. We noticed that some of the radiolarian tests were stained by rose Bengal, suggesting that they had been living in waters overlying Site U1418. A few warm-water subtropical species, such as *Spongaster tetras irregularis* Nigrini, specimens of the pyloniids group (consisting of *Hexapyle, Tetrapyle*, *Octopyle*, and *Pylonium* spp.; Gupta 1996, 2002), *Phormostichoartus carbula* (Harting), *Pterocanium trilobum* (Haeckel), and *Theocorythium trachelium* (Ehrenberg) were also found in the mudline, suggesting the intrusion of warmer water through the subtropical gyres, probably during the warmer El Niño periods into the Gulf of Alaska (Whitney and Welch 2002; Pearcy and Schoener 2012).

3.3 *Radiolarians in the Mudline Hole U1419A*

A total of 1022 Radiolarian specimens made up of 72 species are present in the mudline slides counts (Table 3). Additional species found at this site, relative to the previous sites, are *Collosphaera* sp., *Acrosphaera spinosa, Didymocyrtis tetrathalamus, Tetrapyle octacantha, Phorticium pylonium, Dictyocoryne profunda*, and *Theocorythium trachelium*; these are illustrated in Fig. 5. They represent typical warm-water

Table 1 Mudline radiolarian species from Hole U1417D

Mudline radiolarian species	Counts	Percent
Liosphaera sp.	1	0.20
Actinomma delicatulum	32	6.34
Actinomma leptodermum	19	3.76
Actinomma boreale	12	2.38
Cenosphaera cristata	9	1.78
Hexacontium group	26	5.15
Spongurus pylomaticus	6	1.19
Spongotrochus cf. *glacialis*	6	1.19
Stylacontarium group	18	3.56
Sphaeropyle langii	4	0.79
Lithelius minor	21	4.16
Lithelius nautiloides	4	0.79
Larcopyle butschlii	17	3.37
Stylochlamydium venustum	7	1.39
Stylodictya validispina	52	10.30
Porodiscus sp.	13	2.57
Heliodiscus asteriscus	10	1.98
Tholospyris scaphipes	7	1.39
Cycladophora bicornis	12	2.38
Lamprocyrtis nigriniae	3	0.59
Pterocorys cf. *zancleus*	2	0.40
Cycladophora davisiana davisiana	26	5.15
Spongurus cf. *eliiptica*	15	2.97
Spongurus (?) sp.	21	4.16
Spongopyle osculosa	68	13.47
Acrosphaera spinosa	4	0.79
Pterocanium sp. A + B	11	2.18
Pterocanium korotnevi	13	2.57
Dictyophimus hirundo	4	0.79
Liriospyris reticulata	4	0.79
Anthocyrtidium ophirense	2	0.40
Botryostrobus auritus/australis	6	1.19
Botryostrobus acquilonaris	6	1.19
Giraffospyris angulata	2	0.40
Ceratospyris borealis	22	4.36
Lamprocyclas junonis	2	0.40
Cornutella profunda	2	0.40

(continued)

Table 1 (continued)

Mudline radiolarian species	Counts	Percent
Phomostichoartus corbula	2	0.40
Stichopera pectinata/lagencuna	3	0.59
Pseudodictyophimus sp.	5	0.99
Eucyrtidium hexagonatum	2	0.40
Eucyrtidium sp. aff. *acuminatum*	1	0.20
Botrycyrtis sp.	1	0.20
Siphocampe lineata	1	0.20
Spirocyrtis sp. aff. *scalaris*	1	0.20
Total	505	100

Table 2 Mudline radiolarian species from Hole U1418C

Mudline U1418C radiolarian species	Counts	Percent
Actinomma delicatulum	22	2.21
Actinomma leptodermum	42	4.22
Actinomma medianum	3	0.30
Actinomma boreale	11	1.11
Rhizoplegma ? boreale	6	0.60
Acrosphera spinosa flamabunda	8	0.80
Cenosphaeara cristata	8	0.80
Hexacontium group	31	3.12
Spongurus pylomaticus	8	0.80
Spongurus cf. *elliptica*	15	1.51
Spongurus (?) sp.	23	2.31
Spongodiscus resurgense	30	3.02
Spongotrochus cf. *glacial*	20	2.01
Spongotrochus? venustum	17	1.71
Spongopyle osculosa	67	6.73
Stylochlamydium asteriscus	8	0.80
Stylodictya aculeata	16	1.61
Stylodictya validispina	105	10.55
Porodiscus sp.	14	1.41
Lithelius minor	18	1.81
Lithelius nautiloides	5	0.50
Larcopyle butschlii	12	1.21
Spongaster tetras irregularis	5	0.50
Spongocore pullae/cylendrica	4	0.40
Heliodiscus asteriscus	10	1.01
Stylactactus spp.	23	2.31

(continued)

Table 2 (continued)

Mudline U1418C radiolarian species	Counts	Percent
Spphaeropyle langii	16	1.61
Cladococcus sp.	5	0.50
Pylospira octapyle	5	0.50
Pyloniids (*tetra, hexa, octapyles*)	7	0.70
Didymocyrtis sp.	3	0.30
Pteroscenium pinnatum	2	0.20
Peridium spinipes	3	0.30
Tholospyris scaphipes	16	1.61
Lophopheana sp.	14	1.41
Lamprocyrtis neoheteroporas (?)	3	0.30
Lamprocyrtis nigriniae	8	0.80
Cycladophora bicornis	51	5.13
C. davisiana davisiana	34	3.42
Lamprocyrtis hannai.	11	1.11
Pterocorys minythorax	6	0.60
Pterocorys zancleus gr.	20	2.01
Pterocanium sp.	15	1.51
Pterocanium preatextum gr.	32	3.22
Dictyophimus hirundo	11	1.11
Liriospyris reticulata	7	0.70
Anthocyrtidium ophirense	3	0.30
Theocorythium trachelium	3	0.30
Androcyclas gamphonycha	5	0.50
Bortyostrobus auritus/australis	18	1.81
Bortyostrobus acquilonaris	24	2.41
Giraffospyris angulata	6	0.60
Zygocyrtis productus	4	0.40
Ceratospyris borealis	43	4.32
Cornutella profunda	8	0.80
Phomostichoartus corbula	3	0.30
Stichopera pectinata/lagencuna	20	2.01
Pseudodictyophimus sp.	6	0.60

(continued)

Table 2 (continued)

Mudline U1418C radiolarian species	Counts	Percent
Eucyrtidium hexagonatum	8	0.80
Eucyrtidium sp. aff. *acuminatum*	3	0.30
Botrycyrtis sp.	3	0.30
Siphocampe lineata	4	0.40
Spirocyrtis subscaleris	20	2.01
Bathypyramis sp.	6	0.60
Peripyramis sp.	8	0.80
Total	995	

tropical-subtropical assemblages, which is unusual for the cold subarctic waters of the Gulf of Alaska.

3.4 *Radiolarians in the Mudline, Hole U1421A*

Only 38 species were recorded in the mudline sample from Hole U1421A, and their total counts are limited to 229 in the three slides made. The species count and abundance (%) is presented in Table 4. At this site, additional shallow water species are encountered and they are specifically illustrated in Fig. 6.

4 Conclusion

The mudline radiolarians depict the remnants of standing stock of living radiolarians in the water column over the sites. As they represent the environment of the water mass temperature, salinity, nutrients and the currents, they are indicator of the changes in the water mass characteristics with time. Study records that 45 radiolarian species are encountered in deeper water site U1417 located in the open sea exhibited 46 species, whereas the sites towards the coast but located still in the deep water (U1418-19) have shown enhanced numbers of species between 65 and 72. It could be due to the effect of the coastal upwelled nutrient rich waters during the spring to summer season in the Alaska Basin. The number of species drastically reduced to 38 at site U1421 due to its location on the continental slope-shelf transition, which also be affected by the high energy coastal wave dynamics.

Acknowledgements Author thanks to USIO–IODP and IODP-India for extending invitation, and Dr. S. W. A. Naqvi (the Director CSIR-NIO) for encouragement, permission and extending full financial support to participate in the IODP-341 expedition. Thanks to all the scientific team and crews for retrieving the sediment cores, their sampling and fruitful discussions on board JOIDES

Table 3 Mudline radiolarian species from Hole U1419A

Mudline radiolarian species	Counts	Percent
Cycladophora bicornis	64	6.26
Ceratospyris borealis	52	5.09
Actinomma leptodermum	46	4.50
Lophophaena sp.	38	3.72
Botryostrobus acqullonaris	37	3.62
Stylodictya validispina	36	3.52
Pyloniids (tetra, hexa, octapyle, pylonium)	36	3.52
Larcopyle butschlii	33	3.23
Pterocorys zancleus group	33	3.23
Actinomma delicatalum	31	3.03
Spirocyrtis subscaleris/gyroscaleris	30	2.94
Botryostrobus auritus/australis	29	2.84
Pterocanium preatextum group	28	2.74
Hexacontium group	24	2.35
Stylochlamydium asteriscus	24	2.35
Spongodiscus resurgense	23	2.25
Lithelius minor	22	2.15
Spongopyle osculosa	18	1.76
Spongotrochus ? *venustum*	17	1.66
Lithelius nautiloides	17	1.66
Spongurus cf. *elliptica*	16	1.57
Styiactactus spp.	16	1.57
Stylodictya aculeata	15	1.47
Siphocampe lineata	15	1.47
Spirocyrtis lineata	15	1.47
Sphaeropyle langii	13	1.27
Tholospyris scaphipes	13	1.27
Pterocanium sp.	13	1.27
Spongotrochus cf. *glacialis*	12	1.17
Spongaster tetras	12	1.17
Cenosphaera cristata	11	1.08
Spongurus (?) sp.	11	1.08
Spongodiscus concavus	11	1.08
Liriospyris reticulata	11	1.08
Lamprocyrtis nigriniae	10	0.98
Spongurus pylomaticus	9	0.88
Lamprocyrtis neoheteroporos (?)	9	0.88

(continued)

Table 3 (continued)

Mudline radiolarian species	Counts	Percent
Actinomma boreale	8	0.78
Cladococcus sp.	8	0.78
Lamprocyrtis hannai	8	0.78
Pseudodictyophimus sp.	8	0.78
Collosphaera sp.	7	0.68
Cycladophora davisiana davisiana	7	0.68
Dictyophimus hirundo	7	0.68
Cornutella profunda	7	0.68
Phomostichoartus corbula	7	0.68
Microsphaera spinosa floribunda	6	0.59
Pylospira octapyle	6	0.59
Theocorythium trachelium	6	0.59
Giratfospyris angulata	6	0.59
Zygocyrtis produdus	6	0.59
Eucyrtidium hexagonatum	6	0.59
Actinomma medianum	5	0.49
Heliodiscus asteriscus	5	0.49
Pterocorys minythorax	5	0.49
Anthocyrtidium ophirense	5	0.49
Lithrachnium tentorium	5	0.49
Cubothalus sp.	4	0.39
Spongocore pullae/cylindrica	4	0.39
Botrycyrtis sp.	4	0.39
Bathypyramis	4	0.39
Gondwanaria	4	0.39
Desolenia sp.	3	0.29
Spongaster klingi	3	0.29
Didymocyrtis tetrathalmus	3	0.29
Eucyrtidium sp. aff. *acuminatum*	3	0.29
Carpocanarium	3	0.29
Rhizoplegma ? *boreale*	2	0.20
Stryptacantha	2	0.20
Lipmanella sp.	2	0.20
Liosphaera sp.	1	0.10
Spongoliva sp.	1	0.10
Total	1021	

Table 4 Mudline radiolarian species from Hole U1421A

Mudline radtoiarian species	Counts	Percent
Cycladophora bicornis	25	10.92
Actinomma leptodermum	21	9.1 7
Lophopheana sp.	21	9.17
Stylodictya validispina	16	6.99
Pseudodictyophimus sp.	16	6.99
Pterocorys zancleus group	13	5.68
Botryostrobus auritus/australis	13	5.68
Actinomma delicatulum	3	3.49
Pterocanium sp.	8	3.49
Hexacontium group	7	3.06
Ceratospyris borealis	6	2.62
Spirocyrtis subscaleris/gyroscaleris	6	2.62
Pterocanium preatextum group	6	2.62
Spongotrochus ? *venustum*	6	2.62
Spongodiscus resurgense	6	2.62
Pyloniids (tetra, hexa, octapyle, pylonium)	5	2.18
Giraffospyris angulata	5	2.18
Zygocyrtis productus	5	2.18
Tholospyris scaphipes	4	1.75
Botryostrobus acquilonaris	3	1.31
Spongurus cf. *elliptica*	3	1.31
Eucyrtidium hexagonatum	3	1.31
Heliodiscus asteriscus	3	1.31
Lithelius nautiloides	2	0.87
Spirocyrtis lineata	2	0.87
Lithrachnium tentorium	2	0.87
Cubothalus sp.	2	0.87
Gondwanaria	2	0.87
Larcopyle butschlii	1	0.44
Stylodictya aculeata	1	0.44
Lamprocyrtis nigriniae	1	0.44
C. davisiana davisiana	1	0.44
Comutella profunda	1	0.44
Phomostichoartus corbula	1	0.44
Actinomma medianum	1	0.44
Pterocorys minythorax	1	0.44
Anthocyrtidium ophirense	1	0.44
Carpocanarium	1	0.44
Total	229	

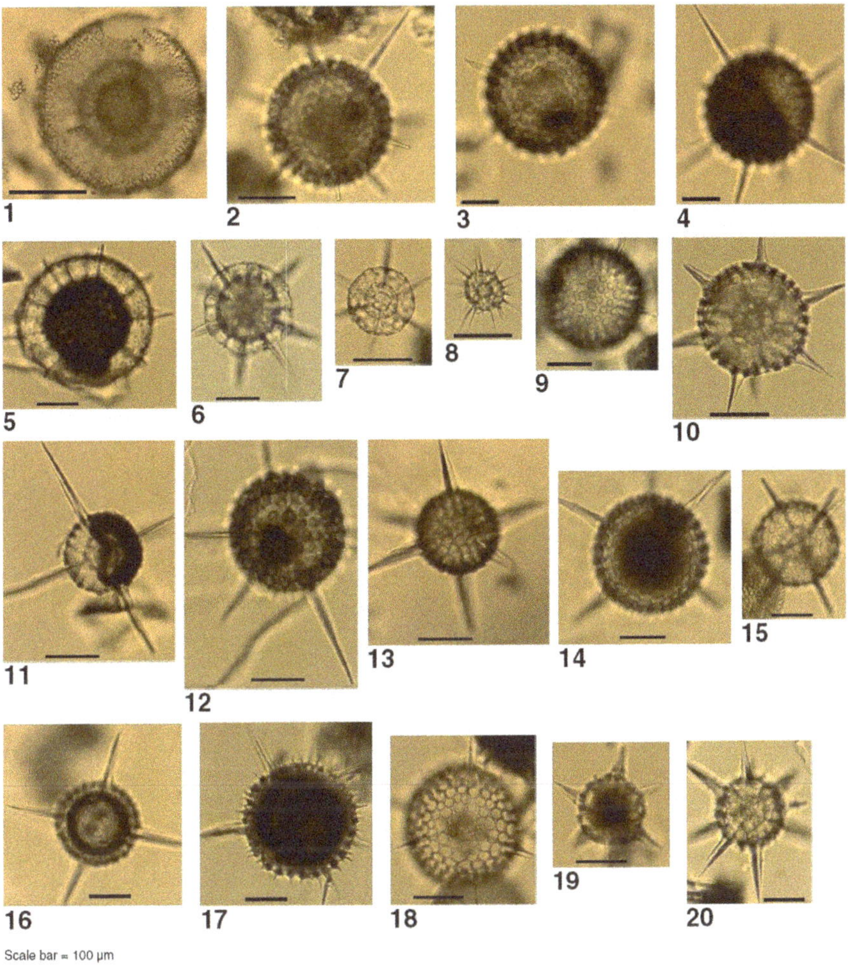

Fig. 2 Photomicrographs of radiolarians from the Hole U1417D mudline sample. **1** *Liosphaera* sp. **2–4** *Actinomma delicatulum* (Dogiel and Reshnetnyak). **5, 6** *Actinomma leptodermum* Jørgensen. **7, 8** *Actinomma boreale* Cleve. **9** *Cenosphaera cristata* Haeckel. **10–20** *Hexacontium* group Westberg-Smith and Riedel

RESOLUTION. Special thanks to Dr. Kenji Matsuzaki for helping with scales set in the radiolarian Figs. 2, 3, 4, 5 and 6.

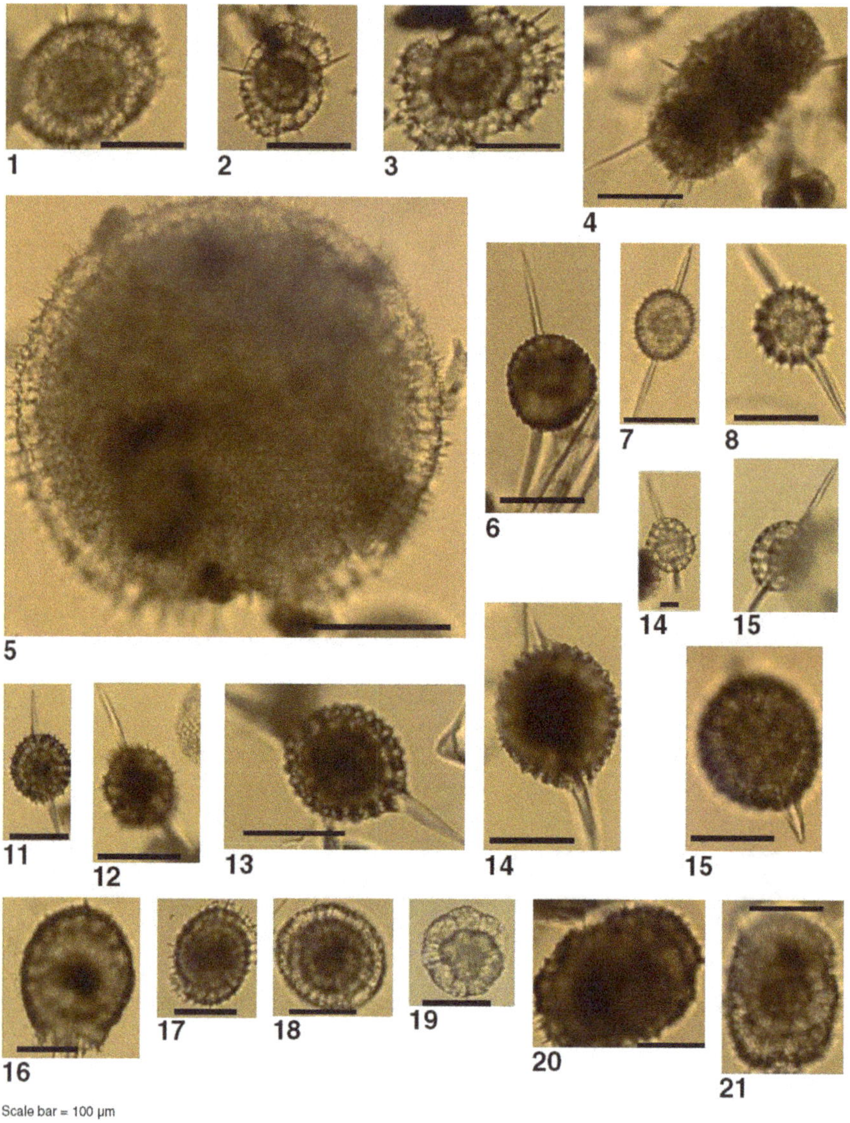

Fig. 3 Photomicrographs of radiolarians from the Hole U1417D mudline sample. **1** *Sphaeropyle* sp.? Dreyer. **2, 3** *Lithelius* spp. **4** *Spongurus pylomaticus* Riedel. **5** *Spongotrochus* cf. *glacial* Popofsky. **6–15** *Stylacontarium* spp. **16** *Sphaeropyle langii* Dreyer. **17, 18** *Lithelius minor* Jørgensen. **19** *Lithelius nautiloides* Popofsky. **20, 21** *Larcopyle butschlii* Dreyer

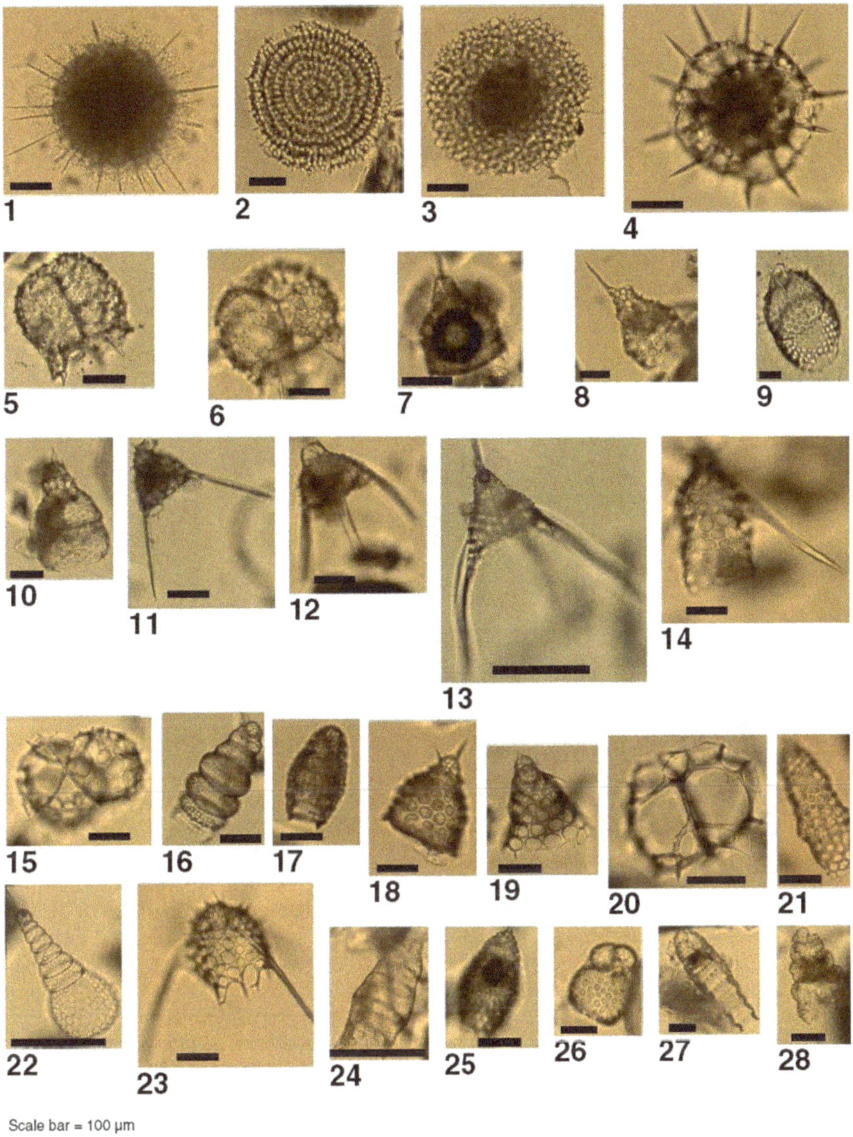

Fig. 4 Photomicrographs of radiolarians from the Hole U1417D mudline sample. **1** *Stylochlamydium venustum* (Bailey). **2** *Stylodictya validispina* Jørgensen. **3** *Porodiscus* sp. Benson. **4** *Heliodiscus asteriscus* Haeckel. **5**, **6** *Tholospyris* cf. *scaphipes* (Haeckel). **7** *Cycladophora bicornis* (Popofsky). **8** *Lamprocyrtis nigriniae* (Caulet). **9**, **10** *Pterocorys* cf. *zancleus* (Mueller). **11.** *Pterocanium* sp. A (*D. infabricatus* Nigrini?). **12** *Pterocanium korotnevi* (Dogiel). **13** *Pterocanium* sp. B. **14** *Dictyophimus hirundo* (Haeckel). **15** *Liriospyris* sp. aff. *reticulata* (Ehrenberg). **16** *Botryostrobus auritus/australis* (Ehrenberg). **17** *Botryostrobus aquilonaris* (Bailey). **18** *Cycladophora bicornis* (Popofsky). **19** *Cycladophora davisiana* Ehrenberg. **20** *Ceratospyris borealis* (Bailey). **21** *Cornutella profunda* Ehrenberg. **22** *Stichopera pectinata/laguncula* Haeckel. **23** *Pseudodictyophimus* sp. **24** *Eucyrtidium hexagonatum* Haeckel. **25** *Botryostrobus* sp. aff. *aquilonaris*. **26** *Botryocyrtis* sp. **27** *Siphocampe lineata* (Ehrenberg). **28** *Botryostrobus* sp

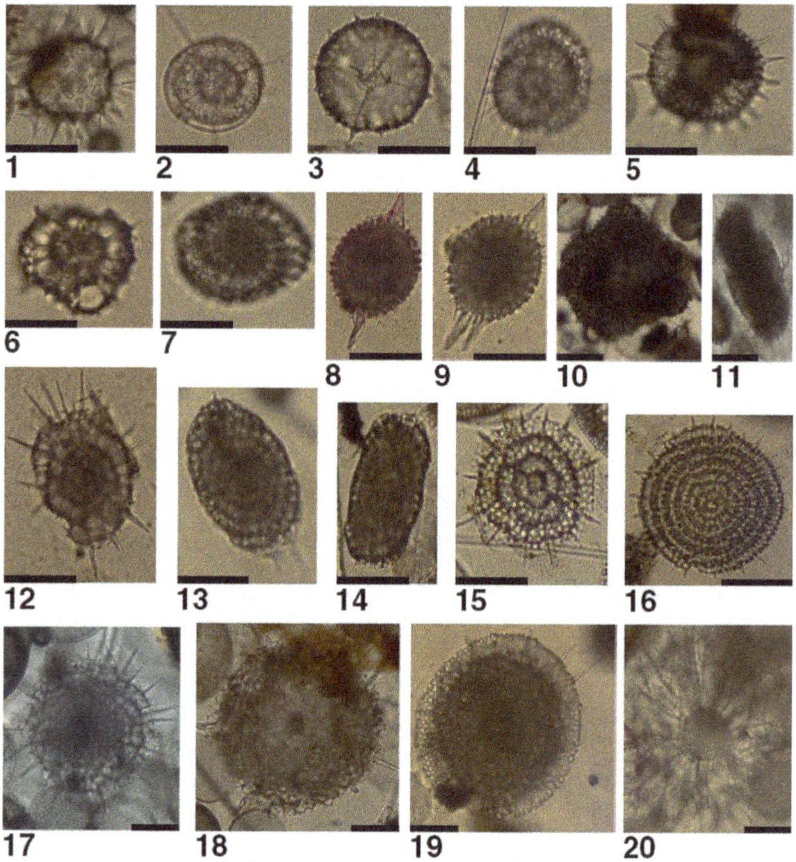

Fig. 5 Photomicrographs of the additional radiolarians encountered at the Sites U1418-21 mud-lines. **1** *Acrosphaera flammabunda* (Haeckel). **2** *Liosphaera* sp. **3** *Hexaconthium* sp. **4** *Dydimocyrtis* sp. **5** *Heliodiscus echiniscus* Haeckel. **6** *Hexapyle* cf. *triangula* Haeckel. **7** *Lithelius nautiloides* Popofsky. **8, 9** Stylosphaeridae. **10** *Spongaster tetras irregularis* Nigrini. **11** *Spongurus pylomaticus* Riedel. **12** *Larcopyle buetschlii* Dreyer. **13, 14** *Larcopyle weddelium* Lazarus, Faust and Popova-Goll. **15** *Stylodictya acuelata* Jørgensen. **16** *Stylodictya validispina* Jørgensen. **17, 19** *Stylochlamydium venustum* Bailey. **18** *Spongotrochus* cf. *glacialis* Popofsky. **20** *Cladococcus* cf. *arborescens* Muller. **21** *Peridium spinipes* Haeckel. **22** *Dimelissa thoracites* (Haeckel). **23** *Eucyrtidium hexagonum* Haeckel. **24** *Botryostrobus aurstrale* (Ehrenberg). **25** Undetermined species. **26** *Pterocanium trilobum* (Haeckel). **27–29** *Lamprocyrtis* sp. **30** *Theocorythium*? sp. **31** *Pterocorys* sp. **32** *Zygocircus productus* (Hertwig). **33** Undetermined species

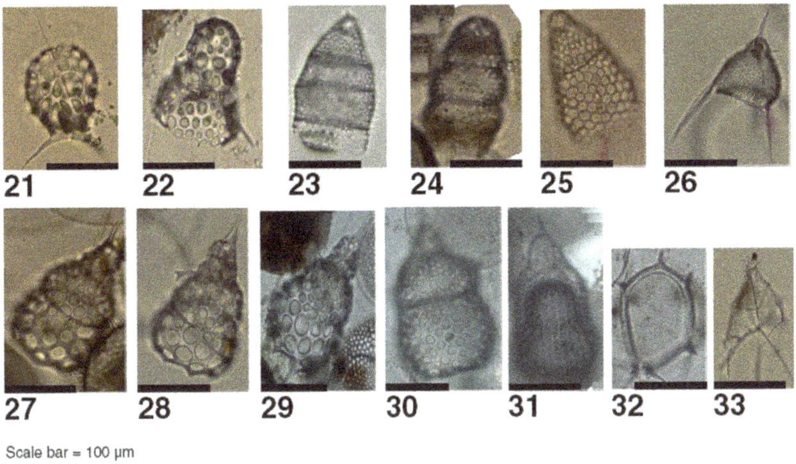

Fig. 5 (continued)

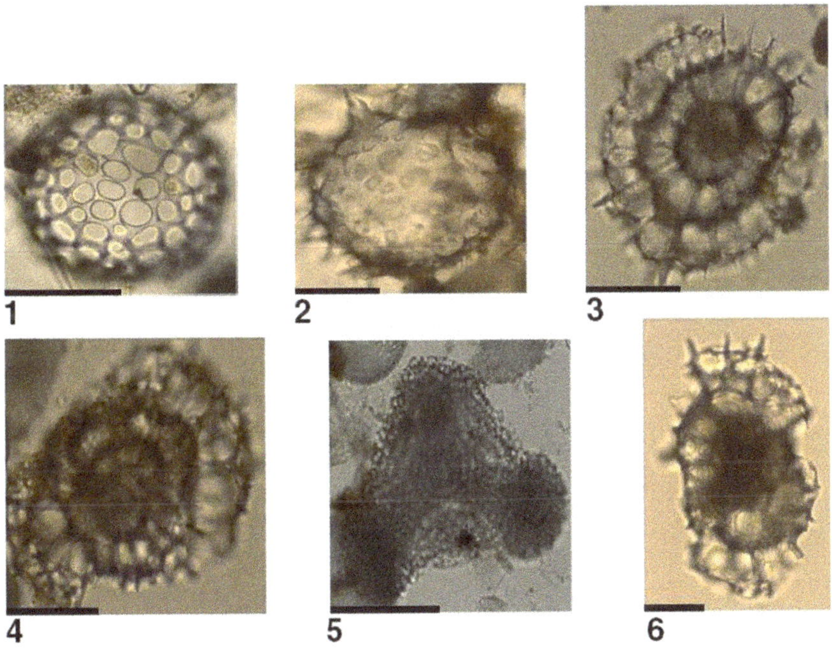

Fig. 6 Photomicrographs of the additional mudline radiolarians from Hole U1419A. Warm water species are in bold. Scale bar = 100 μm. **1** *Collosphaera* sp. Haeckel. **2** *Acrosphaera spinosa* (Haeckel). **3** *Didymocyrtis tetrathalamus* (Haeckel). **4** *Tetrapyle octacantha* Mueller. **5** *Dictyocoryne profunda* Ehrenberg. **6** *Stylachlamydium asteriscus* Haeckel. **7** *Spongotrochus? venustum* (Bailey). **8** *Phorticium pylonium* Haeckel. **9** *Stylodictya aculeata* Jorgensen. **10–12** *Theocorythium trachelium* Ehrenberg. **13** *Anthocyrtidium* sp. (Ehrenberg)

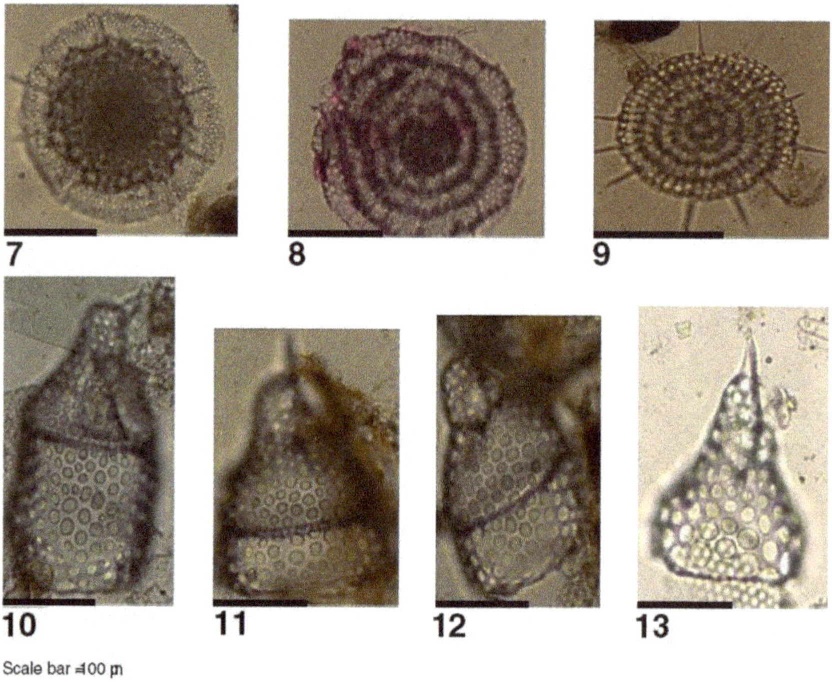

Fig. 6 (continued)

References

Gulick SPS, Jaeger JM, Mix AC, Asahi H, Bahlburg H, Belanger CL, Berbel GBB, Childress L, Cowan E, Drab L, Forwick M, Fukumura A, Ge S, Gupta SM, Kioka A, Konno S, LeVay LJ, Marz C, Matsuzaki KM, McClymont EL, Moy C, Muller J, Nakamura A, Ojima T, Ribeiro FR, Ridgway KD, Romero OE, Slagle AL, Stoner JS, St-Onge G, Suto I, Walczak MD, Worthington LL, Bailey I, Enkelmann E, Reece R, Swartz JM (2015) Mid-Pleistocene climate transition drives net mass loss from rapidly uplifting St Elias Mountains, Alaska. Proc Nat Acad Sci USA 112(49):15042–15047

Gupta SM (1988) Radiolarian zonations and volanic ash layers in two Quaternary sediment cores from the central Indian Ocean Basin. J Palaeontol Soc India 33:59–71

Gupta SM (1996) Quantitative radiolarian assemblages in surface sediments from the Central Indian Basin and their paleomonsoonal significance. J Geol Soc India 47:339–354

Gupta SM (2000) Biostratigraphic analysis of the top layer of sediment cores from the reference and test sites of the INDEX Area. Mar Georesour Geotechnol 18:259–262

Gupta SM (2002) Pyloniid stratigraphy—a new tool to date tropical radiolarian ooze from the central tropical Indian Ocean. Mar Geol 184:85–93

Gupta SM, Fernandes AA (1997) Quaternary radiolarian faunal changes in the central Indian Ocean: inferences to oscillating hydrographic front at 10 degree South of Equator. Curr Sci 72:965–972

Jaeger JM, Gulick SPS, LeVay LJ, Asahi H, Bahlburg H, Belanger CL, Berbel GBB, Childress LB, Cowan EA, Drab L, Forwick M, Fukumura A, Ge S, Gupta SM, Kioka A, Konno S, März CE, Matsuzaki KM, McClymont EL, Mix AC, Moy CM, Müller J, Nakamura A, Ojima T, Ridgway KD, Rodrigues Ribeiro F, Romero OE, Slagle AL, Stoner JS, St-Onge G, Suto I, Walczak MH, Worthington LL (2014) Methods. In Jaeger JM, Gulick SPS, LeVay LJ, The Expedition 341 Scientists (eds), Proceedings of IODP, vol 341. (Integrated Ocean Drilling Program) College Station, TX. https://doi.org/10.2204/iodp.proc.341.102

Johnson DA, Nigrini CA (1980) Radiolarian biogeography in surface sediments of the western Indian Ocean. Mar Micropaleontol 5:111–151. https://doi.org/10.1016/0377-8398(80)90008-0

Nigrini CA, Moore TC (1979) A guide to modern radiolaria. Cushman foundation for foraminiferal research, p S142–N106

Pearcy WG, Schoener A (2012) Changes in the marine biota, coincident with the 1982–83 El-Nino in the North Eastern Pacific Ocean. J Geophys Res (Ocean). https://doi.org/10.1029/JC092ic13p14417

Pisias NG, Roelofs A, Weber M (1997) Radiolarian based transfer function for estimating mean sea surface temperature and seasonal ranges. Paleoceanography 12:365–378

Whitney AF, Welch DW (2002) Impact of the 1997–98 El-Nino and 1999 La-Nina on nutrient supply in the Gulf of Alaska. Prog Oceanogr 54:405–421

Carbon Stable Isotope Source Signature of Bulk Organic Matter in Middle Bengal Fan Sediment Collected During IODP Expedition 354

Supriyo Kumar Das, Ujan Karmakar, Arunabha Dey, Sandip Agrahari and Alf Ekblad

Abstract Middle Bengal Fan in the Bay of Bengal exhibits a unique channel-levee system and are primarily composed of turbidity and hemipelagic sediments. The turbidity sediments are transported and deposited by the Ganga–Brahmaputra (G–B) River system and are considered containing terrestrial organic matter (OM). In contrary, the marine primary production is considered as the source of the OM in hemipelagic sediments. Interestingly, the OM sources in the Middle Bengal Fan sediments have not been extensively studied. Here we show the results of total organic carbon (TOC), total nitrogen (N) and stable isotope ratio of carbon ($\delta^{13}C$) analyses of the bulk OM in the turbidite and hemipelagic sediments from Core U1449A, and compare the results with that of the hemipelagic sediments from Core U1452C. Both the cores have been collected during the International Ocean Discovery Programme (IODP) Expedition 354. The TOC to N ratio (C/N) and the $\delta^{13}C$ composition of the OM indicate marine source in both the turbidite and hemipelagic sediments. This may indicate that the source signature of the terrestrial OM is not efficiently preserved during transport and deposition of the turbidite sediments in the Middle Bengal Fan.

Keywords Sediment · Middle Bengal Fan · IODP Expedition 354 · Stable isotopes of carbon · Organic matter · Sediment

1 Introduction

Accumulation of Bengal Fan sediments started at the beginning of Himalayan uplift in Cenozoic (Curray 1994). The fan sediments primarily consist of turbidites, delivered by the Ganga–Brahmaputra (G–B) River system, and hemipelagic deposits, which is mainly composed of calcareous-clay (Curray and Moore 1974; Kuehl et al. 1989; Weber et al. 1997; Michels et al. 1998; Weber et al. 2018). Galy et al. (2010) and

S. K. Das (✉) · U. Karmakar · A. Dey · S. Agrahari
Department of Geology, Presidency University, College Street 86/1, Kolkata 700073, India
e-mail: sdas.geol@presiuniv.ac.in

A. Ekblad
School of Science and Technology, Örebro University, 70182 Örebro, Sweden

D. K. Pandey et al. (eds.), *Dynamics of the Earth System: Evolution, Processes and Interactions*, Society of Earth Scientists Series,
https://doi.org/10.1007/978-3-030-40659-2_5

Weber et al. (2018) have used carbon stable isotope ratio ($\delta^{13}C$) of the bulk OM as a proxy to interpret the source of the organic matter (OM) in sediments from two separate holes, namely the Deep Sea Drilling Project (DSDP) Leg 22 Site 218 (Galy et al. 2010) and the International Ocean Discovery Programme (IODP) Expedition 354 Site U1452C (Weber et al. 2018), respectively, along the same east-west transect in the Middle Bengal Fan (Fig. 1). The source of the sedimentary OM in turbidity deposits is interpreted as terrigenous because turbidites are primarily transported and deposited by turbidity current in Bengal Fan (Galy et al. 2010) whereas source of the OM in the hemipelagic sediments has been established as marine primary production (Weber et al. 2018). However, Galy et al. (2010) confined their study within the turbidity deposits and Weber et al. (2018) studied only the hemipelagic sediments in the respective cores.

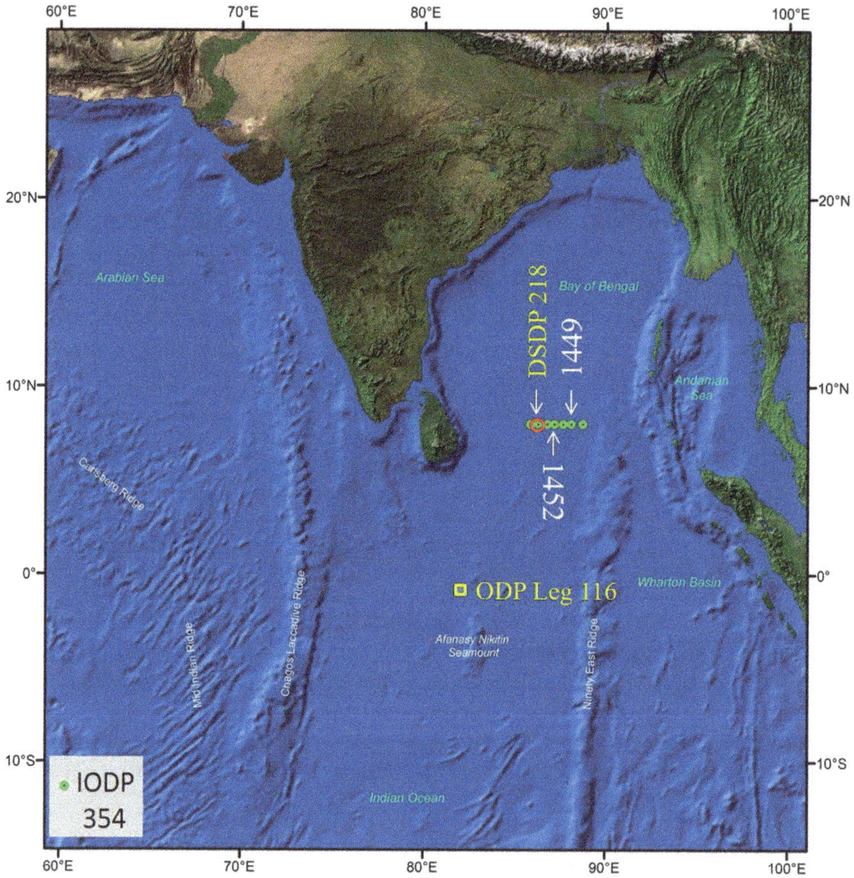

Fig. 1 Map showing the location of Holes U1449 and U1452 at the drilling site of IODP Expedition 354, and the DSDP site 218 in the Bay of Bengal and Ocean Drilling Program Leg 116 in the Central Indian Ocean

The $\delta^{13}C$ of OM has been used to study the source, supply and preservation of the OM in marine sediments (Hedges and Keil 1995; Orginc et al. 2005; Lehmann et al. 2002), but such studies are limited in the Middle Bengal Fan. Galy et al. (2010) have interpreted the shift in the $\delta^{13}C$ of bulk OM towards less depleted values as the C_4 plant expansion in the G–B Basin during Late Miocene (ca. 7.4 Ma). Hein et al. (2017) have linked the variation in the $\delta^{13}C$ and stable isotopic composition of hydrogen (δD) in terrestrial plant wax compounds to monsoon intensity and the paleovegetation changes in the G–B Basin following the last glacial maximum. Weber et al. (2018) have analysed total organic carbon (TOC), total nitrogen (N) and the $\delta^{13}C$ in the uppermost hemipelagic part of the Core U1452C-1H and linked the variation in the geochemical data to the glacial-interglacial scale changes in marine primary production. The objectives of the present study were to identify the source of the OM in the hemipelagic and turbidity deposits in a single core from the Middle Bengal Fan by analysing the TOC, N and $\delta^{13}C$ in the bulk sedimentary OM, and to compare the results with published data. We used the hemipelagic and turbidity sediments in the Core U1449A collected during the IODP Expedition 354 for meeting the above objectives.

2 Site

The IODP Expedition 354 was carried out in the Bay of Bengal at 8° N between 90 East Ridge in the east and the currently active channel-levee system in the west (Fig. 1). Drilling was carried out at 7 locations along a 320 km transect. The site U1449 was part of a six shallow hole transect originally designed to document turbiditic transport and deposition processes in the Middle Bengal Fan at 8° N since the Pliocene. The site is located at 08°0.419′ N, 088°6.599′ E at a water depth of 3653 m. High accumulation of turbidites in the lower 120 m of the hole since ca. 2 Ma was followed by lower accumulation of hemipelagic sediments during the Pleistocene around 1 Ma. Intercalated levee and interlevee deposits later formed until 300,000 years followed by the domination of hemipelagic sedimentation (France-Lanord et al. 2016). The active channel-levee system, which is directly connected to the G–B river system and carries most of the terrestrial OM, is located about 200 km west from the location of U1449 (France-Lanord et al. 2016).

3 Materials and Method

Total 200 samples, including 179 from the Core U1452C-1H (Weber et al. 2018) and 21 from the Core U1449A, were used in the study. The TOC and N in sediments from the Core U1449A were analysed using a Euro EA3024 elemental analyser. The $\delta^{13}C$ in sediments from the Core U1449A was analysed using a Euro EA3024 elemental analyser coupled with Isoprime Isotope Ratio-Mass Spectrometer (IRMS) at Örebro

University. Samples for TOC and $\delta^{13}C$ analyses were treated with HCl acid in order to remove carbonates. Blanks and internal standards were run in between every 8 samples. Results were expressed in conventional δ notation (‰) using standards calibrated against the international Vienna Pee Dee Belemnite (V-PDB) limestone for organic carbon. The results of the $\delta^{13}C$ analysis were calculated using the following equation:

$$\delta(‰) = \left[\left(R_{sample}/R_{standard}\right) - 1\right] \times 1000$$

where R_{sample} for C was the $^{13}C/^{12}C$ ratio of the sample and $R_{standard}$ was the $^{13}C/^{12}C$ ratio of the standard. The precision of the analyses (10 replicated standard samples) was 0.3% for the TOC, 0.03% for the N and 0.06‰ for the $\delta^{13}C$. The reproducibility of duplicate analyses was 0.1‰.

4 Results

4.1 TOC, N, C/N Ratio and $\delta^{13}C$ of Sedimentary OM

The TOC values ranges from 0.4–0.8% (mean 0. 5%; n = 16) in turbidites and 0.4–0.8% (mean 0.6%; n = 5) in hemipelagic sediments of the Core U1449A. The N concentration lies between 0.04 and 0.12% (mean 0.07%) in the turbidites, and between 0.05 and 0.12% (mean 0.08%) in the hemipelagic sediments. The C/N ratio varies from 5.7 to 10.9 (mean 8.3) in the turbidites and from 6 to 10 (mean 7.6) in the hemipelagic sediments. The $\delta^{13}C$ composition ranges from −22 to −20‰ in the turbidites and from −20 to −18‰ in the hemipelagic intervals.

5 Discussion

5.1 Preservation of OM

Stable isotope source signature of the OM, transported via G-B river system and/or produced by phytoplanktons within the marine water column, is expected to survive sinking and sedimentation although typically <1% of the total OM produced in the photic zone reaches to the bottom of the ocean (Lee et al. 2004). However, diagenetic processes during the transport and deposition of the OM may affect the TOC and the $\delta^{13}C$ values. The oxygen concentration of the ocean bottom further determines the extent to which the $\delta^{13}C$ composition could diagenetically be altered. This, coupled with the amount of time the OM is exposed to the oxic environment, determines the level of the $\delta^{13}C$ fractionation (McArthur et al. 1992). Hence, early diagenesis plays a crucial role in the preservation of OM source signature. Microbial processes, both aerobic and anaerobic, mediate the OM decomposition, and could progressively alter the source signature of the bulk OM as different fractions of the OM degrade at different rates under variable substrate conditions (Hedges et al. 1988; Harvey et al. 1995; Meyers and Eadie 1993). In situ bacterial biomass, for example, can convert the biogeochemical signal of the algal-derived OM into bacterial OM up to 5–25% (Harvey et al. 1995). Anaerobic decomposition may further influence the $\delta^{13}C$ by preferential removal of the ^{12}C (Jones et al. 2010). In contrary, Lehmann et al. (2002) have showed a decrease in the $\delta^{13}C$ values due to the selective preservation of ^{13}C depleted OM. Post-depositional microbial decomposition of the organic-carbon-poor (TOC < 1%) Middle Bengal Fan sediments may lead to the decline of the TOC values, and the C/N ratio by selective decomposition of the nitrogen-bearing compounds (Müller and Mathesius 1999). However, lack of correlation ($r^2 = 0.09$; $n = 16$ in turbidites and $r^2 = 0.4$; $n = 5$ in hemipelagic sediments) between the $\delta^{13}C$ and C/N in the Core 1449A imply good preservation of the $\delta^{13}C$ signature in the sedimentary OM record. Consistent with this, the diagenetic effect on the preservation of the elemental and isotopic signals of the bulk OM has been shown negligible in the Core 1452C-1H by Weber et al. (2018).

5.2 *Source of OM*

The C_3 vascular plant-derived OM exhibits higher C/N ratio (≥20) due to low protein content and abundance of cellulose. Low C/N ratio (6–9) in phytoplankton is attributed to the high protein and absence of cellulose (Müller and Mathesius 1999; Prahl et al. 1980; Meyers 1994; Kaushal and Bindford 1999; Schneider and Schlitzer 2003; Martiny et al. 2014). The carbon isotopic composition of bulk OM reflects the dynamics of carbon assimilation during photosynthesis and the isotopic composition of the source carbon. Marine phytoplankton preferentially utilises ^{12}C during photosynthesis (Dean and Stuiver 1993) and exhibits $\delta^{13}C$ values lying between −24 and −18‰ (Khan et al. 2015). Marine sedimentary OM typically shows the $\delta^{13}C$ values lying between −22 and −20‰ (Meyers 1994). Photosynthetic plants using C_3 or Calvin-Benson pathway preferentially incorporate ^{12}C resulting in lighter $\delta^{13}C$ values (average −28‰) of the OM (Meyers and Teranes 2001). In contrast, OM derived from plants using C_4 Hatch-Slack pathway shows heavier $\delta^{13}C$ values (average −14‰) (Meyers and Lallier-Vergès 1999). In general, C_3 dominated terrestrial vegetation derived sedimentary OM shows lower $\delta^{13}C$ values compared to the marine phytoplankton derived OM (Lamb et al. 2006).

The C/N ratio (5–11), $\delta^{13}C$ (−20 to −18‰) and $\delta^{13}C$ to C/N plot (Fig. 2) of the turbidity and hemipelagic sediments in the Core U1449A fall within the range of marine phytoplankton and/or bacterial source of the sedimentary OM (Khan et al. 2015). We do not observe the presence of the terrestrial C_3 plant derived OM in the studied sediments. Interestingly, the $\delta^{13}C$ values of both the turbidity and hemipelagic sediments in the Core U1449A are consistent with the $\delta^{13}C$ values (−17 to −26‰) of particulate OM (POM) in surface waters from the north of subtropical convergence in

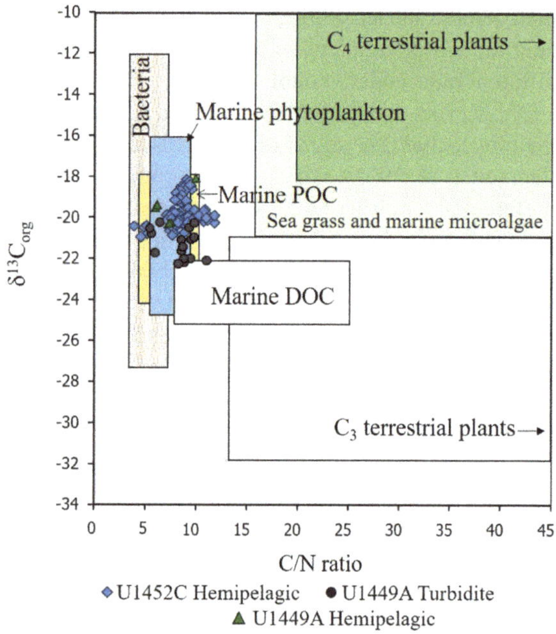

Fig. 2 The $\delta^{13}C$ versus C/N plot of the OM in the sediments from the Core U1449A and Core U1452C-1H (Weber et al. 2018). Boxes indicate typical sources of OM including marine dissolved organic carbon (DOC) and particulate organic carbon (POC) according to Khan et al. (2015)

SW Indian Ocean (Francois et al. 1993). This implies that the terrestrial OM carried by the G–B River system from the G–B Basin to the Middle Bengal Fan does not efficiently preserve the OM source signature. Weber et al. (2018) observe that the $\delta^{13}C$ values in the hemipelagic interval of the Core U1452C-1H lie in between the C_3 and C_4 plant signatures (Fig. 3) reported in Bengal Fan sediments (France-Lanord and Derry 1994; Galy et al. 2010). The $\delta^{13}C$ signature of marine primary production is influenced by a number of factors including the carbon source, the sea surface temperature and hydrogeology. The observed $\delta^{13}C$ values are consistent with the $\delta^{13}C$ values of plankton (ca. −20‰) in the Gulf of Bengal (Fontugne and Duplessy 1978, 1981). Fontugne and Duplessy (1986) have reported heavier $\delta^{13}C$ values (−18 to −23‰) in marine OM during glacial in the Andaman Sea and Gulf of Bengal. The intermediate $\delta^{13}C$ values of OM in the Core U1449A and Core 1452C-1H may further be attributed to the heavier (ca. −7.5‰) dissolved CO_2 composition in the warm north Indian Ocean water that is considered as the primary source of carbon to photosynthetic marine phytoplankton in the north of Indian Ocean (Francois et al. 1993).

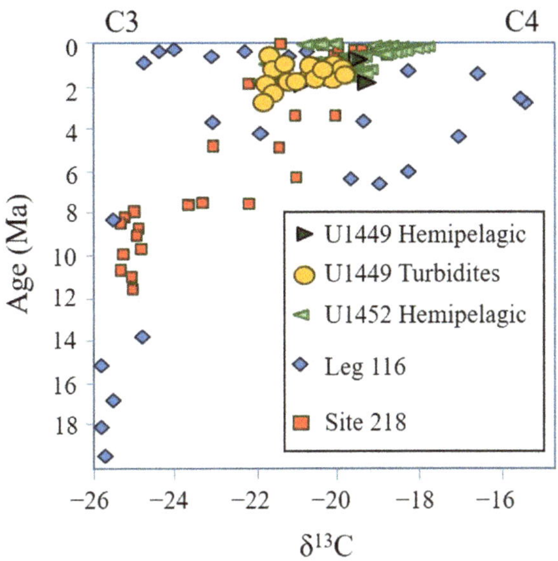

Fig. 3 The $\delta^{13}C$ composition of OM in the turbidites and hemipelagic intervals in the Core U 1449A and hemipelagic interval in the Core U1452C-1H (Weber et al. 2018), and in the Bengal Fan sediments reported by France-Lanord and Derry (1994) and Galy et al. (2010). Note that $\delta^{13}C$ values fall between the C_3 and C_4 plant source signature of the OM

6 Conclusion

Carbon stable isotope ratio of sediments from the Core U1449A shows that the terrestrial OM source signature is not preserved in the turbidity sediments, and indicates marine source of the OM in both the turbidites and hemipelagic sediments of the Middle Bengal Fan. The heavier $\delta^{13}C$ signal may also be attributed to a heavier source of the carbon and/or alteration of stable isotope signature during transport and deposition of the OM.

Acknowledgements The authors gratefully acknowledge the financial help of the University Grants Commission (UGC) India and IODP–India (NCAOR/IODP/20.15/15(III)) to S.K.D for the study.

References

Curray JR (1994) Sediment volume and mass beneath the Bay of Bengal. Earth Planet Sci Lett 125:371–383

Curray JR, Moore DG (1974) Sedimentary and tectonic processes in the Bengal deep-sea fan and geosyncline. In: Burk CA, Drake CL (eds) The geology of continental margins. Springer, Berlin, Heidelberg

Dean WE, Stuiver M (1993) Stable carbon and oxygen isotope studies of the sediments of Elk Lake, Minnesota. In: Bradbury JP, Dean WE (eds) Elk Lake, Minnesota: evidence for rapid climate change in the North-Central United States, vol 276. Geological Society of America Special Paper, pp 63–180

Fontugne M, Duplessy JC (1978) Carbon isotope ratio of marine plankton related to surface water masses. Earth Planet Sci Lett 41:365–371

Fontugne MR, Duplessy JC (1981) Organic-carbon isotopic fractionation by marine plankton in the temperature range −1 to 31 °C. Oceanologia Acta 4:85–90

Fontugne MR, Duplessy JC (1986) Variations of the monsoon regime during the upper quaternary: evidence from carbon isotopic record of organic matter in North Indian Ocean sediment cores. Palaeogeogr Palaeoclimat Palaeoecol 56:69–88

France-Lanord C, Derry LA (1994) $\delta^{13}C$ of organic carbon in the Bengal Fan: source evolution and transport of C_3 and C_4 plant carbon to marine sediments. Geochim Cosmochim Acta 58:4809–4814

France-Lanord C, Derry LA (1997) Organic carbon burial forcing of the carbon cycle from Himalayan erosion. Nature 390:65–67

France-Lanord C, Spiess V, Klaus A, Schwenk T, The Expedition 354 Scientists (2016) Bengal Fan. In: Proceedings of the international Ocean Discovery Program, vol 354, College Station, TX (International Ocean Discovery Program)

Francois R, Altabet MA, Goericke R, McCorkle DC, Brunet C, Poisson A (1993) Changes in the $\delta^{13}C$ of surface water particulate organic matter across the subtropical convergence in the SW Indian Ocean. Global Biogeochem Cycles 7:627–644

Galy V, Eglinton TI (2011) Protracted storage of biospheric carbon in the Ganges–Brahmaputra basin. Nat Geosci 4:843–847

Galy A, France-Lanord C (1999) Weathering processes in the Ganges-Brahmaputra basin and the riverine alkalinity budget. Chem Geol 159:31–60

Galy V, France-Lanord C, Beyssac O, Faure P, Kudrass H, Palhol F (2007) Efficient organic carbon burial in the Bengal fan sustained by the Himalayan erosional system. Nature 450:407–410

Galy V, France-Lanord C, Peucker-Ehrenbrink B, Huyghe P (2010) Sr-Nd-Os evidence for a stable erosion regime in the Himalaya during the past 12 Myr. Earth Planet Sci Lett 290:474–480

Galy V, Eglinton T, France-Lanord C, Sylva S (2011) The provenance of vegetation and environmental signatures encoded in vascular plant biomarkers carried by the Ganges-Brahmaputra rivers. Earth Planet Sci Lett 304:1–12

Harvey HR, Tuttle JH, Bell JT (1995) Kinetics of phytoplankton decay during simulated sedimentation: Changes in biochemical composition and microbial activity under oxic and anoxic conditions. Geochim Cosmochim Acta 59:3367–3377

Hedges JI, Keil RG (1995) Sedimentary OM preservation: an assessment and speculative synthesis. Mar Chem 49:81–115

Hedges JI, Clark WA, Quay PD, Richey JE, Devl U, De Santo U (1986) Compositions and fluxes of particulate organic material in the Amazon River. Limnol Oceanogr 31:717–738

Hedges JI, Clark WA, Cowie GL (1988) Fluxes and reactivities of organic matter in a coastal marine bay. Limnol Oceanogr 33:1137–1152

Hein CJ, Galy V, Galy A, France-Lanord C, Kudrass H, Schwenk T (2017) Post-glacial climate forcing of surface processes in the Ganges-Brahmaputra river basin and implications for carbon sequestration. Earth Planet Sci Lett 478:89–101

Jones MC, Peteet DM, Sambrotto R (2010) Late-glacial and Holocene $\delta^{15}N$ and $\delta^{13}C$ variation from a Kenai Peninsula, Alaska peatland. Palaeogeogr Palaeoclimatol Palaeoecol 293:132–143

Kaushal S, Bindford MW (1999) Relationship between C:N ratios of lake sediments, organic matter sources, and historical deforestation of Lake Pleasant, Massachusetts, USA. J Paleolimnol 22:439–442

Khan NS, Vane CH, Horton BP (2015) Stable carbon isotope and C/N geochemistry of coastal wetland sediments as a sea-level indicator. In: Handbook of sea-level research. Wiley, pp 295–311

Kuehl SA, Hariu TM, Moore WS (1989) Shelf sedimentation off the Ganges-Brahmaputra river system: evidence for sediment bypassing to the Bengal fan. Geology 17:1132–1135

Lamb AL, Wilson GP, Leng MJ (2006) A review of coastal palaeoclimate and relative sea-level reconstructions using $\delta^{13}C$ and C/N ratios in organic material. Earth Sci Rev 75:29–57

Lehmann MF, Bernasconi SM, Barbieri A, McKenzie JA (2002) Preservation of organic matter and alteration of its carbon and nitrogen isotope composition during simulated and in situ early sedimentary diagenesis. Geochim Cosmochim Acta 66:3573–3584

Lee C, Wakeham S, Arnosti C (2004) Particulate organic matter in the sea: the Composition Conundrum. Ambio 33:565–575

Martiny AC, Vrugt JA, Lomas MW (2014) Concentrations and ratios of particulate organic carbon, nitrogen, and phosphorus in the global ocean. Sci Data 1:140048

McArthur JM, Tyson RV, Thomson J, Mattey D (1992) Early diagenesis of marine organic matter: alteration of the carbon isotopic composition. Mar Geol 105:51–61

Meyers PA (1994) Preservation of elemental and isotopic source identification of sedimentary organic matter. Chem Geol 144:289–302

Meyers PA, Eadie BJ (1993) Sources, degradation and recycling of organic matter associated with sinking particles in Lake Michigan. Org Geochem 20:47–56

Meyers PA, Lallier-Vergès E (1999) Lacustrine sedimentary organic matter records of Late Quaternary paleoclimates. J Paleolimnol 21:345–372

Meyers PA, Teranes JL (2001) Sedimentary organic matter. In: Last WM and Smol JP (eds) Tracking environmental change using lake sediments. Vol. 2, Physical and geochemical methods. Kluwer Academic Publishers, Dordrecht, The Netherlands, pp 239–269

Michels KH, Kudrass HR, Hübscher C, Suckow A, Wiedicke M (1998) The submarine delta of the Ganges–Brahmaputra: cyclone-dominated sedimentation patterns. Mar Geol 149:133–154

Müller A, Mathesius U (1999) The palaeoenvironments of coastal lagoons in the southern Baltic Sea. The application of sedimentary C_{org}/N ratios as source indicators of organic matter. Palaeogeogr Palaeoclimatol Palaeoecol 145:1–16

Orginc N, Fontolan G, Faganeli J, Covelli S (2005) Carbon and nitrogen isotope compositions of organic matter in coastal marine sediments (the Gulf of Trieste, N Adriatic Sea): indicators of sources and preservation. Mar Chem 95:163–181

Ortiz JE, Torres T, Delgado A, Julia R, Lucini M, Llamas FJ, Reyes E, Soler V, Valle M (2004) The palaeoenvironmental and palaeohydrological evolution of Padul Peat Bog (Granada, Spain) over one million years, from elemental, isotopic and molecular organic geochemical proxies. Org Geochem 35:1243–1260

Prahl FG, Bennett JT, Carpenter R (1980) The early diagenesis of aliphatic hydrocarbons and organic matter in sedimentary particulates from Dabob Bay, Washington. Geochim Cosmochim Acta 44:1967–1976

Rao Z, Li Y, Zhang J, Jia G, Chen F (2016) Investigating the long-term palaeoclimatic controls on the δD and δ^{18}O of precipitation during the Holocene in the Indian and East Asian monsoonal regions. Earth Sci Rev 159:292–305

Schelske CL, Hodell DA (1995) Using carbon isotopes of bulk sedimentary organic matter to reconstruct the history of nutrient loading and eutrophication in Lake Erie. Limnol Oceanogr 40:918–929

Schneider B, Schlitzer R (2003) Depth-dependent elemental compositions of particulate organic matter (POM) in the ocean. Global Biogeochem Cycles 17:1032

Weber ME, Wiedicke MH, Kudrass HR, Hübscher C, Erlenkeuser H (1997) Active growth of the Bengal Fan during sea-level rise and highstand. Geology 25:315–318

Weber ME, Lantzsch H, Dekens P, Das SK, Reilly B, Martoc YM, Meyer-Jacob C, Agrahari S, Ekblad A, Titschack J, Holmes B, Wolfgramm P (2018) 200,000 years of monsoonal history recorded on the lower Bengal Fan—strong response to insolation forcing. Global Planet Change 166:107–119

Geochemistry of Marine Carbonates from Hole 1394, off the Coast of Montserrat, IODP Expedition-340; Implications on Provenance, Paleoenvironment and Lesser Antilles Arc Migration

K. S. V. Subramanyam, V. Balaram, C. Manikyamba, Parijat Roy, A. K. Krishna, S. S. Sawant, Ch. Narshimha and The IODP Expedition 340 Scientists

Abstract The marine carbonate sediment samples are subsampled and collected from various cores spanning from top depths 11.67 m (core depth below sea floor; CSF) and 11.45 m (CSF) up to the depth of 168.57 m (CSF) respectively of U1394 A and B holes drilled to south east of Montserrat Volcanic Island during IODP Expedition 340. These thin horizons are characterised for major, minor, trace, rare earth elements and total carbon to decipher their mode of origin, provenance and implications on paleo-environment. The carbonate sediments show variation in major oxide contents in SiO_2 (11.4–35.5 wt%), CaO (22.4–44.7 wt%), $CaCO_3$ (3.6–76.6 wt%), Fe_2O_{3T} (1.1–6.0 wt%), MgO (2.9–4.0 wt%) with very low K_2O, TiO_2 and P_2O_5 contents. Their Mg/Ca, reciprocals (Ca/Mg) and Sr/Ca ratios classify these sediments as dolomitic limestones. These sediments depict very low total organic carbon (TOC) contents compared to total inorganic carbon (T_{ic}) indicating their lithogenic origin. The PAAS normalised rare earth element and yttrium (REY) distribution patterns show slightly depleted LREE abundances relative to HREE with $(La/Yb)_{SN}$ range between 0.3 and 0.7 indicating low degree of fractionation with varying $\sum$REE contents (35–105 ppm). Subchrondritic to superchondritic Y/Ho (25.7–31.6), Ce/Ce* (0.2–0.5) and Pr/Pr* (1.0–1.3) indicate low magnitude negative anomalies, slight negative to positive anomalies of Eu (Eu/Eu*; 0.9–1.6) point towards minimal terrigenous input, varied nature of sediments and plagioclase addition. The studied sediments mimic the seawater REY distribution patterns indicating REY are sourced

K. S. V. Subramanyam (✉) · V. Balaram · C. Manikyamba · A. K. Krishna · S. S. Sawant
CSIR-National Geophysical Research Institute, Hyderabad 500007, India

P. Roy
National Center for Polar and Ocean Research, Vasco da Gama, Goa 403804, India

Ch. Narshimha
Department of Applied Geochemistry, Osmania University, Hyderabad 500007, India

The IODP Expedition 340 Scientists
Integrated Ocean Drilling Programe (IODP) at Texas A & M University, College Station, USA

D. K. Pandey et al. (eds.), *Dynamics of the Earth System: Evolution, Processes and Interactions*, Society of Earth Scientists Series,
https://doi.org/10.1007/978-3-030-40659-2_6

from the seawater. The major, minor, immobile element ratios and redox sensitive element proxies indicate that the studied carbonate sediments are derived from an intermediate to felsic source. These sediments deposited in a low saline, oxygen rich open ocean shallow water environment akin to an active continent margin affinity. It can also be stated that later to their dissolution in extinct arc zones (between the old and young arc) led to the re-deposition in favourable sites in the marine sediment column during middle Miocene period or later due to the submarine slope failures and debris avalanches.

Keywords IODP Expedition-340 · Lesser Antilles arc · Marine carbonates · Rare earth elements · Montserrat volcanic island

1 Introduction

The Integrated Ocean Drilling Program (IODP) undertook Expedition 340 with an objective to obtain drill cores from several sites of Lesser Antilles volcanic arc region (Fig. 1). These cores are used to recover the information to establish records of volcanic eruptions, landslides and their impact on the marine sedimentary record. The drill site U 1394 A and B are first among all and located to ~24 km southeast of Montserrat volcanic island (Fig. 1). The site U 1394 is at a water depth of 1115 m and consists of hemipelagic muds, turbidite sequences, mafic volcaniclastics, volcaniclastic sands, tephra layers and carbonate muds as main lithologies (Fig. 2).

The Lesser Antilles arc is an oceanic island arc located at the eastern boundary of Caribbean Plate under which the American plate is subducting (Macdonald et al. 2000). The rate of subduction is reported to be 2 cm/year (Feuillet et al. 2002 and references therein) with a highly curved subduction zone geometry which is tectonically bounded by two strike-slip faults towards north and south. Although, several minor and major events of volcanism are reported from Lesser Antilles, the recent report of a 2.36 Ma Plinian eruption from Guadeloupe volcanic island, with volcanic explosivity index ~6 is known to be the largest documented volcanic event in this region since time (Palmer et al. 2016). The Lesser Antilles arc, with active volcanism in the Caribbean Sea since 40 Ma (Soufriere Hills Volcano of Montserrat Volcanic Island) divided into two arcs to the north of Martinique (Germa et al. 2011). The east and west diverged two volcanic axes converges to single at Martinique island. The eastern segment also called as the Limestone Caribbees is inactive and older than the young (~7 My ago) active western segment start with the volcanic chain islands from north to south Montserrat, Guadeloupe, Dominica, Martinique, St. Lucia, St. Vincent and Grenada (Nagle et al. 1976; Briden et al. 1979). The eastern segment consists of carbonate platform type sediment islands (such as Angulia, St. Martin, Barbuda, Antigua etc., of eastern older arc) rested on the volcanic basement (Murray et al. 2016; Palmer et al. 2016; Garmon et al. 2017). Carbonate sediments mixed with pelagic and hemipelagic muds, clays are omnipresent in every marine stratigraphic

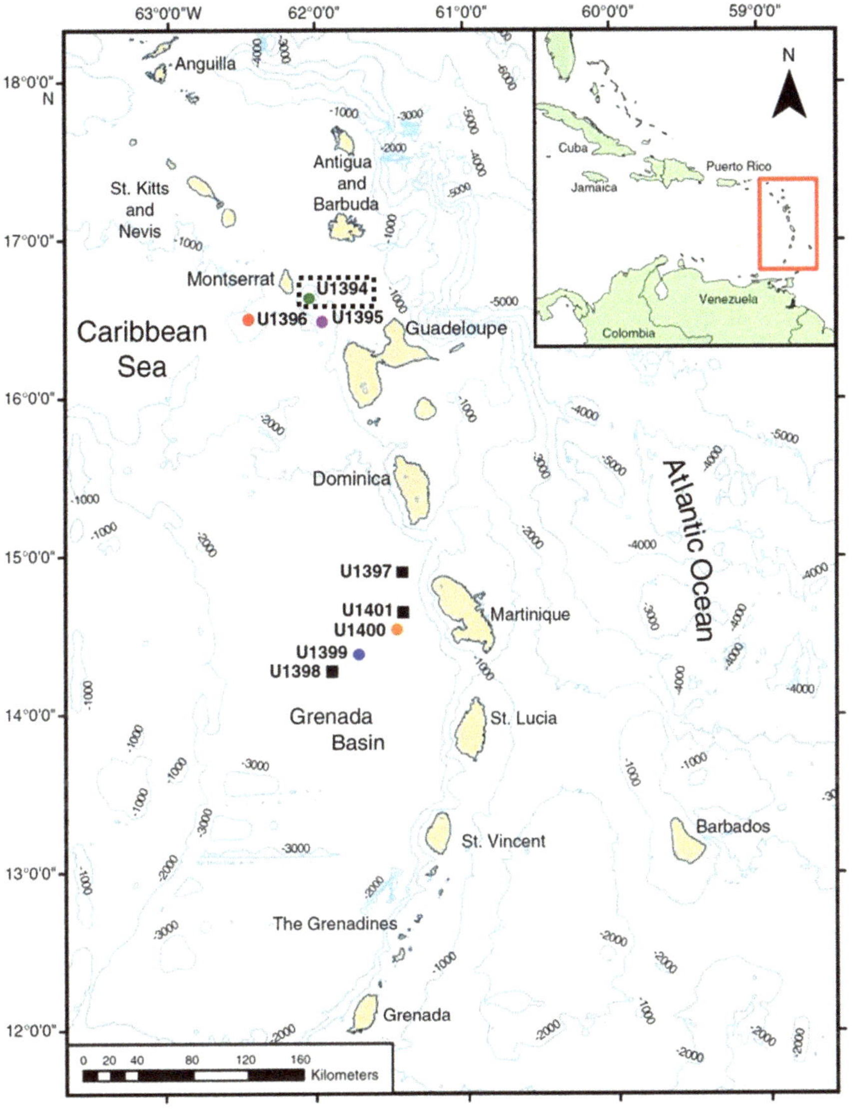

Fig. 1 Map of the study area (inset: red box is the location of Lesser Antilles Island Arc) showing volcanic islands and IODP Expedition 340 drilling sites indicated by solid colour circles and squares. Studied borehole U1394 A and B are shown in the dotted line box. Modified after Le Friant et al. (2013)

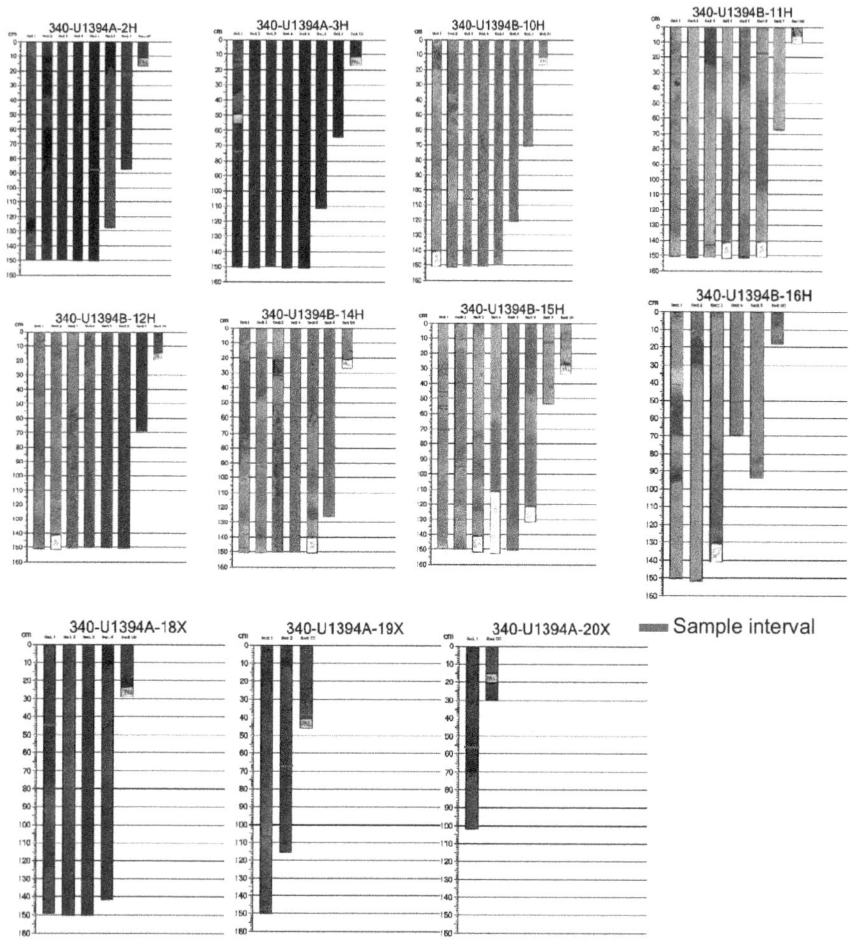

Fig. 2 Core sections of U1354 A and B showing lithological horizons and sub-sample locations for the present study

column in modern as well as old settings. Such sediment horizons occurring in shallow marine conditions in a volcanic/intraoceanic arc tectonic regime in convergent margins record key information on the ocean red-ox conditions, sea-level changes, terrestrial sediment input, paleodepocenters, depositional environments and diagenetic changes etc. (Dorobek 2008; Hawkins et al. 1984). Although, such studies are not much in existence, they throw light on the geochemical and tectonic implications on the shallow marine carbonate systems in a convergent margin tectonic setting. Present study, deals with such shallow marine carbonate system encountered in holes U1394 A and B drill cores pertaining to the Soufriere Hills Volcano of Montserrat Volcanic Island along the Lesser Antilles Arc system to constrain the influence of siliciclastic and volcano-sedimentary inputs from active volcanism through major, trace and rare earth element (REE) geochemistry, total organic carbon content etc. Our studies indicate that these sediment horizons having varying carbonate contents are deposited in an oxygen rich environment under active continental margin tectonic setting.

2 Materials and Methods

The samples are collected from cores of holes U1394 A and B at different depth intervals (Table 1). Based on the shipboard Carbonate and XRD data, the carbonate horizons are identified and sub sampled at IODP Gulf coast repository, Texas A & M University, College Station, USA, at depths between top depth 11.67 m core depth below sea floor A (hereafter CSF) and 11.45 m CSF-B up to the bottom depth of 168.57 m CSF of U1394 A and B holes. These samples are dried and further homogenised in an agate mortar and pestle until fineness. The sample quantities are very less, as they represent the working half cores and all the geochemical analytical methods are destructive in nature, care has been taken to provide sufficient amount of sample for each analysis. Thus the samples made to fine fraction of size <0.075 mm were stored in glass bottles labelled with all the details as per IODP sample curation protocol are subjected to carbonate, total carbon on board RV Joides Resolution while major, minor oxides, loss on ignition, trace and REE determinations are conducted at CSIR-National Geophysical Research Institute at Hyderabad, India. For carbonate analyses, a 5 cm^3 plug sample was taken and was freeze-dried for ~24 h and pulverised manually using an agate mortar and pestle during sample preparation. Inorganic carbon concentrations were determined using a CO_2 coulometer available on board. 2 M HCl was added to around ~10 mg of dried, ground sediment and the liberated CO_2 was titrated to an end point determination using an on board photodetection cell. $CaCO_3$ weight percentage was calculated from the obtained inorganic carbon content, assuming that all evolved CO_2 was derived from calcium carbonate as follows: $CaCO_3$ (wt%) = 8.33 × IC (wt%), where IC = inorganic carbon. As this approach does not discriminate among carbonate minerals, all the evolved CO_2 is assumed to derived from $CaCO_3$.

Total carbon is determined using a ThermoElectron Corporation Flash EA 1112 CHNS elemental analyser available on board calibrated using the Thermo Soil Reference Material NC standard (338 40025 Lot N12A) using ~10–20 mg of dried, ground sediment. Total organic carbon content was estimated as the difference between inorganic carbon and total carbon (for further details refer to Expedition 340 Summary, Methods 2013).

The loss on ignition (%LOI) was determined separately using a high temperature muffle furnace and pure silica crucibles at 1000 °C for 60 min in muffle furnace. Sediment samples and standards (International reference materials such as GSR-5 (Shale, Chinese Geochemical Standard Reference Samples), JDO-1 (dolomite, JGS,

Table 1 Precession and accuracy of JLs-1 (certified reference material from GSJ) used in the present study

Analyte	Certified values	Obtained values	STDV	% RSD
SiO_2 (wt.%)	0.15	0.15	0.01	0.08
Al_2O_3	0.02	0.02	0.004	0.27
Fe_2O_3	0.01	0.01	0.07	1.31
MnO	0.00	0.00	0.00	0.00
MgO	0.63	0.63	0.05	0.07
CaO	55.01	55.01	2.22	0.04
Na_2O	0.00	0.00	0.00	0.00
K_2O	0.11	0.11	0.03	0.22
TiO_2	0.00	0.00	0.00	0.00
P_2O_5	0.04	0.04	0.01	0.30
Total	55.86	55.97	–	–
LOI	43.73	–	–	–
Sc (ppm)	0.03	0.03	0.01	0.16
V	3.6	3.87	0.31	0.08
Cr	3.0	3.12	0.10	0.03
Co	0.1	0.08	0.002	0.02
Ni	0.3	0.33	0.03	0.08
Cu	0.3	0.28	0.01	0.04
Zn	2.9	3.01	0.07	0.02
Ga[a]	0.1	0.19	0.09	0.46
Rb	0.2	0.21	0.01	0.04
Sr	296	292.4	5.95	0.02
Y	0.22	0.23	0.01	0.02
Zr	4.19	4.29	0.06	0.01
Nb	1.30	1.41	0.12	0.08
Cs	0.02	0.08	0.001	0.01
Ba	476	475.9	4.09	0.01
La	0.15	0.15	0.002	0.02
Ce	0.93	0.92	0.01	0.01
Pr	0.04	0.04	0.0004	0.01
Nd	0.14	0.14	0.003	0.02
Sm	0.16	0.17	0.003	0.02
Eu	0.01	0.01	0.0001	0.02
Gd	0.04	0.04	0.003	0.09
Tb	0.004	0.01	0.0028	0.51

(continued)

Table 1 (continued)

Analyte	Certified values	Obtained values	STDV	% RSD
Dy	0.03	0.03	0.0006	0.02
Ho	0.005	0.005	0.00005	0.01
Er	0.01	0.01	0.0002	0.02
Tm	0.002	0.003	0.0017	0.55
Yb	0.02	0.02	0.0004	0.02
Lu	0.03	0.03	0.0003	0.01
Hf	0.10	0.10	0.001	0.01
Ta	0.01	0.02	0.001	0.04
Pb	1.21	1.18	0.024	0.02
Th	0.03	0.03	0.001	0.05
U	1.75	1.70	0.04	0.02

Note Average calculated for five number of observations
STDV standard deviation; *RSD* relative standard deviation
[a]Certified value (CV) not available; hence 0.1 is taken arbitrarily. CV are taken from Georem website georem.mpch-mainz.gwdg.de/

Japan), JLS-1 (limestone, GSJ, Japan) were prepared as pressed pellets Major (SiO_2, Al_2O_3, Fe_2O_{3T}, MgO, CaO, K_2O, Na_2O) and minor elements (TiO_2, MnO and P_2O_5) concentrations were determined by X-ray fluorescence spectrometry (XRF; Phillips MAGIX PRO Model 2440), on pressed pellets prepared from powdered whole-rock samples following the method of Krishna et al. (2007) at CSIR-NGRI, Hyderabad, India. Trace element contents and REE were determined by high resolution inductively coupled plasma mass spectrometry (HR-ICP-MS) using international standard reference materials JDO-1 (dolomite, JGS, Japan), JLS-1 (limestone, GSJ, Japan). Satyanarayanan et al. (2018) is followed for analytical parameters and instrument calibration methods, for trace and REE determination. Sample digestion procedures are as given in Subramanyam et al. (2013). Analytical precision for major, minor oxides and trace elements should be less than 5% relative standard deviation (RSD) where as in the present case it is less than 2% (Table 1).

Total description of samples from cores Exp-340; Hole U1394 A and B, $CaCO_3$, total carbon (TC), total organic carbon (TOC) and total inorganic carbon (TIC) data sets are presented in Table 2. Major and minor oxide composition of carbonate samples are shown in Table 3. Trace and REE data are normalised to the PAAS compositions after Taylor and MacLennan (1985) presented in Table 4. For curator procedures, sample depth calculations, stratigraphic correlations, XRD, TOC, TIC, TC and carbonate analysis methods please see the Chapter Methods in Le Friant et al. (2013).

Table 2 Total description of samples from cores Exp-340; Hole U1394 A & B. $CaCO_3$, TC, TOC and TIC data sets

Hole	Core	Type	Section	A/W	1	2	3	4	5	6	7	8	9
A	2	H	5	W	87	88	11.67	11.459	3.657	0.46	0.439	0.02	37
A	2	H	7	W	22	23	13.79	13.515	7.53	1.06	0.904	0.16	36
A	3	H	1	W	37	38	14.67	14.67	65.89	8.84	7.91	0.93	31
A	3	H	1	W	73	74	22.53	22.53	34.503	4.57	4.142	0.43	35
B	10	H	2	W	46	47	74.455	74.446	32.362	4.35	3.885	0.46	30
B	10	H	3	W	106	107	76.56	76.541	64.474	7.83	7.74	0.09	39
B	11	H	1	W	92	93	82.915	82.915	60.692	7.62	7.286	0.33	27
B	11	H	6	W	17	18	89.69	89.46	76.694	9.89	9.207	0.68	25
B	12	H	1	W	97	98	92.47	92.47	53.687	7.61	6.445	1.17	23
B	12	H	3	W	129	130	95.79	95.674	21.241	2.67	2.55	0.12	38
B	14	H	5	W	50	51	116.5	116.48	38.235	5.2	4.59	0.61	22
B	14	H	6	W	79	80	118.29	118.265	70.78	9.16	8.497	0.66	24
B	15	H	1	W	71	72	119.71	119.71	61.659	8.37	7.402	0.97	40
B	15	H	2	W	95	96	121.45	121.422	51.504	6.97	6.183	0.79	29
B	15	H	7	W	12	13	127.94	127.776	62.408	8.23	7.492	0.74	28
B	16	H	4	W	50	51	133.4	133.386	69.056	8.82	8.29	0.53	26
A	18	X	1	W	44	45	149.24	149.24	36.602	4.9	4.394	0.51	21
A	19	X	2	W	67	68	160.57	160.57	54.92	7.65	6.593	1.06	33
A	19	X	CC	W	38	39	161.43	161.43	66.432	9.07	7.975	1.1	32
A	20	X	1	W	57	58	168.57	168.57	32.104	4.37	3.854	0.52	34

1. Top offset on section (cm); 2. Bottom offset on section (cm); 3. Top depth CSF-A (m); 4. Top depth CSF-B (m); 5. Calcium carbonate (wt%); 6. Total carbon (TC; wt%); 7. Inorganic carbon (TIC; wt.%); 8. Organic carbon (TOC; wt%) by difference (CHNS-COUL); 9. Sample codes used hereafter

Table 3 Major and minor oxide composition of carbonate samples (wt.%)

Sample	37	36	31	35	30	39	27	25	23	38	22
SiO_2	34.86	22.83	46.18	16.32	56.85	15.56	12.91	21.05	33.97	25.33	23.79
Al_2O_3	10.41	6.49	12.36	5.21	15.57	4.7	4.2	5.48	8.94	8.51	6.96
Fe_2O_3	4.73	2.46	4.51	1.97	8.21	1.82	1.71	2.16	3.87	3.26	2.94
MnO	0.19	0.17	0.19	0.19	0.28	0.1	0.09	0.17	0.16	0.12	0.18
MgO	2.99	2.88	2.57	3.92	2.78	3.1	3.51	3.37	3.27	3.07	3.09
CaO	24.71	35.53	19.28	37.13	12.7	42.33	40.68	37.04	24.7	33.06	34.19
Na_2O	2.07	1.52	2.45	1.68	3.32	1.49	1.68	1.19	2.22	1.79	1.6
K_2O	0.78	0.73	0.77	0.33	0.91	0.38	0.24	0.56	0.8	0.73	0.73
TiO_2	0.61	0.28	0.45	0.22	0.85	0.22	0.21	0.28	0.38	0.39	0.33
P2O5	0.11	0.14	0.06	0.07	0.08	0.11	0.08	0.09	0.08	0.13	0.16
Sum	81.5	73.0	88.8	67.0	101.5	69.8	65.3	71.4	78.4	76.4	74.0
LIO%	16.8	26.1	10.6	31.7	0.4	27.8	34.5	27.5	21.4	12.9	22.9
Total	98.3	99.1	99.4	98.7	101.9	97.6	99.8	98.9	99.8	89.3	96.9
Mg/Ca	0.10	0.07	0.11	0.09	0.18	0.06	0.07	0.08	0.11	0.08	0.08
Ca/Mg	9.8	14.6	8.9	11.2	5.4	16.2	13.7	13.0	8.9	12.8	13.1
CIA	27.4	14.7	35.5	11.7	47.9	9.6	9.0	12.4	24.4	19.3	16.0
(Na + Ca)/(Na + Ca + K)	0.97	0.98	0.97	0.98	0.98	0.99	0.99	0.99	0.97	0.94	0.96
Si/Si + Al	0.74	0.75	0.77	0.83	0.77	0.72	0.73	0.77	0.72	0.76	0.77
Sr/Ca	0.0016	0.003	0.011	0.0011	0.0045	0.0059	0.0059	0.007	0.017	0.0089	0.0083
Fe/Mn	18.1	14.7	21.8	13.6	11.5	12.9	17.2	11.9	30.2	26.5	21.4

(continued)

Table 3 (continued)

Sample	24	40	29	28	26	21	33	32	34
SiO_2	23.38	19.53	45.62	11.38	17.11	26.97	35.53	19.97	15.51
Al_2O_3	4.17	6.27	15.77	2.99	5.77	8.35	10.32	5.99	4.4
Fe_2O_3	1.81	2.17	11.03	1.05	2	3.4	6.03	2.22	1.74
MnO	0.12	0.13	0.33	0.08	0.14	0.17	0.25	0.14	0.1
MgO	3.01	3.13	4.73	3.6	3.89	3.55	4.01	3.25	3.74
CaO	33.17	39.05	15.23	44.65	37.29	31.06	22.35	37.68	40.6
Na_2O	2.22	1.35	2.73	1.6	1.77	1.55	2.5	1.28	1.23
K_2O	0.76	0.66	0.54	0.24	0.3	0.77	0.56	0.54	0.28
TiO_2	0.13	0.25	1.23	0.13	0.22	0.43	0.6	0.28	0.23
P2O5	0.07	0.12	0.1	0.08	0.07	0.1	0.1	0.08	0.08
Sum	68.8	72.7	97.3	65.8	68.5	76.4	82.2	71.4	67.9
LIO%	26.4	27.9	2.6	34.1	28.9	23.4	14.6	27.8	33.4
Total	95.2	100.6	99.9	99.9	97.4	99.8	96.8	99.2	101.3
Mg/Ca	0.08	0.07	0.26	0.07	0.09	0.10	0.15	0.07	0.08
Ca/Mg	13.1	14.8	3.8	14.7	11.4	10.4	6.6	13.7	12.9
CIA	10.3	13.2	46.0	6.0	12.8	20.0	28.9	13.2	9.5
(Na + Ca)/(Na + Ca + K)	0.98	0.97	0.99	0.99	0.98	0.97	0.98	0.99	0.98
Si/Si + Al	0.75	0.75	0.76	0.73	0.76	0.75	0.72	0.75	0.73
Sr/Ca	0.0077	0.0114	0.0056	0.007	0.0098	0.006	0.0077	0.0068	0.004
Fe/Mn	14.3	21.8	15.7	9.4	13.1	22.5	24.5	16.4	15.1

Calculations based on
CIA = $[(Al_2O_3)/(Al_2O_3 + CaO + Na_2O) * 100]$ (Harnois 1988); Eu/Eu* = [EuN/(SQRT SmN + GdN)] (Bau and Dulski 1996); Ce/Ce* = $[3Ce/(2La + Nd)]_{SN}$ (Bau and Dulski 1996); Pr/Pr* = $[Pr/(0.5Ce + 0.5Nd)]_{SN}$ (Bau and Dulski 1996)

Table 4 Trace and rare earth elements concentrations (ppm) for carbonaceous sediments of U 1384 A & B; Exp. 340

Sample	37	36	31	35	30	39	27	25	23	38	22
Sc	23.7	9.7	6.2	12.9	8.0	5.9	6.2	3.9	8.8	13.5	11.7
V	234	87	45	116	59	45	43	24	59	126	77
Cr	48	27	31	30	39	23	40	42	43	29	48
Co	69	40	6	101	11	6	6	4	7	16	19
Ni	21	13	22	14	23	18	14	14	20	15	19
Cu	60	33	34	37	45	29	31	35	40	33	34
Zn	152	112	140	214	133	370	62	144	101	143	131
Ga	17.1	13.4	6.1	15.3	11.3	5.9	6.7	3.2	8.0	13.7	9.6
Rb	10.4	14.2	16.4	18.9	15.7	4.8	4.7	5.0	15.7	10.4	8.2
Sr	365	644	1865	250	1191	1565	1718	2178	1869	808	1139
Zr	63.4	65.1	19.3	83.5	43.5	17.6	10.8	11.6	29.3	18.5	26.9
Nb	4.1	2.5	2.6	3.1	2.3	1.2	1.1	0.9	2.7	2.6	2.1
Cs	0.3	0.5	1.2	0.6	0.7	0.6	0.3	0.2	1.1	0.4	0.3
Ba	221	250	167	295	270	325	94	74	188	238	152
Hf	1.9	1.9	0.7	2.3	1.4	0.6	0.5	0.5	1.0	0.9	1.0
Ta	0.3	1.2	0.2	0.3	0.1	0.1	0.1	0.1	0.1	0.2	0.3
Pb	9.2	7.3	7.3	7.7	9.7	6.7	7.7	7.7	8.7	7.1	23.2
Th	1.9	2.7	2.1	3.2	1.5	0.9	0.6	0.7	1.7	1.9	1.1
U	1.2	1.4	3.0	6.0	2.3	3.9	2.2	1.3	2.9	1.6	1.2
La	8.7	10.3	11.3	12.2	9.6	5.2	5.6	7.9	10.0	9.7	7.2
Ce	17.6	18.7	16.9	23.2	16.1	8.2	7.8	8.4	16.4	18.0	13.2
Pr	2.5	2.4	2.6	3.0	2.4	1.3	1.4	1.8	2.5	2.6	1.8
Nd	10.3	9.4	10.3	11.6	9.4	5.1	5.7	7.7	9.9	10.4	7.2
Sm	2.6	2.2	2.3	2.6	2.2	1.3	1.4	1.7	2.3	2.5	1.7
Eu	1.0	0.8	0.6	0.9	0.7	0.4	0.5	0.4	0.6	0.8	0.6
Gd	3.0	2.3	2.3	2.7	2.3	1.4	1.5	1.8	2.3	2.8	1.8
Tb	0.5	0.4	0.4	0.5	0.4	0.2	0.3	0.3	0.4	0.5	0.3
Dy	3.0	2.3	2.1	2.8	2.2	1.4	1.5	1.7	2.3	2.8	1.9

(continued)

Table 4 (continued)

Sample	37	36	31	35	30	39	27	25	23	38	22
Y	16.1	13.6	14.0	16.1	13.4	8.7	9.3	11.1	13.4	15.8	11.4
Ho	0.6	0.5	0.4	0.6	0.5	0.3	0.3	0.4	0.5	0.6	0.4
Er	1.8	1.5	1.3	1.8	1.3	0.9	0.9	1.0	1.3	1.7	1.2
Tm	0.3	0.3	0.2	0.3	0.2	0.1	0.2	0.2	0.2	0.3	0.2
Yb	1.9	1.6	1.2	2.0	1.5	0.9	0.9	0.9	1.3	1.9	1.3
Lu	0.3	0.3	0.2	0.3	0.2	0.2	0.2	0.2	0.2	0.3	0.2
$\sum$REE + Y	54.2	66.6	66.0	80.7	62.5	35.4	37.4	45.4	63.6	70.6	50.4
Y/Ho	25.7	26.6	31.2	26.6	27.7	28.7	29.9	31.6	28.0	26.6	28.0
V/Cr	4.9	3.2	1.4	3.8	1.5	1.9	1.1	0.6	1.4	4.3	1.6
Ni/Co	0.3	0.3	3.6	0.1	2.1	2.8	2.3	3.9	2.7	0.9	1.0
Cu/Zn	0.4	0.3	0.2	0.2	0.3	0.1	0.5	0.2	0.4	0.2	0.3
Uau	0.5	0.5	2.2	4.9	1.8	3.7	2.0	1.0	2.3	1.0	0.8
U/Th	0.62	0.50	1.38	1.85	1.54	4.56	3.63	1.72	1.73	0.84	1.09
Th/U	1.61	1.98	0.72	0.54	0.65	0.22	0.28	0.58	0.58	1.19	0.92
Zr/Sc	2.7	6.7	3.1	6.5	5.4	3.0	1.7	3.0	3.3	1.4	2.3
Th/Sc	0.1	0.3	0.3	0.3	0.2	0.1	0.1	0.2	0.2	0.1	0.1
La/Sc	0.4	1.1	1.8	0.9	1.2	0.9	0.9	2.0	1.1	0.7	0.6
Cr/Th	25.5	10.2	14.6	9.4	25.7	26.7	64.4	57.0	26.1	15.4	43.7
LREE	65.7	9.5	10.7	11.3	8.9	4.7	5.1	7.5	9.4	8.9	6.7
Er/Nd	0.2	0.2	0.1	0.2	0.1	0.2	0.2	0.1	0.1	0.2	0.2
(V/V + Ni)	0.9	0.9	0.7	0.9	0.7	0.7	0.8	0.6	0.7	0.9	0.8
$(La/Yb)_{SN}$	0.33	0.46	0.69	0.44	0.48	0.43	0.44	0.62	0.55	0.38	0.40
$(Eu/Eu^*)_{SN}$	0.83	0.74	0.54	0.78	0.64	0.55	0.58	0.46	0.56	0.72	0.61
$(Ce/Ce^*)_{SN}$	0.85	0.84	0.69	0.87	0.76	0.71	0.62	0.48	0.74	0.81	0.81
$(Pr/Pr^*)_{SN}$	1.02	1.04	1.11	1.03	1.07	1.09	1.13	1.20	1.08	1.04	1.05
$(Nd/Yb)_{SN}$	5.32	5.71	8.51	5.68	6.46	5.74	6.11	8.23	7.31	5.57	5.40
$(Dy/Yb)_{SN}$	1.52	1.41	1.72	1.38	1.49	1.56	1.57	1.79	1.67	1.50	1.41

(continued)

Table 4 (continued)

Sample	24	40	29	28	26	21	33	32	34
Sc	5.4	6.4	5.0	7.2	6.3	13.6	8.1	5.6	11.7
V	40	45	16	47	42	64	55	38	72
Cr	43	28	38	43	41	47	29	23	27
Co	5	7	3	7	15	7	7	5	9
Ni	21	24	18	18	17	17	21	15	18
Cu	37	41	48	44	38	42	40	36	42
Zn	184	258	127	257	119	122	115	122	201
Ga	5.0	5.6	6.0	6.5	5.0	11.2	7.9	5.1	11.3
Rb	8.9	12.9	16.9	14.8	4.4	24.0	19.8	5.1	17.5
Sr	2064	1813	1622	1836	2477	1066	1816	2060	1121
Zr	21.2	22.9	44.4	26.3	11.0	47.5	38.8	13.8	41.0
Nb	1.7	2.2	2.0	2.7	1.0	4.2	3.4	1.2	3.1
Cs	0.6	0.9	0.7	1.0	0.2	1.5	1.4	0.3	1.0
Ba	140	152	174	147	78	223	180	107	188
Hf	0.7	0.8	1.7	0.9	0.5	1.8	1.2	0.5	1.6
Ta	0.1	0.2	0.1	0.2	0.1	0.2	0.2	0.1	0.2
Pb	7.8	10.5	8.3	11.6	8.2	9.8	8.1	8.1	8.0
Th	1.1	1.9	1.6	1.7	0.7	2.5	2.7	0.9	2.4
U	2.9	3.4	4.4	4.6	2.9	2.0	3.8	2.9	1.8
La	8.1	10.2	9.8	13.3	8.2	14.0	11.4	9.6	10.8
Ce	11.4	14.9	16.4	19.9	8.6	25.6	19.7	10.0	20.8
Pr	1.9	2.4	2.6	3.2	2.0	3.7	2.8	2.2	3.0
Nd	7.8	9.6	11.1	12.7	8.4	15.3	11.0	9.4	12.0
Sm	1.7	2.2	2.7	2.9	1.9	3.7	2.4	2.1	3.0
Eu	0.4	0.5	0.6	0.7	0.5	0.9	0.6	0.6	0.8
Gd	1.8	2.3	3.0	3.0	2.1	4.0	2.5	2.3	3.2
Tb	0.3	0.4	0.5	0.5	0.3	0.7	0.4	0.4	0.6
Dy	1.6	2.2	3.1	2.7	1.9	4.1	2.3	2.2	3.3

(continued)

Table 4 (continued)

Sample	24	40	29	28	26	21	33	32	34
Y	10.5	13.4	19.8	16.7	12.3	23.5	13.2	13.9	19.5
Ho	0.4	0.5	0.7	0.6	0.4	0.9	0.5	0.5	0.7
Er	1.0	1.3	2.0	1.5	1.1	2.5	1.3	1.3	2.1
Tm	0.2	0.2	0.3	0.3	0.2	0.4	0.2	0.2	0.4
Yb	1.0	1.3	2.1	1.5	1.1	2.5	1.4	1.3	2.1
Lu	0.2	0.2	0.4	0.2	0.2	0.4	0.2	0.2	0.3
$\sum$REE + Y	48.1	61.6	75.0	79.7	49.3	102.2	69.9	56.1	82.6
Y/Ho	29.8	28.8	28.9	29.9	30.1	27.5	28.1	30.5	27.8
V/Cr	0.9	1.6	0.4	1.1	1.0	1.3	1.9	1.7	2.7
Ni/Co	4.3	3.6	5.6	2.6	1.1	2.4	2.8	2.8	2.0
Cu/Zn	0.2	0.2	0.4	0.2	0.3	0.3	0.3	0.3	0.2
Uau	2.5	2.8	3.9	4.0	2.6	1.1	2.9	2.6	1.1
U/Th	2.52	1.78	2.84	2.67	4.14	0.78	1.38	3.25	0.78
Th/U	0.40	0.56	0.35	0.38	0.24	1.28	0.72	0.31	1.28
Zr/Sc	3.9	3.6	8.8	3.6	1.7	3.5	4.8	2.5	3.5
Th/Sc	0.2	0.3	0.3	0.2	0.1	0.2	0.3	0.2	0.2
La/Sc	1.5	1.6	2.0	1.8	1.3	1.0	1.4	1.7	0.9
Cr/Th	37.7	14.4	24.1	25.2	59.0	18.8	10.7	24.9	11.3
LREE	7.7	9.6	9.3	12.6	7.7	13.0	10.8	9.0	10.0
Er/Nd	0.1	0.1	0.2	0.1	0.1	0.2	0.1	0.1	0.2
(V/V + Ni)	0.7	0.7	0.5	0.7	0.7	0.8	0.7	0.7	0.8
$(La/Yb)_{SN}$	0.62	0.60	0.34	0.66	0.56	0.41	0.62	0.56	0.37
$(Eu/Eu^*)_{SN}$	0.48	0.53	0.48	0.59	0.56	0.69	0.56	0.57	0.67
$(Ce/Ce^*)_{SN}$	0.64	0.67	0.71	0.68	0.46	0.79	0.78	0.47	0.83
$(Pr/Pr^*)_{SN}$	1.14	1.11	1.07	1.11	1.20	1.04	1.07	1.20	1.05
$(Nd/Yb)_{SN}$	8.05	7.65	5.21	8.54	7.77	6.03	8.08	7.46	5.61
$(Dy/Yb)_{SN}$	1.69	1.73	1.45	1.84	1.77	1.60	1.69	1.75	1.53

3 Stratigraphy and Nature of Sediments

The hole U 1394 A was drilled at latitude: 16°38.4259′ N, longitude: 62°02.2822′ W at a water depth 1114.9 (m), Seafloor depth (DRF in m): 1126.3, Cored interval (m): 244.5 having recovery 23% and generated 27 nos. of total cores. The hole U1394 B is drilled 20 m north of hole A at latitude: 16°38.4375′ N, longitude: 62°02.2819′ W at water depth of 1114.2 (m), penetration depth DRF (m):181.4 (m), cored interval (m): 181.4 having recovery 78% and generated 21 nos. of total cores. Both the holes are located off the tip of southeast coast of Montserrat volcanic island at an aerial distance of 24 km which have deposits of Soufriere Hills Volcano. Both the holes were drilled to target the debris avalanche deposits (DAD) to constrain the Soufriere Hills Volcanism of Montserrat volcanic Island. The subsampled points locations on the various cores are presented in Fig. 2.

4 Results

4.1 Geochemistry

4.1.1 Major and Minor Oxides

Major oxide geochemistry of 17 sediment samples of U1394 A and B cores between top depth 11.6 m (CSF-A) to 168.5 m (CSF-A) indicate highly variable SiO_2 contents (11.4–35.5 wt%), Al_2O_3 (2.9–10.4 wt%), Fe_2O_3 (1.1–6.0 wt%), MnO (0.1–0.3 wt%), MgO (2.8–4.0 wt%), CaO (24.7–44.6 wt%), total alkalies (1.5–3.0 wt%), TiO_2 (0.2–0.6 wt%), P_2O_5 (0.08–0.16 wt%) and loss on ignition range between (LOI; 13.4–33.6 wt%). These sediments also vary widely in Mg/Ca (0.06–0.15 wt%, except 0.26 in one sample) and the reciprocal Ca/Mg (3.8–14.8 wt%, except 16.2 in one sample) but very consistent in ratios such as (Na + Ca)/(Na + Ca + K) 0.96–0.99, (Si/Si + Al) 0.72–0.83, Sr/Ca (0.001–0.01) and Fe/Mn (9.4–30.2) classify them as dolomitic limestones as per the chemical criteria described by Todd (1966). The major oxide relationships of with other major, minor and trace elements are presented in Table 3 and Fig. 3.

4.1.2 Trace Elements

The immobile elements contents (ppm) such as Th, Sc, Ti, Zr, Nb and Y are in the range of 0.6–3.2 (avg. 1.7; PAAS-14.6), 3.9–23.7 (avg. 9.0; PAAS-16), 779–7373 (avg. 2314; PAAS-4916), 10.8–83.5 (avg. 32.8; PAAS-201), 0.9–4.2 (avg. 2.3; PAAS-15.4) and 8.7–23.5 (avg. 14.4; PAAS-33, Table 4) respectively. The large ion lithofile elements such as Ba, Rb, Cs, K, Sr and U show abundances (ppm)

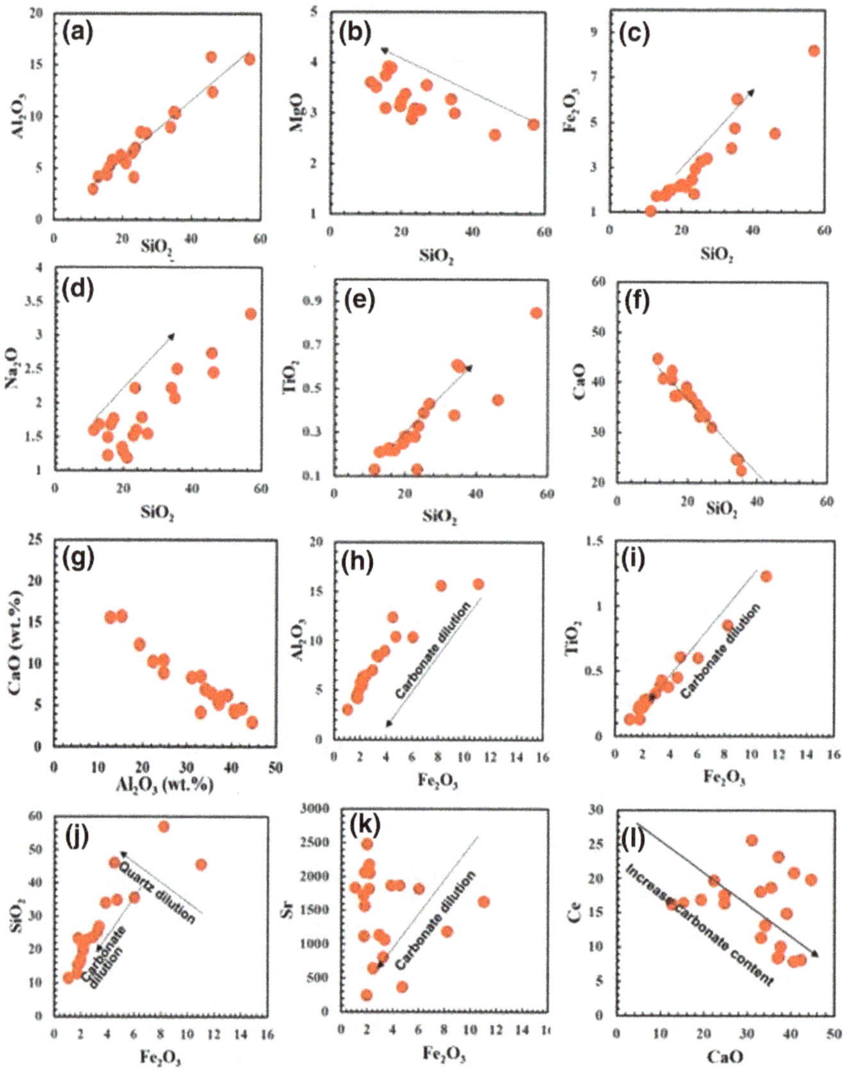

Fig. 3 Major element and selected trace element cross plot relationships of studied marine carbonates of U 1394 A and B

in the range of 73.9–325.2 (avg. 183.2, PAAS-551), 4.4–24.0 (avg. 12.4, PAAS-163), 0.2–1.5 (avg. 0.7, PAAS-6.0), 0.2–0.7 (avg. 0.48, PAAS-3.18), 250–2477 (avg. 1473; PAAS-136) and 1.2–6.0 (avg. 2.8; PAAS-2.9, Table 4). The PAAS normalised spidergram of selected trace elements show depletion of V, Cr, Ni, Cu, Sc, Y, Rb, Ba, Nb, Pb and Th whereas the Mn, Zn and Sr show enrichment characteristics (Fig. 4a).

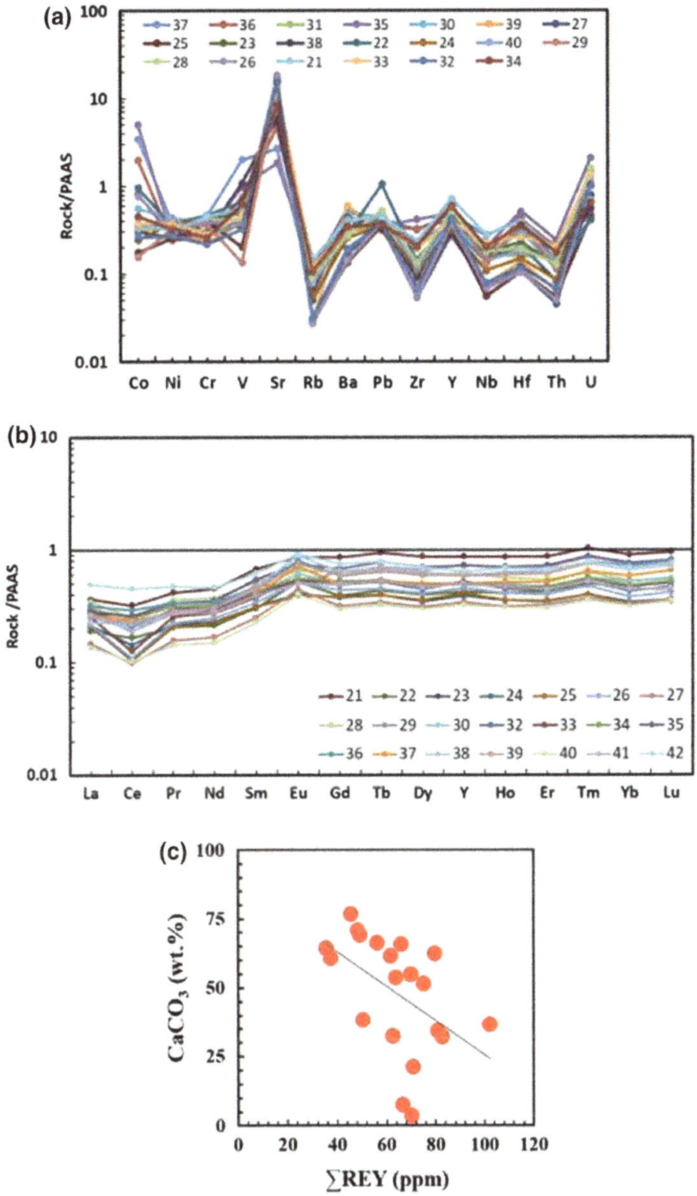

Fig. 4 **a** PAAS normalised selected trace elements distribution patterns of the studied samples showing depletion of V, Cr, Ni, Cu, Sc, Y, Rb, Ba, Nb, Pb and Th whereas the Mn, Zn and Sr show enrichment characteristics. **b** PAAS normalised selected rare earth element distribution patterns showing slightly depleted LREE relative to HREE with Ce negative and Eu positive anomalies. **c** Plot of ∑REY versus $CaCO_3$ showing negative relationship indicating the REY are sourced from seawater

4.1.3 Rare Earth Elements

Rare earth elements along with yttrium abundances, Ce, Eu, La anomalies and other elemental ratios are presented in Table 4. Total REE concentrations ($\sum$REY) range from 35 to 102 (ppm) in the studied sediments. The PAAS normalised REY distribution patterns are smooth and coherent with progressive enrichment in the heavy REE (Fig. 4b). This is due to the formation of trivalent REE and mono- and di-carbonate ion complexes in seawater, thus the filling of 4f electron shell along the series (Tostevin et al. 2016 and references there in). The studied samples show varying abundances of rare earth elements such as Ce in the range of 7.8–25.6 ppm (avg. 15.6, PAAS Ce-82), La 5.2–14.0 ppm (avg. 9.7, PAAS 38.8), Eu 0.4–1.0 ppm (avg. 0.6, PAAS 1.14), Y 8.7–23.5 (avg. 14.4, PAAS 33), $\sum$REY 35–102 (avg. 63.7, PAAS 216). The inverse relationship observed between $CaCO_3$ and $\sum$REY clearly indicate the role of seawater as a major source of REY in these samples (Fig. 4c).

5 Discussion

The major element ratios such as SiO_2/Al_2O_3, Al_2O_3/Na_2O and K_2O/Na_2O are excellent proxies for the interpretation of sediment recycling and detrital input (Zaid 2013; Madhavaraju et al. 2016; Zaid et al. 2015; Tapia-Fernandez et al. 2017). If sediment recycling and detrital inputs are high in the depositional system SiO_2/Al_2O_3 ratio is >6, similarly, the Al_2O_3/Na_2O ratio is >5 and K_2O/Na_2O ratio is >1. The studied samples show SiO_2/Al_2O_3 (2.9–5.6), Al_2O_3/Na_2O (1.9–5.8, >5 only one sample) and K_2O/Na_2O ratios (0.1–0.5) indicating no sediment recycling or detrital input in the studied samples. Further, Roser et al. (1996) have identified SiO_2/Al_2O_3 ratio of sediments as a measure of maturity of sedimentary rocks where the ratio more than 5 is treated as resulted from mature chemical components of the rocks. SiO_2/Al_2O_3 ratio of the studied samples as mentioned above is very consistent except one sample recorded 5.6 indicating their low maturity. This is further supported by low K_2O/Na_2O and very consistent and low Al_2O_3/TiO_2 ratios. The distinct negative relationship obtained between SiO_2 and CaO indicate distinct origins of the two elements in the studied samples (Fig. 3). Very clear carbonate dilution trends are obtained from the positive relationships of Fe_2O_3 when plotted against Al_2O_3, TiO_2, CaO and SiO_2 which is also resulted in high Sr and Ca in these samples (Table 4; Sawant et al. 2017). SiO_2 when plotted against Fe_2O_3, Na_2O and TiO_2 show positive relationship suggesting that these sediments are derived from stable constituents of rocks showing low to moderate sediment maturity level (Fig. 3).

Trace and rare earth element (REE) concentrations are shown in Table 4. The PAAS normalised (Taylor and McLennan 1985) trace element distribution patterns (Fig. 4a) show extreme depletion in V, Sc, Y, Rb, Nb. Cr, Ni, Cu, Ba, Pb and Th whereas enrichment is observed in case of Mn, Zn and Sr relative to PAAS abundances and on the spidergram (Fig. 4a). The trace element pairs such as Zr-Hf, Nb-Ta, Cr-Ni, Ti-V and are compatible elements and behave identically in magmatic systems whereas in alteration-weathering-sedimentary systems these igneous pairs may be fractionated which are influenced by several weathering and depositional (marine/freshwater) conditions (Veizer 1983). The studied marine sediments show positive covariation between Zr-Hf and Nb-Ta (Fig. 5a and b) indicating a significant role played by felsic or intermediated sources which are weakly weathered

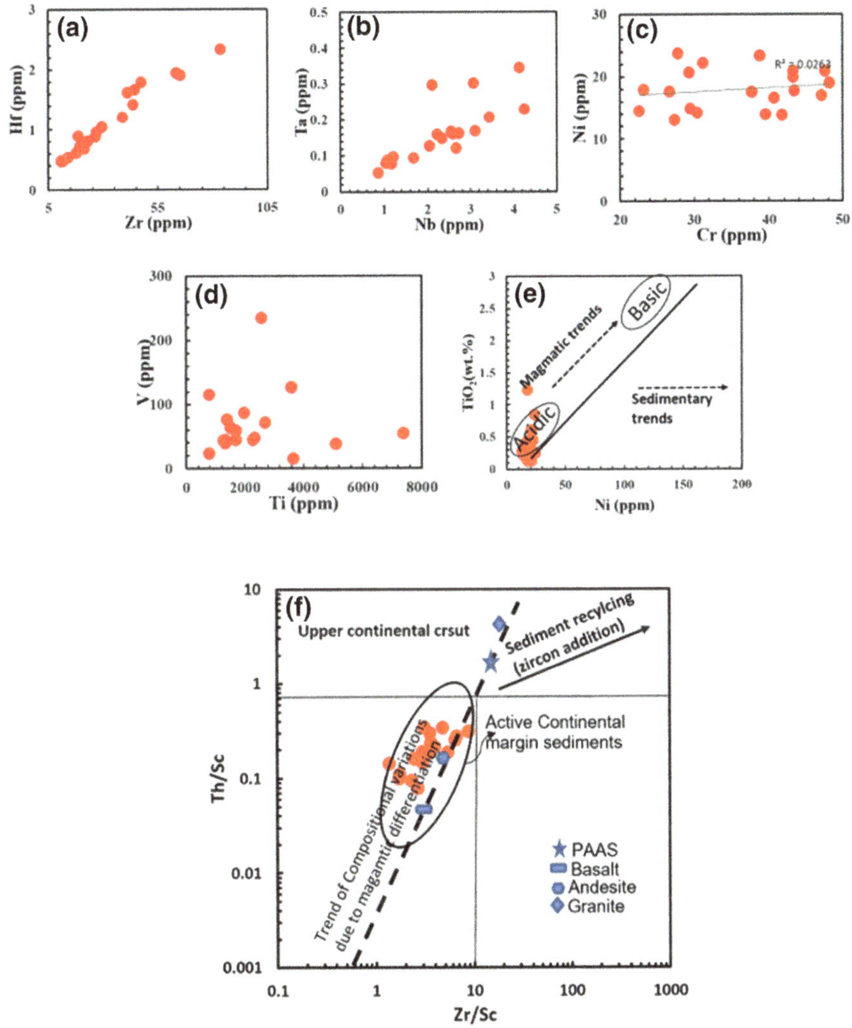

Fig. 5 **a** and **b** Relationship between felsic element pairs Zr versus Hf and Nb versus Ta showing positive covariance indicating their source rocks are felsic to intermediate. **c** and **d** Where negative correlations between mafic element pairs Ni versus Cr and Ti versus V confirms the above felsic-intermediate source for the carbonaceous sediments in the present study. **e** Zr/Sc versus Th/Sc (after Floyd et al. 1989) indicating their derivation from rocks of magmatic differentiation, wherein all the studied samples following a typical trend of active continental margin tectonic affinity

and remain coupled till their deposition, whereas Cr-Ni and Ti-V which are mostly controlled by mafic-ultramafic sources are not correlated in present case (Fig. 5c and d). These element pairs might have been highly fractionated and decoupled before during their transportation to the site of deposition (Brand and Veizer 1980; Veizer 1983). Floyd et al. (1989) described a relationship between two most abundant end members of felsic (TiO_2) and mafic (Ni) systems to verify the provenance from which the sediments derived. The studied sediment samples having low Ni and low TiO_2 contents plot along the typical magmatic trend further indicating their derivation from an acidic source (Fig. 5e). Th/Sc versus Zr/SC relationship after McLennan et al. (1993) indicating that the sediments are derived from varied compositions due to igneous differentiation that is only possible in an active continental margin setting (Fig. 5f). The studied samples follow the trend of Th/Sc (incompatible/compatible elements in igneous chemical differentiation system) and clustered below the PAAS and UCC indicating their derivation from a differentiated felsic igneous source.

Rare earth element (REE + Y) geochemistry in sedimentary systems offer proxies for weathering, depositional, diagenesis, source rock, oceanic paleoredox conditions. Yttrium behaves similar to REE due to their coherent geochemical properties, based on the ionic radii. Yttrium is placed between Dy and Ho in the REE plots which behaves more like a HREE (Nozaki et al. 1997; Allwood et al. 2010). Thus Y/Ho ratios indicate anomalous nature of sea water and marine sediments (Bau and Dulski 1996; Quinn et al. 2004). In addition to this the LREE/HREE, Eu/Eu*, Ce/Ce*, Pr/Pr* ratios as well as Er/Nd characteristics are discussed for the studied samples which offer excellent clues on the genesis of carbonate fraction as well as source of terrigenous contamination if any (Greaves et al. 1999; Taylor and McLennan 1985). The pure carbonate facies record very low abundance of REY whereas the admixed terrigenous material along with seawater influence will increase the REE abundances (Piper 1974; Bellanca et al. 1997). Total REY ($\sum$REY) contents in the studied samples range from 35 to 105 ppm. PAAS (Post Archean Australian Shale) normalised REY distribution patterns show overall less fractionation (La/Yb_N; 0.3–0.7) with slight LREE depletion relative to flat HREE with slight negative to positive Ce (Ce/Ce* 0.2–1.1; most of the samples record <1) and positive Eu (Eu/Eu* 0.9–1.6; most of the samples record >1) anomalies. The studied samples exhibit modern sea water like REE distribution patterns with slight LREE depletion relative to HREE except positive Y-anomalies.

5.1 Europium Anomalies (Eu/Eu)*

The $(Eu/Eu^*)_{SN}$ values are presented in Table 4. The shale normalised Eu anomalies range between 0.9 and 1.6 indicating their varied nature in the sediment column. Michard et al. (1983) and German et al. (1993) attributed Eu anomalies due to the effect of hydrothermal solutions, intense diagenesis of sediments (Murray et al. 1991a; MacRae et al. 1992), variations in plagioclase contents (Nath et al. 1992) whereas Elderfield (1988) speculated about the aeolian input into the sediment column up to 50 m. Hence, it is imperative to verify the Eu relationship with immobile trace elements such as Zr, Y and Th and found the positive correlations indicating non-diagenetic origin of Eu (Fig. 6a–c). Further, the regional influence of aeolian input and hydrothermal effects on Eu positive anomalies have been negated by the poor correlation between the Eu when plotted against Na/Al and K/Al (Murray et al. 1991b; Fig. 6d and e). Thus, the influence of plagioclase addition in the sediments played a role for positive Eu anomalies in the studied samples.

5.2 Cerium Anomalies (Ce/Ce)*

Ce/Ce* ratios are calculated using the technique of Bau and Dulski (1996; modified after Webb and Kamber 2000; $3Ce/(2La + Nd)_{SN}$; Table 4). Trivalent cerium from seawater can undergo oxidation to the highly insoluble tetravalent state. This is mainly a change in the ionic state of Ce as a function of oxidation state of the medium. It's scavenging from seawater and adsorption to the particulate matter is considered as distinctive depletion of Ce in well-oxygenated seawater. Nagarajan et al. (2011) compiled the reasons of anomalous cerium values in the marine system due to the following process 1. Biologically mediated oxidation of Ce (III) to Ce (IV) in coastal and open ocean (Sholkovitz et al. 1994), 2. Ce (III) to Ce (IV) redox shift across anoxic/oxic interfaces in the water columns of marine basins (German and Elderfield 1989; Sholkovitz et al. 1992) and 3. Due the development of largest Ce anomalies in rivers with the highest pH (Goldstein and Jacobsen 1988). Hence, Ce/Ce* parameter is very helpful to determine the redox conditions prevalent in the water column at the time of the REE uptake by these phases (Elderfield and Greaves 1982; German and Elderfield 1990; Bellanca et al. 1997; Moller and Bau 1993; Balaram et al. 2012 and references there in). Ce/Ce* values of the normal seawater range from <0.1 to 0.4 (Elderfield and Greaves 1982; Wang et al. 1986; Piepgras and Jacobsen 1992) while in average shale Ce/Ce* is 1 (Haskin and Haskin 1966; Murray et al. 1991a). A low magnitude Ce-negative anomalies (Ce/Ce* = 0.2–1.1; 04 samples show $\geq$1.0) are observed for the studied samples on the PAAS normalised REE patterns (Fig. 4b) indicating inherited sea water compositions (Tostevin et al. 2016).

5.3 Er/Nd Ratios

REE enrichment or depletion (LREE relative to HREE or vice versa) depends on the redox potentials of the depositional system as well as their abundance in sea water. Fractionation of REE in marine sediments is also possible due to burial and diagenesis. Hence, the Er(HREE)/Nd(LREE) ratios are indicators of diagenetic changes, seawater contribution and the redox process operative during deposition. Er/Nd ratio

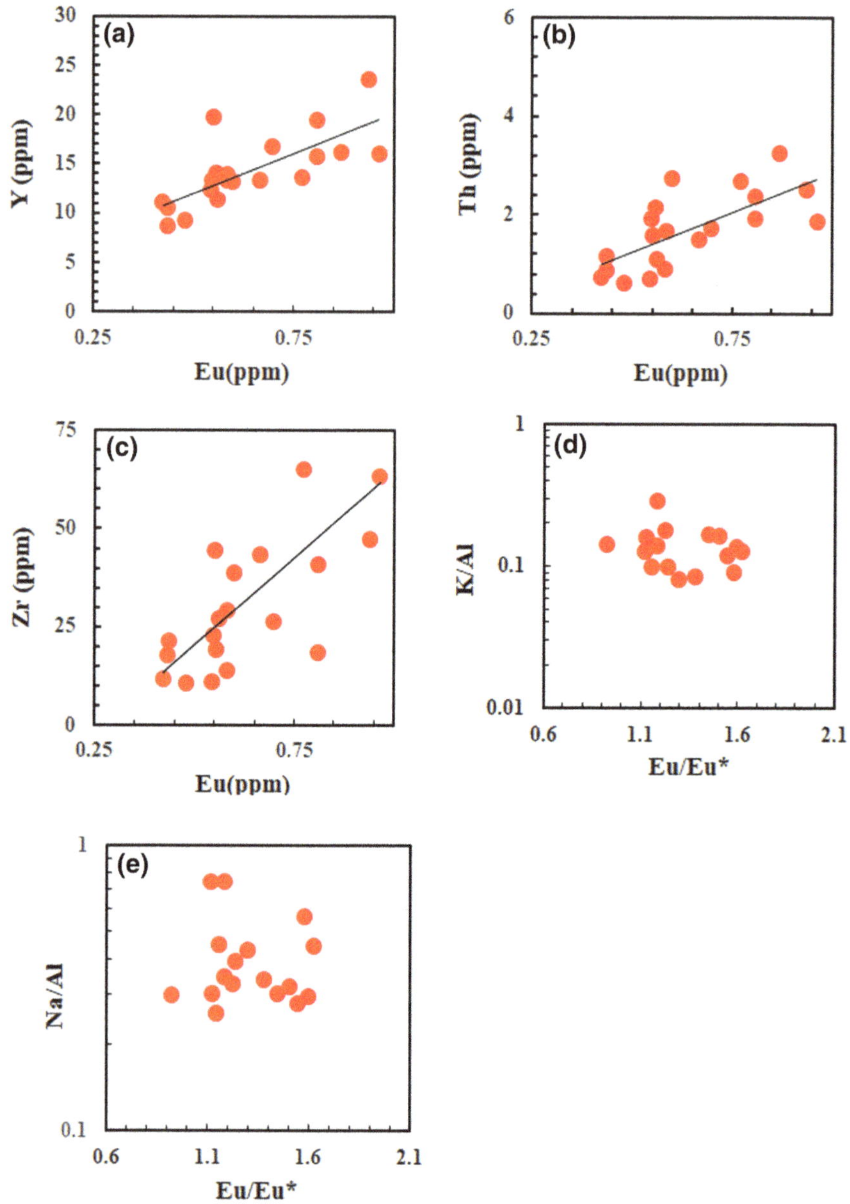

Fig. 6 **a–c** Eu (ppm) plotted against immobile trace elements Zr, Th, Zr display positive relationship indicating non-diagenetic origin of Eu (Murray et al. 1991a; MacRae et al. 1992; Nath et al. 1992). **d** and **e** are showing no correlations between Eu and major element ratios such as Na/Al and K/Al, indicating thus the influence of aeolian input is negated (Murray et al. 1991b)

of normal seawater is 0.27 (de Baar et al. 1988) whereas, the high Er/Nd ratios >0.1 of the limestones reflect the seawater signature retained by the marine carbonates (Bellanca et al. 1997). The Er/Nd ratios tend to decrease below 0.1 in case of addition of detrital input in the marine carbonate system. This is due to the preferential absorption of Nd over Er on to the particle surfaces (De Baar et al. 1988; German and Elderfield 1989; Bellanca et al. 1997; Nagarajan et al. 2011). In the studied samples Er/Nd ratios range between 0.12 and 0.18 representing the pristine nature of carbonates with very little detrital input into a few samples which show Er/Nd ~ 0.12. Another important LREE/HREE ratio namely the $(Nd/Yb)_{SN}$ ratios are good indicators of contamination. The studied samples show $(Nd/Yb)_{SN}$ ratios in the range of 5–8, which are well below the shale values (PAAS-11.4) but much above the ratios for 50 m water depth of normal ocean waters ~0.205–0.497 (Zhang and Nozaki 1996; De Baar et al. 1985). The LREE depletion characteristics are responsible for such high ratios in the present samples. Relative HREE enrichment when compared to LREE prompted us to see the HREE/HREE ratios such as the $(Dy/Yb)_{SN}$ which are well constrained with the studied samples showing 1.4–1.8; whereas PAAS values are 1.6 and the modern seawater (~0.8–1.1) ratios are slightly low. Shields and Stille (2002) reported similar range of ratios recorded for Early Cambrian phosphorites (1.1–2.0) of South China and the same ratios are as low as 0.8–1.0 as reported by Hein et al. (1999), Siegmund (1995) and Kawabe et al. (1991), which are similar to the lower range of near surface modern and open Pacific seawater (Zhang and Nozaki 1996; Nothdurft 2001; Webb and Kamber 2000).

5.4 *Age of the Carbonates*

The present study deals with samples recovered from south eastern part of Montserrat volcanic island (MVI) from the drilled cores of U-1394 A and B holes which intersected debris of mass wasting events of Soufriere Hills volcano of Monserrat Island located along the Bouillante-Montserrat Graben. Young et al. (1998) studied the heightened volcano-seismic activity of Soufriere Hills volcano during 18th July 1995 to next three years till 1998. The authors further reported that the phreatic explosions have continually erupted juvenile andesitic lavas built a lava dome during November 1995. The collapse of eastern flank occurred followed by an explosive eruption on 17th September 1996 sustained till June 1997. The volcano has vented pyroclastic material on 22 and 23 March, 2012 just after we completed the drilling of U1396 site located towards southwest of MVI. The Sofriere Hills pyroclast flows of quantity 2×10^6 m^3, which occupied the shallow shelf area around the MVI mostly as proximal lobes and rest as distal turbidites (Trofimovs et al. 2006). Trofimovs et al. (2010) argued about the large submarine slope failure of carbonate platform and rapid raise of sea level in Lesser Antilles around Soufriere Hills volcano of MVI during the last 26 kyr as well as the age the occurrence of mass flow was 14 ka. Further, along the south eastern segment of MVI, the Bouillante-Montserrat Graben where numerous piston cores, vibracore samples are studied prior to and during IODP Exp. 340, resulted in a better understanding of the dynamic evolution of volcanic island, mass flow deposits, erosional surfaces, Tsunami genesis and failure of carbonate platforms due to seismogenesis. Wall-Palmer et al. (2014 and references there in) dated the sediments intersected in hole U1396 C located towards the southwest of MVI using a combination of parameters such as high-resolution oxygen isotope stratigraphy, AMS radiocarbon dating, biostratigraphy of foraminifera, calcareous nanofossils, clast componentry and established five major events since 250 ka to present. Further, the authors established laterally correlatable stratigraphic horizons across the drill holes on the south west and south east of Montserrat (U 1396 A, B and C with U 1394 A and B) and identified high intensity erosive density-currents that travelled at least 33 km offshore of Montserrat having heterogeneously distributed in space. Coussens et al. (2016) characterised three important age peaks (0.93–0.90, 0.81–0.76 and 0.19–0.12 Ma) mass wasting (flank collapse) and high volcanic instability that are coinciding with Soufriere Hills volcanic activity. The authors have shown excellent lateral stratigraphic correlations up to 15 m depth. Debris Avalanche Deposits (DAD) 1, 2, 4, 8 are located to the southeastern part of Monserrat (Wall-Palmer et al. 2014; Coussens et al. 2016). However, earlier workers used shallow cores of various expeditions to constrain the evolutionary history and volcanism around MVI but the carbonate horizons which are very important in terms of their location along with the packet of hemipelagic and pelagic muds are not studied. Trofimovs et al. (2013) alternatively argued two possible hypotheses for the collapse of carbonate platforms. 1. Subjected to mass wasting of 12–14 ka non-volcanic carbonate shelf of Antigua and Rodanda reached till Montserrat. 2. The Sofriere Hills flank collapse related DAD 1 which also have triggered the collapse of carbonate shelf.

Coussens et al. (2016) studied DAD 2 from U 1394B drill hole intersected at depth of around 11.0–19.0 m (CSF) assigned an age of 130 ka and these data sets are limited to 15 m. The samples from depths 11.6 to 14.6 m (CSF) studied in the present investigation characterise the DAD 2 having very little carbonate content and dominated by siliceous sediments. Wall-Palmer et al. (2014) also studied the cores up to 5.9 m and constrained age up to 246 ka events having done the lateral correlations up to 4.75 m (mbsf) only. Since the studied samples are recovered from depths starting from 11.6 up to 168.5 m (CSF) upper part samples (11.6–14.6 m) can be speculated for age correlations since the drilled cores JR123-5V, JR123-6V and U-1394B are very closely spaced. Hence, it is speculated that the samples in the present study are recovered from the buried DADs 2, 4, 8 located along the Bouillante-Montserrat Graben (Figs. 1 and 11 in Wall-Palmer et al. 2014).

Deposit boundary ages are determined using marine isotope stage (MIS) (Coussens et al. 2016) which have been studied at core depths 28.31, 35.93 and 43.84 m (CSF) from U 1395 B drill cores indicate ages of 191, 243, 300 ka respectively. These ages can be well correlated with U 1394 B which is very closely spaced with U 1395 B drill hole at a slightly shallower depth i.e., 24.1, 31.6 and 39.6 m (CSF; upper surface of Deposit 2 is located 4.2 m shallower in U 1394 B than in U 1395 B). These age boundaries do not reflect any significant changes in the studied geochemical parameters. Thus above aspects combined with the results of present study indicate episodic failures of platformal carbonates linked to the volcanic event of MVI.

5.5 Depth Wise Variation in Geochemical Parameters

It can be interpreted from the depth profiles of studied geochemical parameters of carbonates that high Al and Fe contents are compensating the $CaCO_3$ and TOC contents at carbonate barren zones from 22 to 78 m (CSF; Fig. 7). From 78 to 168 m (CSF) small numerous carbonate horizons show low Al, Fe and V/Cr ratios

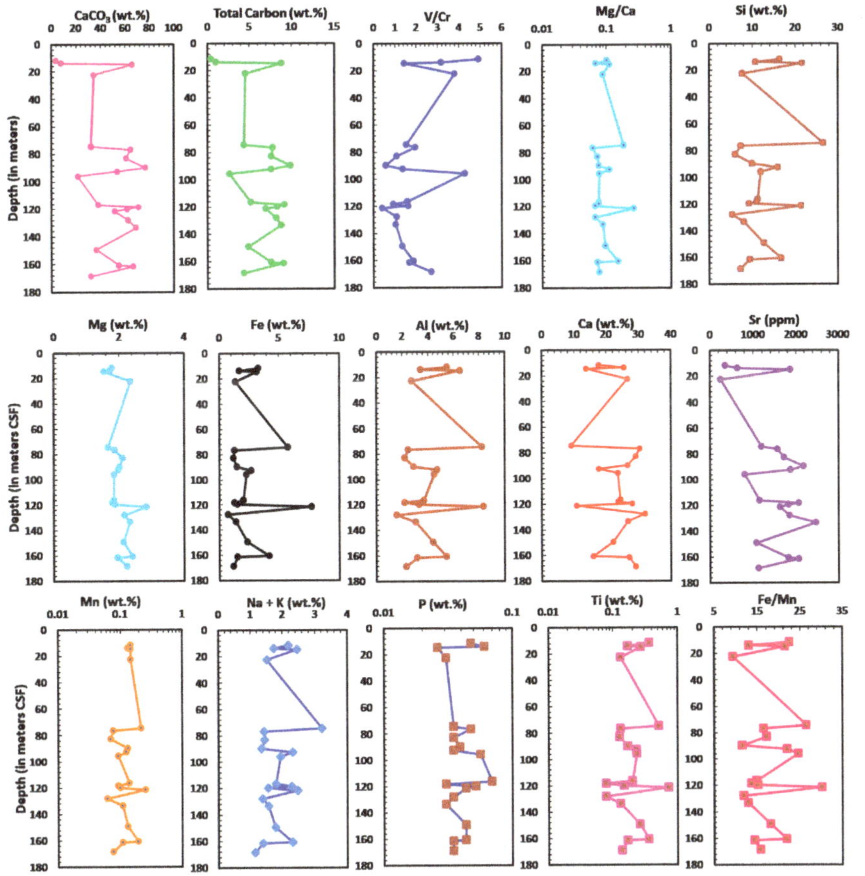

Fig. 7 Down the depth profiles of various geochemical parameters showing three major peaks w.r.t. red-ox sensitive elements (Sr, Mg, Fe, V/Cr, Fe/Mn and Mg)

with moderate Mg contents. An increase in Sr (250–2477 ppm) from 20 to 98 m (CSF) along with concomitant increase in Al_2O_3 and Fe, depletion in Mg, Ca and V/Cr indicate influence of redox conditions below the Deposit 2 assign an age older than 130 ka as well as increase in dolomitisation downward (Sr/Ca ratios discussed elsewhere). Mg content ranges from 2.7 to 4.7 (wt%) having high values at 120 m (CSF) show near consistent downward trends, where Fe, Al also show their peaks at 120 m (CSF) but V/Cr record a very low ratio indicating a redox influence during >1 Ma (Fig. 7).

5.6 Source Weathering and Alteration Characteristics

Goldberg and Humayun (2010) concluded that CIA should not be used for carbonate rich sediments having >30% carbonate content which should be treated by acetic acid prior to analysis to remove the carbonate content. Having very small sample size, the authors could not perform such pre-treatment to eliminate the $CaCO_3$, but as an alternative to this the CIW values are calculated (similar to CIA except K) which are in the range of 34.7–3.4 taking CaO corrected for Ca in apatite only, thus removing the carbonate fraction from non-silicate part (whole anhydrous Ca from anhydrous CaO = anhydrous CaO/56.077, this many moles of oxide is calculated for apatite Ca by moles of oxide Ca − (moles of oxide P × (10/3); Harnois 1988; Fedo et al. 1995). These values when plotted on the A-CN-K trilinear diagram after Nesbitt and Young (1982; Fig. 8a), the studied samples occupying the CN apex indicating the clinopyroxene and hornblende weathering in the provenance; they show a parallel trend with the A-CN line inconsistent with the trends published by Nesbitt and Young (1982) indicating multiple factors involved in the weathering and suggest more influence on the composition of the rock than metamorphism and diagenetic process (McLennan et al. 1993). Further, the ideal weathering trend would be parallel to the A-CN side, whereas the rocks which have not suffered strong weathering occupy the left bottom apex of the diagram below the plagioclase-K-feldspar joining line (Hu et al. 2015). Diagenetic alteration of carbonates increases Mn at the extent of Sr (Nagarajan et al. 2011). In the studied samples such relationship is obscure indicating that the samples have not undergone significant diagenetic alteration (Fig. 8b).

5.7 Contamination and Diagenetic Alterations

Lack of correlation between CaO and $\sum$REY (Fig. 8c) indicate that the REE are derived from either sea water or from the mildly admixed terrigenous material. Such anomalies on the REY distribution diagram are relative to adjacent element abundances whereas Pr/Pr* versus Ce/Ce* relationship indicate the studied samples plot in the modern sea water field with negative Ce anomaly (Fig. 8d). Similar REE

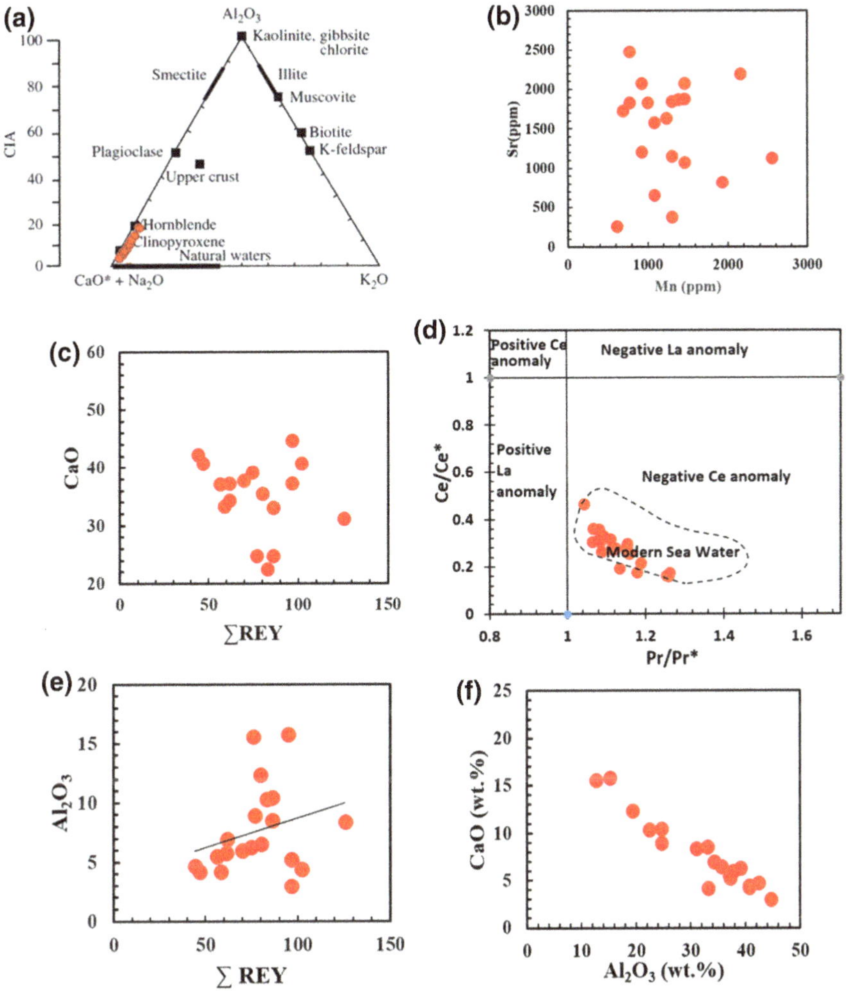

Fig. 8 **a** Ternary plot of molecular proportions of [$Al_2O_3/(Al_2O_3 + CaO + Na_2O)$] with the chemical index of weathering (CIW) calculated after (Harnois 1988; Fedo et al. 1995) plotted on alteration scale shown on the left after Nesbit and Young (1982). Also plotted are selected idealised igneous sedimentary minerals. **b** Sr versus Mn; **c** ∑REY versus CaO; **d** plot of Ce/Ce* versus Pr/Pr* after Bau and Dulski (1996) to identify La and Ce anomalies of carbonaceous sediments indicating negative Ce and positive La anomalies and occupying the zone of modern seawater (upper 50 m of water column) for which the data collected from different locations of the Pacific and Indian Oceans (Nagarajan et al. 2011 and references there in); **e** ∑REY versus Al_2O_3 relationships showing no correlations; **f** negative correlations between Al_2O_3 and CaO indicating that the carbonates are not effected by terrigenous input

distribution patterns are described for carbonate sediments of late Devonian Reefal carbonates by Nothdurft et al. (2004).

The Y/Ho ratios (20.7–31.5; average 28.0; Chondritic Y/Ho = 28; Sun and McDonough 1989; Kamber and Webb 2001) of the studied samples are subchondritic to superchondritic. Y/Ho ratio of modern sea water is strongly superchondritic (40–90; Bau 1996; Global ocean waters 90–162; Johannesson et al. 2006). Hence, the studied samples mimic the Y/Ho ratios of volcanic ash or minimal terrestrial sediment inputs (Y/Ho = 28; Bau and Dulski 1996; Nothdurft et al. 2004). Feeble positive correlation of $\sum$REY with Al_2O_3 (Fig. 8e), negative correlation of $\sum$REE and CaO (Fig. 8c) and very significant negative correlation between Al_2O_3 with CaO (Fig. 8f) along with limited variation in Y/Ho ratios indicate that the carbonate sediments are not affected by terrigenous input.

5.8 *Temperate (Cool Water) Versus Tropical (Warm Water) Carbonates*

Winefield et al. (1996) have discriminated the carbonates precipitated from temperate (cooler waters) and tropical (warm waters) seawaters of both ancient and modern sedimentary basins based on elemental geochemistry such as Sr, Mg, Mn and Na contents. Using 600 bulk samples studied from temperate Cenozoic limestones of New Zealand, the authors reported that low Mg and Sr contents and relatively higher Na, Fe and Mn contents than tropical (warm water) carbonates which are very consistent with the temperature effects on the carbonate dissolution. Moreover, such studies are rare other than few Tasmanian examples and eastern Iran (Azizi et al. 2014).

The samples in the present study have recorded average values (ppm) for Mn 1278; Sr 1433; Na 13814; Mg 20339 and Fe 24161 much higher in abundance then the average contents of New Zealand Cenozoic limestones. These five elements are of utmost importance to discriminate between the tropical and temperate carbonates (Winefield et al. 1996). The ratio Sr/Na and Mn contents in carbonates very well discriminate the carbonate sediments of temperate (cool water) zones from their tropical (warm water) counterparts of both modern and ancient occurrences (Rao 1991; Winefield et al. 1996). The tropical (hot water) carbonate sediments of modern settings will have low Mn and High Sr/Na ratios (3–5), in contrary to this the modern temperate (cool water) carbonates will have high Mn and low Sr/Na ratios (~1). In the studied samples the Mn content and Sr/Na ratio are less than 1, (0.06–0.25 ppm; avg. 0.128) and (0.02–0.2; avg. 0.11) respectively. On the binary discrimination of Sr-Na and Sr-Mg and Sr-Mn (Fig. 9) for the studied samples plot along with the Cenozoic New Zealand carbonates (but on the higher side of the Mn, Mg and Na) indicating their genesis in a temperate cooler waters but in actual condition the MVI (Lesser Antilles) is located in the zone of tropics which is a known latitudinal tectonic setting. Here we report a deviation from the general observation

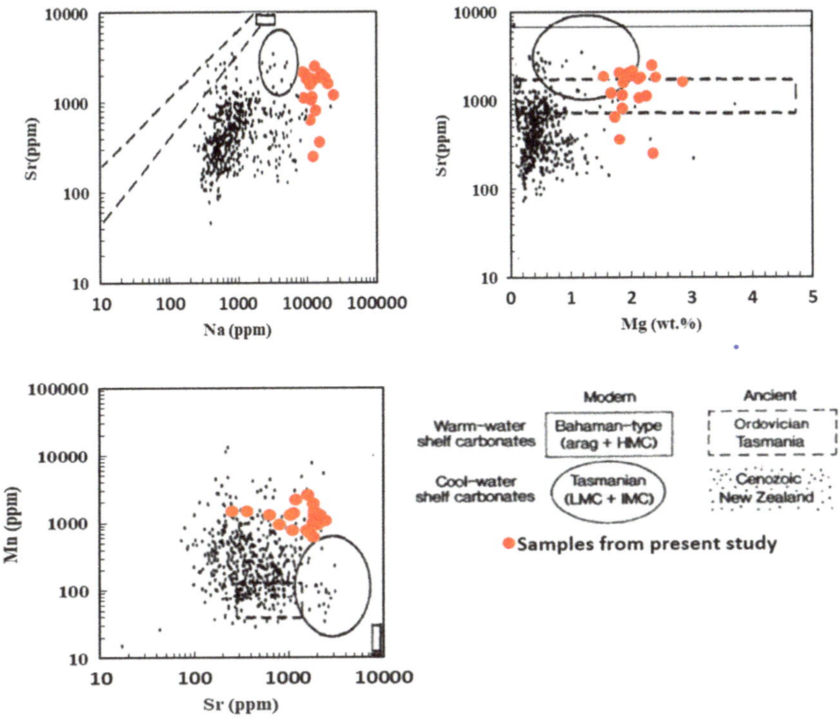

Fig. 9 Binary plots with fields demarcated for modern warm-water, Tasmanian cool water shelf carbonates as well as Ordovician Tasmanian subtropical shelf carbonates along with the carbonates of under present study (red solid circles; data and figures after Winefield et al. 1996) plotted along with the Cenozoic New Zealand carbonates of temperate waters for the Sr-Mn and Sr-Na relationships. On the Sr-Mg binary plot the same samples plot along the tropical (warm water) fields with overlap

of Winefield et al. (1996), Rao (1991), Hood et al. (2004) and Azizi et al. (2014) on the temperate limestones, although the geochemical parameters and ratios as mentioned above satisfy the discrimination. The authors also suggested that the regional generalisations can be made limited to the skeletal differences, diagenetic process involved etc., Based on the above and analysed data sets of the studied samples, the net effect of volcanogenesis, carbonate dilution and dissolution play an important role in discriminating the carbonates of tropical origins. The failure of carbonate platforms and later events might have also modified the chemical character of the studied samples. Boudon et al. (2007) reported that the MVI has recorded more than 47 volcanic events where as 15 of them have occurred during the last 12 kyrs. These observations have been fairly matching with the DAD 2 where we have seen an age of 130 ka and mass flow of debris are observed at 14 ka due to submarine landslides and due to slope failures during rapid sea level changes (Trofimovs et al. 2010 and references there in). Bowles and Fleischer (1985) reported enhanced Amazon and Orinoco sediment influx to the eastern Caribbean during interglacial periods and at times of higher sea levels. Although, the authors speculated about the sediment dispersals and terrigenous dilution produced the reversals of carbonate trends as well as high accumulation rates of sediments during glacial periods around Aves Ridge (located south west to the MVI), the high amounts of cool water mixing into the Lesser Antilles and the possibility of increased production of cool water carbonates caused by upwelling cannot be ruled out. Further studies on these samples are warranted using stable isotope systematics to unravel the growth and destruction of shelf carbonates of Lesser Antilles to further substantiate the present findings.

5.9 Th/U Ratios

Th/U whole-rock ratios of the studied samples (0.3–2.0; predominantly most of the samples show <1.0) suggest that these are much below the upper continental crust igneous values (3.5–4.0; McLennan et al. 1993). During weathering these values will increase much further whereas the low values as obtained in the present case are indicative of active continent margin values (see the Th vs. Th/U relationship, Fig. 10a). U enrichment in sedimentary process lead to lower ratios of Th/U. McLennan et al. (1993) suggested that low Th/U (U enrichment) in sedimentary processes commonly seen in active margin sediments also reflects geochemically depleted mantle sources of the arc provenance (also refer to Hofer et al. 2013). The high Th contents are generally added to sediments from continental sources as detrital material which is also a measure of sediment maturity. U enrichment is generally believed to be due to its binding or trapping with organic materials in the marine system. Thus low Th/U values also indicate immature or less mature nature of the studied samples.

5.10 Paleo-redox Characteristics

The REE such as Ce and Eu (LREE), the trace elements and their ratios like V, Cr, Ni, Co, Cu, Zn, U, Th etc., are also considered as good proxies for paleo-redox characteristics of marine sediments since these cationic metals possess more than one oxidation state by losing electrons (except Zn and Th have single oxidation states as II and IV respectively). These elements are also named redox sensitive elements (ROS) in Marine Geochemistry (Jones and Manning 1994; Cullers 2002; Kato et al. 2002; Swanner et al. 2014; Goldberg and Humayun 2016; Armstrong-Altrin et al. 2015). The redox mechanism of aquatic systems operates based on the available oxidizable and reducible metallic compounds. In other words, it is the balance between the atmospheric oxygen supply and the oxygen consumption by microbial decomposition of organic matter (Drever 1997).

Ni/Co ratio is a strong indicator of paleo-redox conditions of aquatic systems. Ni/Co <5 and >5 are considered to represent oxic and anoxic conditions respectively. The studied samples consistently show the ratio in the range of 0.1–4.3 (except

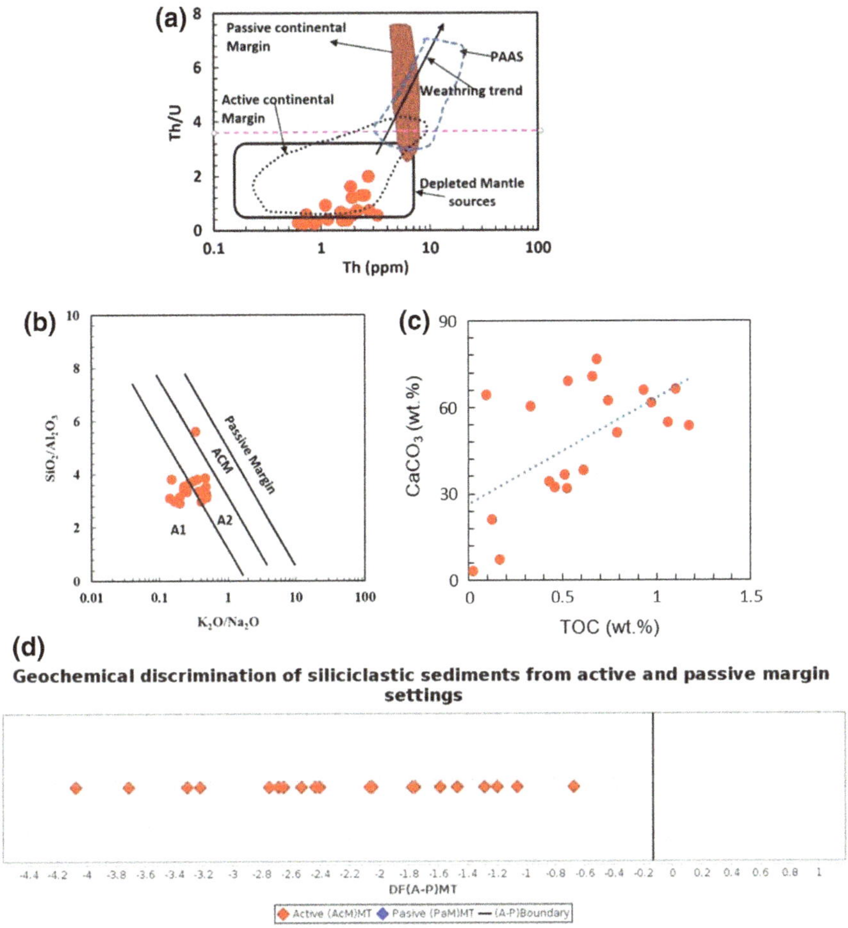

Fig. 10 **a** Plot of Th/U versus Th (after McLennan et al. 1993) indicating the carbonaceous sediments occupy the region of active continental margin sediments also indicating that these sediments are unaltered and unweathered since these are well below the Th/U of upper crustal igneous rocks (3.5–4) which is characteristic of PAAS (shales). **b** K_2O/Na_2O versus SiO_2/Al_2O_3 major oxide based tectonic discrimination diagram after Roser and Korsch (1986) discriminating the marine carbonates into A1 to A2 fields (A1 = arc setting, basaltic and andesitic detritus; A2 = evolved arc setting, felsitic-plutonic detritus). **c** TOC (wt%) versus $CaCO_3$ (wt%) showing a slightly weak positive anomaly indicating oxic depositional environment. **d** Active and passive margin discrimination diagram based on log transformed major and selected trace element compositional data effectively discriminating the studied marine carbonates with active continental margin affinity

one sample showing 5.6). Similarly, the V/Cr ratio is a good indicator of paleoredox conditions. V/Cr ratio more than >4.5 is considered as anoxic, whereas <2 is considered as oxic condition (Jones and Manning 1994). The studied samples show a range of V/Cr ratios from 0.4 to 4.9 (in which 5 samples show V/Cr > 2.0). Further, Goldberg and Humayun (2016) reported Cu/Zn ratio as a good indicator of paleo-redox conditions. If this ratio is >1, it is considered as anoxic, when it is <1 represent oxic conditions. All the studied samples are in the range of 0.1–0.5 confirming to the prevalence of oxic conditions during their deposition. Authigenic uranium (U_{au}; non-detrital fraction derived from $U_{au} = U_{Tot}$-Th/3) precipitates under reducing conditions being carried as uranyl carbonates in solution (Langmuir 1978). It is the difference between the total uranium (measured) and detrital uranium. Detrital uranium can be taken as Th/3, since the Th/U ratio in normal mudstone is 3; where the U is considered to be derived entirely of detrital origin is 3 (Wignall and Myers 1988). The studied samples show a range of ratios 0.5–4.9 of U_{au} suggesting oxic depositional conditions wherein the values <5, between 5–12 and >12 suggest oxic and suboxic to anoxic depositional conditions respectively.

5.11 Total Carbon and Carbonate Content

The studied samples show highly variable $CaCO_3$ contents between 3.6 and 76.6 (wt%). These sediments depict total carbon (TC) contents vary in a wide range from 0.46 to 9.89 (wt%) whereas the inorganic carbon (T_{ic}) shows similar variation from 0.43 to 9.2 (wt%). Total organic carbon is generally very low when compared to the inorganic carbon (Frimmel 2009; Hofer et al. 2013) as obtained in the present studies (0.02–1.17 wt%) which is a general case for many marine carbonate sediments indicating the samples are lithogenic. The positive relationship between TOC (wt%) versus $CaCO_3$ (wt%) (Fig. 10c) for the carbonates indicate their deposition in oxic environment and at a high rate of sedimentation (Nioti et al. 2013).

5.12 Shifting of Platformal Carbonates to Deep Ocean: Evidence from Ce and La (Pr/Pr) Anomalies*

The slight relative depletion of LREE compared to HREE on shale normalised patterns slight positive La, Eu, Gd and Y anomalies, Ce negative anomalies, superchondritic Y/Ho ratios are proxies for retention of seawater characteristics in sediments (Zhang and Nozaki 1996; De Baar et al. 1985; Zhang and Nozaki 1998; Bau 1996; Elderfield and Greaves 1982; Piepgras and Jacobsen 1992). Although, some deviations to the above mentioned criteria observed in the later studies due to addition of detrital material from adjacent continents, addition of biogenic sedimentation to the ocean basins, changes in Eh, Ph conditions of the seawater which modify the sea water composition and sediments due to the above contamination that have operated at depth, varying temperatures, salinity conditions as well as authigenic removal (scavenging) of rare earths and early diagenesis and chemical alteration of sediments (Nagarajan et al. 2011 and references there in). The above mentioned conditions may dilute the original seawater characteristics retained by the marine carbonates resulting in a non-seawater like patterns of REE due to the addition contaminants such as Fe–Mn oxyhydroxides, phosphatic sediments, non carbonates, siliceous sediments etc., The REE distribution patterns of studied carbonate sediments have grossly retained the seawater characteristics except lack of yttrium-positive anomaly whereas Y/Ho values are near chondritic (range of present study 20.7–31.5; average 28.0; Chondritic Y/Ho = 28; Sun and McDonough 1989; Kamber and Webb 2001). Lawrence and Kamber (2006) suggested that the Y/Ho ratios vary as a function of salinity of sea water. Similar Y/Ho values and lack of Y anomalies in most of the carbonate sediments, in spite of no shale contamination obtained by Frimmel (2009). The authors attributed such lacking of Y positive anomaly as a primary feature of precipitating water. Further, Lawrence et al. (2006) reported that the freshwater influence and sediment load admixed by rivers on the marine sediments may generate such low magnitude Y anomalies. While considering these aspects along with the PAAS normalised Ce/Ce* versus Pr/Pr* discrimination diagram (Fig. 7d; modified after Bau and Dulski 1996) which discriminates the studied carbonate samples into a tight cluster having negative Ce and positive La anomalies which are in perfect agreement with modern open oceanic surface waters. The demarcated area shown by a field enclosed in Fig. 7d is an average of modern seawater samples (upper 50 m of water column) for which the data collected from different locations of the Pacific and Indian Oceans (Nagarajan et al. 2011 and references there in). This observation gave a chance of speculation about the origin of the carbonates in an open ocean setting in between the younger and older arc towards north-western margin (located on the overriding plate) of Lesser Antilles volcanic arc as a carbonate platform which are also known as intra arc carbonates. These carbonate platforms due to the Lesser Antilles volcanism and associated tectonics slumped onto the ocean floor in the form of a collapsed carbonate platform as a debris avalanche deposit. Doborek (2008) explained the growth of such thin shallow marine carbonate platforms in an active volcanic front such as the present case, which are effected by differential tectonic subsidence, uplift, relative sea-level changes and paleoclimate variations. The present day examples are Grenadine Islands, Lesser Antilles, where intra-arc carbonate facies form a submerged and incipiently drowned carbonate platform (the Grenadine Bank) that is currently at 20–40 m water depths (D'Anglejan and Mountjoy 1973; Dey and Smith 1989). Soja (1996) described such worldwide platformal carbonates related to the island-arc settings, their dissolution in extinct arc zones and deposition in favourable sites in the marine sediment column. Further, Soja (1996) explained the mechanism by which the carbonate platforms descend to deep marine conditions developed during the intermittent phase of volcanism.

5.13 Tectonic Setting

Carbonaceous sediments may be accumulated at various sites in passive and active margin setting (Dorobek 2008). The Lesser Antilles volcanic arc is an oceanic island arc with active margin conditions wherein American plate is being subducted below the Caribbean plate given rise to a chain of volcanic islands in an arcuate shape (Le Friant et al. 2013; Macdonald et al. 2000). Based on the geochemical characteristics of the studied sediments, the tectonic setting in which their deposition took place is although known to be an active continental margin setting, a verification has been made by using major oxide ratios K_2O/Na_2O versus SiO_2/Al_2O_3 proposed by Roser and Korsch (1986; Fig. 10b). These discrimination diagrams have been used by many workers for texturally dissimilar rocks (coarse grained sandstones to med grained greywackes to fine grained shales/mudstones) which are derived from complex active margins including material derived from continental margin magmatic arcs (and deposited in a variety of basin settings including trench, fore-arc, intra-arc, and back-arc). On the other side when these discrimination diagrams were used for modern sediments which are chemically distinctive such as trailing-edge and forearc muds, but compositions of continental margin arc, strike-slip, and back-arc muds overlapped considerably. The studied carbonaceous samples when plotted on the K_2O/Na_2O versus SiO_2/Al_2O_3 diagram occupy the field of A1–A2 (A1 = arc setting, basaltic and andesitic detritus; A2 = evolved arc setting, felsitic-plutonic detritus) very similar to the results obtained by various other geochemical ratios of the present study (Fig. 10b). Further, an online computer programme known as APM Disc (Active and Passive Margin Discrimination diagrams) for the discrimination of active and passive margins was written in Java using the framework ZK developed and published by Verma and Altrin (2016) is used for the studied samples. The compositional data of major oxides and trace elements (DF A-P MT) for the studied samples have been processed using the excel sheets and discrimination diagrams are obtained. Figure 9c shows that all the studied samples effectively discriminated with active margin affinity. Since the drilled core samples are collected from Lesser Antilles arc (a known active margin tectonic setting), the samples with <30 wt% SiO_2 are also included during data processing (out of 20 samples 14 have SiO_2 contents less than <30%) for APM Disc programme, as the authors (Verma and Altrin 2016, who also supported the idea to include samples of <30% SiO_2 during personnel communication) have considered only samples having >30% SiO_2 contents under this discrimination scheme.

6 Conclusions

1. The studied marine carbonates of drilled core samples from the off coast of Montserrat volcanic island of Lesser Antilles arc retain seawater REE characteristics showing slight LREE depletion relative to HREE, negative Ce and positive Eu-anomalies.
2. Various geochemical proxies and total organic carbon contents indicate that the rare earth elements and yttrium are sourced from seawater whereas carbonates are lithogenic.
3. These marine carbonates are deposited in an open ocean, shallow water predominantly oxic environment devoid of any terrigenous or detrital contamination.
4. These sediments are immature to slightly mature and derived from intermediate to felsic nearby provenance.
5. Our studies indicate that the carbonates are platform type formed in a shallow seawater environment subsequently have undergone dissolution prior to their deposition into the present deeper marine sediment column located on the overriding plate of the Lesser Antilles subduction system.

Acknowledgements The authors thank Dr. Virendra Mani Tiwari, Director, CSIR-NGRI and Dr. M. Ravichandran, Director, NCPOR, Goa for permitting to publish this work. KSVS acknowledges the funding provided by IODP (India) and Ministry of Earth Sciences, New Delhi for his participation in IODP Expedition-340. Dr. D. K. Pandey, Programming Officer, IODP (India) is specially thanked for his kind help and motivation during the preparation of this manuscript. This research is based on shipboard data and core samples provided by IODP Gulf Coast repository, TAMU and CSIR-NGRI is acknowledged for providing the geochemical data. The authors are grateful to the Editors of the special issue and anonymous reviewers for their constructively critical and thought provoking review which has greatly enhanced the content of the manuscript.

References

Allwood AC, Kamber BS, Walter MR, Burch IW, Kanik I (2010) Trace elements record depositional history of an Early Archean stromatolitic carbonate platform. Chem Geol 270:148–163

Armstrong-Altrin JS, Machain-Castillo ML, Rosales-Hoz L, Carranza-Edwards A, Sanchez-Cabeza JA, Ruíz-Fernández AC (2015) Provenance and depositional history of continental slope sediments in the Southwestern Gulf of Mexico unravelled by geochemical analysis. Cont Shelf Res 95:15–26

Azizi SHH, Shabestariv GM, Khazaei A (2014) Petrography and geochemistry of Paleocene-Eocene limestones in the Ching-dar syncline, eastern Iran. Geosci Front 5:429–438

Balaram V, Banakar VK, Subramanyam KSV, Roy P, Satyanarayanan M, Mohan MR, Sawant SS (2012) Yttrium and rare earth element contents in seamount cobalt crusts in the Indian Ocean. Curr Sci 103(11):1334–1338

Bau M (1996) Controls on fractionation of isovalent trace elements in magmatic and aqueous systems: evidence from Y/Ho, Zr/Hf, and lanthanide tetrad effect. Contrib Mineral Petrol 123:323–333

Bau M, Dulski P (1996) Distribution of yttrium and rare-earth elements in the Penge and Kuruman iron-formations, Transvaal Supergroup, South Africa. Precambrian Res 79:37–55
Bellanca A, Masetti D, Neri R (1997) Rare earth elements in limestone/marlstone couplets from the Albian-Cenimanian Cismon section (Venetian region, northern Italy): assessing REE sensitivity to environmental changes. Chem Geol 141:141–152
Boudon G, Le Friant A, Komorowski J-C, Deplus C, Semet MP (2007) Volcano flank instability in the Lesser Antilles arc: diversity of scale, processes, and temporal recurrence. J Geophys Res Solid Earth 112:B08205
Bowles FA, Fleischer P (1985) Orinoco and Amazon River sediment input to the eastern Caribbean Basin. Mar Geol 68:53–72
Brand U, Veizer J (1980) Chemical diagenesis of a multicomponent carbonate system. 1. Trace elements. J Sediment Res 50(4):1219–1236
Briden JC, Rex DC, Faller AM, Tomblin JF (1979) K-Ar geochronology and palaeomagnetism of volcanic rocks in the Lesser Antilles island arc. Philos Trans R Soc Lond 291(1383):485–528
Cullers RL (2002) Implications of elemental concentrations for provenance, redox conditions, and metamorphic studies of shales and limestones near Pueblo, CO, USA. Chem Geol 191:305–327
Coussens M et al (2016) The relationship between eruptive activity, flank collapse, and sea level at volcanic islands: a long-term (>1 Ma) record offshore Montserrat, Lesser Antilles. Geochem Geophys Geosyst 17:2591–2611. https://doi.org/10.1002/2015GC006053
D'Anglejan B, Mountjoy EW (1973) Submerged reefs of the eastern Grenadines shelf margin. Geol Soc Am Bull 84:2445–2454
De Baar HJW, Bacon MP, Brewer PG (1985) Rare earth elements in the Pacific and Atlantic oceans. Geochim Cosmochim Acta 49:1943–1959
De Baar HJW, German CR, Elderfield H, Van Gaans P (1988) Rare earth element distributions in anoxic waters of the Cariaco Trench. Geochim Cosmochim Acta 52:1203–1219
Dey S, Smith L (1989) Carbonate and volcanic sediment distribution patterns on the Grenadines Bank, Lesser Antilles Island Arc, Eastern Caribbean. Bull Can Petrol Geol 37:18–30
Dorobek SL (2008) Carbonate-platform facies in volcanic-arc settings: characteristics and controls on deposition and stratigraphic development. In: Draut AE, Clift PD, Scholl DW (eds) Formation and applications of the sedimentary record in arc collision zones. Geol Soc Spec Pap 436, pp 55–90. https://doi.org/10.1130/2008.2436(04)
Drever JI (1997) The geochemistry of natural waters: surface and groundwater environments, 3rd edn. Prentice-Hall, Upper Saddle River, NJ, 436 pp
Elderfield H (1988) The oceanic chemistry of the rare-earth elements. Philos Trans R Soc Lond 325:105–126
Elderfield H, Greaves MJ (1982) The rare earth elements in seawater. Nature 296:214–219
Expedition 340 summary (2013) In: Le Friant A, Ishizuka O, Stroncik NA, the Expedition 340 Scientists (eds) Proceedings of the IODP, 340: Tokyo (Integrated Ocean Drilling Program Management International, Inc.). https://doi.org/10.2204/iodp.proc.340.101.2013
Fedo CM, Wayne Nesbitt H, Young GM (1995) Unravelling the effects of potassium metasomatism in sedimentary rocks and paleosols, with implications for paleoweathering conditions and provenance. Geology 23(10):921–924
Feuillet N, Manighetti I, Tapponnier P, Jacques E (2002) Arc parallel extension and localization of volcanic complexes in Guadeloupe, Lesser Antilles. J Geophys Res Solid Earth 107(B12):ETG-3
Floyd PA, Winchester JA, Park RG (1989) Geochemistry and tectonic setting of Lewisian clastic metasediments from the Early Proterozoic Loch Maree Group of Gairloch, NW Scotland. Precambrian Res 45(1–3):203–214
Frimmel HE (2009) Trace element distribution in Neoproterozoic carbonates as palaeoenvironmental indicator. Chem Geol 258(3–4):338–353
Garmon WT, Allen CD, Groom KM (2017) Geologic and tectonic background of the Lesser Antilles. In: Allen CD (ed) Landscapes and landforms of the Lesser Antilles. World geomorphological landscapes. Springer, Cham. https://doi.org/10.1007/978-3-319-55787-8_2

Germa A, Quidelleur X, Labanieh S, Chauvel C, Lahitte P (2011) The volcanic evolution of Martinique Island: Insights from K-Ar dating into the Lesser Antilles arc migration since the Oligocene. J Volcanol Geotherm Res 208(3–4):122–135

German CR, Elderfield H (1989) Rare earth elements in Saanich Inlet, British Columbia, a seasonally anoxic basin. Geochim Cosmochim Acta 53:2561–2571

German CR, Higgs NC, Thomson J, Mills R, Elderfield H, Blusztajn J, Fleer AP, Bacon MP (1993) A geochemical study of metalliferous sediment from the TAG Hydrothermal Mound, 26° 08′ N, Mid-Atlantic Ridge. J Geophys Res Solid Earth 98(B6):9683–9692

Goldberg K, Humayun M (2010) The applicability of the Chemical Index of Alteration as a paleoclimatic indicator: an example from the Permian of the Paraná Basin, Brazil. Palaeogeogr Palaeoclimatol Palaeoecol 293(1–2):175–183

Goldberg K, Humayun M (2016) Geochemical paleoredox indicators in organic-rich shales of the Irati Formation, Permian of the Paraná Basin, southern Brazil. Braz J Geol 46(3):377–393

Goldstein SJ, Jacobsen SB (1988) Rare earth elements in river waters. Earth Planet Sci Lett 89:35–47

Greaves MJ, Elderfield H, Sholkovitz ER (1999) Aeolian sources of rare earth elements to the Western Pacific Ocean. Mar Chem 68:31–38

Harnois L (1988) The CIW index: a new chemical index of weathering. Sediment Geol 55:319–322

Haskin MA, Haskin LA (1966) Rare earths in European shales, a redetermination. Science 154:507–509

Hawkins JW, Bloomer SH, Evans CA, Melchior JT (1984) Evolution of intra-oceanic arc-trench systems. Tectonophysics 102:175–205

Hein JR, Fan D, Ye J, Liu T, Yeh H-W (1999) Composition and origin of Early Cambrian Tiantaishan phosphorite-Mn carbonate ores, Shaanxi Province, China. Ore Geol Rev 15:95–134

Hofer G, Wagreich M, Neuhuber S (2013) Geochemistry of fine-grained sediments of the upper Cretaceous to Paleogene Gosau Group (Austria, Slovakia): implications for paleoenvironmental and provenance studies. Geosci Front 4(4):449–468

Hood SD, Nelson CS, Kamp PJJ (2004) Discriminating cool-water from warm-water carbonates and their diagenetic environments using element geochemistry: the Oligocene Tikorangi Formation (Taranaki Basin) and the dolomite effect. NZJ Geol Geophy 47(4):857–869. https://doi.org/10.1080/00288306.2004.9515093

Hu J, Li Q, Fang N, Yang J, Ge D (2015) Geochemistry characteristics of the Low Permian sedimentary rocks from central uplift zone, Qiangtang Basin, Tibet: insights into source-area weathering, provenance, recycling, and tectonic setting. Arab J Geosci 8(8):5373–5388

Johannesson KH, Doyle LH, Alejandra C (2006) Do Archean chemical sediments record ancient seawater rare earth element patterns? Geochim Cosmochim Acta 70(2006):871–890

Jones B, Manning DA (1994) Comparison of geochemical indices used for the interpretation of palaeoredox conditions in ancient mudstones. Chem Geol 111(1–4):111–129

Kamber BS, Webb GE (2001) The geochemistry of late Archaean microbial carbonate: implications for ocean chemistry and continental erosion history. Geochim Cosmochim Acta 65:2509–2525

Kato Y, Nakao K, Isozaki Y (2002) Geochemistry of Late Permian Triassic pelagic cherts from southwest Japan: implications for an oceanic redox change. Chem Geol 182:15–34

Kawabe I, Kitahara Y, Naito K (1991) Non-chondritic Yttrium/Holmium ratio and lanthanide tetrad effect observed in pre-Cenozoic limestones. Geochem J 25:31–44

Krishna AK, Murthy NN, Govil PK (2007) Multielement analysis of soils by wavelength-dispersive X-ray fluorescence spectrometer. At Spectrosc 28(6):202

Langmuir D (1978) Uranium solution-mineral equilibria at low temperatures with applications to sedimentary ore deposits. Geochim Cosmochim Acta 42:547–569

Lawrence MG, Kamber BS (2006) The behaviour of the rare earth elements during estuarine mixing-revisited. Mar Chem 100:147–167

Lawrence MG, Greig A, Collerson KD, Kamber BS (2006) Rare earth element and yttrium variability in south east Queensland waterways. Aquat Geochem 12:39–72. https://doi.org/10.1007/s10498-005-4471-8

Le Friant A, Ishizuka O, Stroncik NA (2013) The Expedition 340 Scientists. In: Proceedings of Integrated Ocean Drilling Program, 340. https://doi.org/10.2204/iodp.proc.340.101.2013

Macdonald R, Hawkesworth CJ, Heath E (2000) The Lesser Antilles volcanic chain: a study in arc magmatism. Earth Sci Rev 49:1–76

MacRae ND, Nesbitt HW, Kronberg BI (1992) Development of a positive Eu anomaly during diagenesis. Earth Planet Sci Lett 109(3–4):585–591

Madhavaraju J, Tom M, Lee YI, Balaram V, Ramasamy S, Carranza-Edwards A, Ramachandran A (2016) Provenance and tectonic settings of sands from Puerto Peñasco, Desemboque and Bahia Kino beaches, gulf of California, Sonora, México. J S Am Earth Sci 71:262–275

McLennan SM, Hemming S, McDaniel DK, Hanson GN (1993) Geochemical approaches to sedimentation, Provenance and Tectonics. In: Johnsson MJ, Basu A (eds) Processes controlling the composition of clastic sediments. Geol Soc Spec Pap 285, pp 21–40. https://doi.org/10.1130/spe284-p21

Michard A, Albarede F, Michard G, Minster JF, Charlou JL (1983) Rare-earth elements and uranium in high-temperature solutions from East Pacific Rise hydrothermal vent field (13° N). Nature 303(5920):795

Moller P, Bau M (1993) Rare-earth patterns with positive cerium anomaly in alkaline waters from Lake Van, Turkey. Earth Planet Sci Lett 117:671–676. https://doi.org/10.1016/0012-821x(93)90110-u

Murray RW, Buchholtz MR, Brumsack HJ (1991a) Rare earth elements in Japan Sea sediments and diagenetic behavior of Ce/Ce*, results from ODP leg 127. Geochim Cosmochim Acta 55:2453–2466

Murray RW, Buchholtz Brink MR, Brink MR, Gerlach DC, Russ GP III, Jones DL (1991b) Rare earth, major and trace elements in chert from the Franciscan complex and Monterey group, California: assessing REE sources to fine grained marine sediments. Geochim Cosmochim Acta 55:1875–1895

Murray NA, Muratli JM, Hartwell AM, Manners H, Megowan MR, Goñi M, Palmer M, McManus J (2016) Data report: dissolved minor element compositions, sediment major and minor element concentrations, and reactive iron and manganese data from the Lesser Antilles volcanic arc region, IODP Expedition 340 Sites U1394, U1395, U1396, U1399, and U1400. In: Le Friant A, Ishizuka O, Stroncik NA, the Expedition 340 Scientists (eds) Proceedings of the Integrated Ocean Drilling Program, 340: Tokyo (Integrated Ocean Drilling Program Management International, Inc.). https://doi.org/10.2204/iodp.proc.340.207.2016

Nagarajan R, Madhavaraju J, Armstrong-Altrin JS, Nagendra R (2011) Geochemistry of Neoproterozoic limestones of the Shahabad formation, Bhima basin, Karnataka, southern India. Geosci J 15(1):9–25

Nagle F, Stipp JJ, Fisher DE (1976) K-Ar geochronology of the Limestone Caribbees and Martinique, Lesser Antilles, West Indies. Earth Planet Sci Lett 29(2):401–412

Nath BN, Roelandts I, Sudhakar M, Plueger WL (1992) Rare earth element patterns of the Central Indian Basin sediments related to their lithology. Geophys Res Lett 19:1197–1200

Nesbitt HW, Young G (1982) Early Proterozoic climates and plate motions inferred from major element chemistry of lutites. Nature 299(5885). https://doi.org/10.1038/299715a0

Nioti D, Maravelis A, Tserolas P, Zelildiis A (2013) TOC and $CaCO_3$ content in Oligocene shelf deposits on Lemnos Island and their relation with depositional conditions. In: Bull Geol Soc Greece XLVII, proceedings of the 13th international congress, Chania, Sept 2013

Nothdurft LD (2001) Rare earth element geochemistry of Late Devonian reefal carbonates, Canning Basin, Western Australia: a proxy for ancient seawater chemistry. Honours thesis, Queensland University of Technology, Brisbane, 103 pp

Nothdurft LD, Webb GE, Kamber BS (2004) Rare earth element geochemistry of Late Devonian reefal carbonates, Canning Basin, Western Australia: confirmation of a seawater proxy in ancient limestones. Geochim Cosmochim Acta 68:263–283

Nozaki Y, Zhang J, Amakawa H (1997) The fractionation between Y and Ho in the marine environment. Earth Planet Sci Lett 148:329–340

Palmer MR, Hatter SJ, Gernon TM, Taylor RN, Cassidy M, Johnson P, Le Friant A, Ishizuka O (2016) Discovery of a large 2.4 Ma Plinian eruption of Basse-Terre, Guadeloupe, from the marine sediment record. Geology 44(2):123–126

Piepgras DJ, Jacobsen SB (1992) The behaviour of rare earth elements in seawater: precise determination of ferromanganese crusts. Geochim Cosmochim Acta 56:1851–1862

Piper DZ (1974) Rare earth elements in the sedimentary cycle: a summary. Chem Geol 14:285–304

Quinn KA, Byrne RH, Schijf J (2004) Comparative scavenging of yttrium and the rare earth elements in seawater: competitive influences of solution and surface chemistry. Aquat Geochem 10:59–80

Rao CP (1991) Geochemical differences between subtropical (Ordovician), temperate (Recent and Pleistocene) and subpolar (Permian) carbonates, Tasmania, Australia. Carbonate Evaporite 6:83–106

Roser BP, Korsch RJ (1986) Determination of tectonic setting of sandstone-mudstone suites using content and ratio. J Geol 94(5):635–650

Roser BP, Cooper RA, Nathan S, Tulloch AJ (1996) Reconnaissance sandstone geochemistry, provenance, and tectonic setting of the lower Paleozoic terranes of the West Coast and Nelson, New Zealand. N Z J Geol Geophys 39:1–16. https://doi.org/10.1080/00288306.1996.9514690

Satyanarayanan M, Balaram V, Sawant SS, Subramanyam KSV, Vamsi Krishna G, Dasaram B, Manikyamba C (2018) Rapid determination of REEs, PGEs, and other trace elements in geological and environmental materials by high resolution inductively coupled plasma mass spectrometry. Atom Spectrosc 39(1):1–15

Sawant SS, Vijaya Kumar K, Balaram V, Subba Rao DV, Rao KS, Tiwari RP (2017) Geochemistry and genesis of craton-derived sediments from active continental margins: insights from the Mizoram Foreland Basin, NE India. Chem Geol 470:13–32

Shields G, Stille P (2002) Diagenetic constraints on the use of cerium anomalies as palaeoseawater redox proxies: an isotopic and REE study of Cambrian phosphorites. Chem Geol 175:29–48

Sholkovitz ER, Shaw TJ, Schneider DL (1992) The geochemistry of rare earth elements in the seasonally anoxic water column and pore waters of Chesapeake Bay. Geochim Cosmochim Acta 56:3389–3402

Sholkovitz ER, Landing WM, Lewis BL (1994) Ocean particle chemistry: the fractionation of rare earth elements between suspended particles and seawater. Geochim Cosmochim Acta 58(6):1567–1579

Siegmund H (1995) Fazies und Genese unterkambrischer Phosphorite und mariner Sedimente der Yangtze Plattform, Südchina. Berliner geowissenschaftliche Ahandlungen, Reihe A 173:114

Soja CM (1996) Island-arc carbonates: characterization and recognition in the ancient geologic record. Earth Sci Rev 41(1–2):31–65

Subramanyam KSV, Balaram V, Reddy UVB, Satyanarayanan M, Sawant SS, Roy Parijat (2013) Problems involved in using improper calibration CRMs in geochemical analyses—a case study on mafic rocks of Boggulakonda Pluton, east of Cuddapah Basin, India. MAPAN 28(1):1–9

Sun SS, McDonough WF (1989) Chemical and isotopic systematics of oceanic basalts: implications for mantle composition and processes. Geol Soc London Spec Publ 42(1):313–345

Swanner ED, Planavsky NJ, Lalonde SV, Robbins LJ, Bekker A, Rouxel OJ et al (2014) Cobalt and marine redox evolution. Earth Planet Sci Lett 390:253–263. https://doi.org/10.1016/j.epsl.2014.01.001

Tapia-Fernandez HJ, Armstrong-Altrin JS, Selvaraj K (2017) Geochemistry and U-Pb geochronology of detrital zircons in the Brujas beach sands, Campeche, southwestern Gulf of Mexico, Mexico. J S Am Earth Sci 76:346–361

Taylor SR, McLennan SM (1985) The continental crust: its composition and evolution. Blackwell, Oxford, UK, 312 pp. ISBN-13: 978-0632011483

Todd TW (1966) Petrographic classification of carbonate rocks. J Sediment Petrol 36(2):317–340

Tostevin R, Shields GA, Tarbuck GM, He T, Clarkson MO, Wood RA (2016) Effective use of cerium anomalies as a redox proxy in carbonate-dominated marine settings. Chem Geol 438:146–162

Trofimovs J, Amy L, Boudon G, Deplus C, Doyle E, Fournier N, Hart MB, Komorowski JC, Le Friant A, Lock EJ, Pudsey C (2006) Submarine pyroclastic deposits formed at the Soufrière

Hills volcano, Montserrat (1995–2003): what happens when pyroclastic flows enter the ocean? Geology 34(7):549–552

Trofimovs J et al (2010) Evidence for carbonate platform failure during rapid sea-level rise; ca 14 000 year old bioclastic flow deposits in the Lesser Antilles. Sedimentology 57(3):735–759

Trofimovs J, Talling PJ, Fisher JK, Sparks RSJ, Watt SFL, Hart MB, Smart CW, Le Friant A, Cassidy M, Moreton SG, Leng MJ (2013) Timing, origin and emplacement dynamics of mass flows offshore of SE Montserrat in the last 110 ka: implications for landslide and tsunami hazards, eruption history, and volcanic island evolution. Geochem Geophys Geosyst 14(2):385–406

Veizer J (1983) Trace elements and isotopes in sedimentary carbonates. Rev Mineral Geochem 11(1):265–299

Verma SP, Armstrong-Altrin JS (2016) Geochemical discrimination of siliciclastic sediments from active and passive margin settings. Sediment Geol 332:1–12. https://doi.org/10.1016/j.sedgeo.2015.11.011

Wall-Palmer D, Coussens M, Talling PJ, Jutzeler M, Cassidy M, Marchant I, Palmer MR, Watt SFL, Smart CW, Fisher JK, Hart MB (2014) Late Pleistocene stratigraphy of IODP Site U1396 and compiled chronology offshore of south and south west Montserrat, Lesser Antilles. Geochem Geophys Geosyst 15(7):3000–3020

Wang YL, Liu YG, Schmitt RA (1986) Rare earth element geochemistry of south Atlantic deep sediments, Ce anomaly change at ~ 54 My. Geochim Cosmochim Acta 50:1337–1355

Webb GE, Kamber BS (2000) Rare earth elements in Holocene reefal microbialites: a new shallow seawater proxy. Geochim Cosmochim Acta 64:1557–1565

Wignall PB, Myers KJ (1988) Interpreting benthic oxygen levels in mudrocks: a new approach. Geology 16(5):452–455

Winefield PR, Nelson CS, Hodder APW (1996) Discriminating temperate carbonates and their diagenetic environments using bulk elemental geochemistry: a reconnaissance study based on New Zealand Cenozoic limestones. Carbonate Evaporite 11:19–31

Young SR, Steven RJS, Aspinall WP, Lynch LL, Miller AD, Robertsson REA, Shepard JB (1998) Overview of the eruption of Soufriere Hills volcano, Montserrat, 18 July 1995 to December 1997. Geophys Res Lett 25(18):3389–3392

Zaid SM (2013) Provenance, diagenesis, tectonic setting and reservoir quality of the sandstones of the Kareem Formation, Gulf of Suez, Egypt. J Afr Earth Sci 85:31–52

Zaid SM, El-Badry O, Ramadan F, Mohamed M (2015) Petrography and geochemistry of pharaonic sandstone monuments in tall san Al Hagr, Al Sharqiya governorate, Egypt: Implications for provenance and tectonic setting. Turk J Earth Sci 24(4):344–364

Zhang J, Nozaki Y (1996) Rare earth elements and yttrium in seawater: ICP-MS determinations in the East Caroline, Coral Sea, and South Fiji basins of the western South Pacific Ocean. Geochim Cosmochim Acta 60:4631–4644

Zhang J, Nozaki Y (1998) Behaviour of rare earth elements in seawater at the ocean margin: a study along the slopes of the Sagami and Nankai troughs near Japan. Geochim Cosmochim Acta 62(8):1307–1317

Morpho-taxonomy of Corycaeid Cyclopoids from Lakshadweep Sea, South Eastern Arabian Sea—A Part of the Indian Ocean

R. Radhika and S. Bijoy Nandan

Abstract Classical taxonomy, the science of naming and classifying organisms accurately plays an important role in biodiversity conservation. Taxonomy has been endorsed as an important gadget for the execution and monitoring of the Convention on Biological Diversity which in turn has significantly reformed the climate in the field of taxonomic research. This paper endeavor to accomplish the goal of strengthening capacity building (taxonomists) as a part of Global Taxonomy Initiative of convention on biological diversity and also due to rapidly dwindling cyclopoid copepod taxonomic expertise in the national scenario. Despite the vastness of the volumes of seas and oceans in this world, the echelon of work done on the taxonomical facet of marine cyclopoid copepods are surprisingly fewer. Since the taxonomy of marine cyclopoids along the southwestern part of Indian Ocean have not been attempted much in the Indian scenario, the present study grips much significance as it becomes a pioneering effort on the morphotaxonomy of Corycaeid copepods (Order: Cyclopoida) from Lakshadweep Sea and associated islands, South Eastern Arabian Sea—a part of Indian Ocean. Sampled small Corycaeidae group of copepods in high abundance from five lagoons (Kavaratti, Kalpeni, Minicoy, Agatti and Bangaram) and open ocean zones of Minicoy of the Lakshadweep archipelago during premonsoon, monsoon and post monsoon seasons during 2013–2016 period thus handing over the morphotaxonomic descriptions of twenty-three species under five genera (*Corycaeus*, *Urocorycaeus*, *Onychocorycaeus*, *Ditrichocorycaeus* and *Farranula*) belonging to the family Corycaeidae.

Keywords Marine cyclopoids · Corycaeidae · Lakshadweep · Indian Ocean

1 Introduction

The family Corycaeidae (Dana 1852) now including five genera, viz, *Corycaeus*, *Onychocorycaeus*, *Urocorycaeus*, *Ditrichocorycaeus* and *Farranula* are marine

R. Radhika · S. Bijoy Nandan (✉)
Department of Marine Biology, Microbiology and Biochemistry, Cochin University of Science and Technology, Cochin 682016, India

D. K. Pandey et al. (eds.), *Dynamics of the Earth System: Evolution, Processes and Interactions*, Society of Earth Scientists Series,
https://doi.org/10.1007/978-3-030-40659-2_7

pelagic copepods occurring typically in epipelagic zone of tropical to temperate seas (Motoda 1963; Boxshall and Halsey 2004; Wi and Soh 2013a). Boxshall and Halsey (2004) through a major phylogenetic revision established all the subgenera mentioned above as valid genera. Yet, appropriate morphological criteria for all generic status could not be presented for the same. Corycaeid group of copepods which are easily recognized by their peculiar body structure and large paired eyes seem to be very useful indicator forms of warm ocean currents. The major research topics have spotlighted on their community structure and biology, thus providing new insights into the largely unknown ecological importance of this copepod family (Robin et al. 2013; Radhika et al. 2014a). Apart from ecological issues, the taxonomy of the Corycaeidae family has been the subject of progressively more detailed studies over the past two decades. Nevertheless, despite these efforts, the taxonomy of many species under this family is still inadequate and have not been described in ample detail. Moreover, the species belonging to this family could be distinguished only by a very few characters, since many of the congenic species are morphologically very similar. These include micro structures like ornamentation of endopodal segments of antenna, spination of exopodal segments of the swimming legs and length-width ratio of the concerned parts, which in turn require a level of detail, which is not, by and large, adopted in most taxonomic descriptions.

2 Materials and Methods

Conducted field sampling invariably during the early morning hours by hired boats in different Lakshadweep islands (lagoon stations of Kavaratti, Kalpeni, Minicoy, Agatti and Bangaram) from 2013 to 2016 during premonsoon (spring intermonsoon), monsoon (fall inter monsoon) and post monsoon (winter monsoon) seasons for the collection of mesozooplankton (Figs. 1 and 2). Sampling from open ocean zones (eighteen stations located between 8°12′–8°24′ N and 72°54′–73°12′ E) in Minicoy Island was also undertaken during premonsoon (spring intermonsoon) on board the Fishery Oceanographic Research Vessel (FORV) Sagar Sampada (Cruise #338) in April 2015 (Fig. 3). Zooplankton samples were collected using a modified WP (working Party) plankton net (mesh size 200-μm) having a mouth area of 0.28 m^2 (Jagadeesan et al. 2013). Carried out horizontal tows just below the water surface, at a fixed speed of 1 Knot for 10 min, with the net attached to a calibrated flow meter (General Oceanics model number −2030 R, 2012). The average depth of the collection site is 6 m with a maximum depth of about 16 m. Immediately after sampling, those for morphological examination were preserved in 4% buffered formaldehyde. Used Magnesium chloride (7–10%) as narcotizing agent (Steedman 1976; Anon 1968; Omori and Ikeda 1984; Harris et al. 2000) in high-quality airtight bottles labeled with the date, time, location of the station and season. Sorting, analysis and species level identification of Corycaeidae copepods was done under Stereo Microscope in the Ecology lab of the department using standard identification keys (Boxshall and Halsey 2004; Dahl 1912; Farran 1911; Kasturirangan 1963;

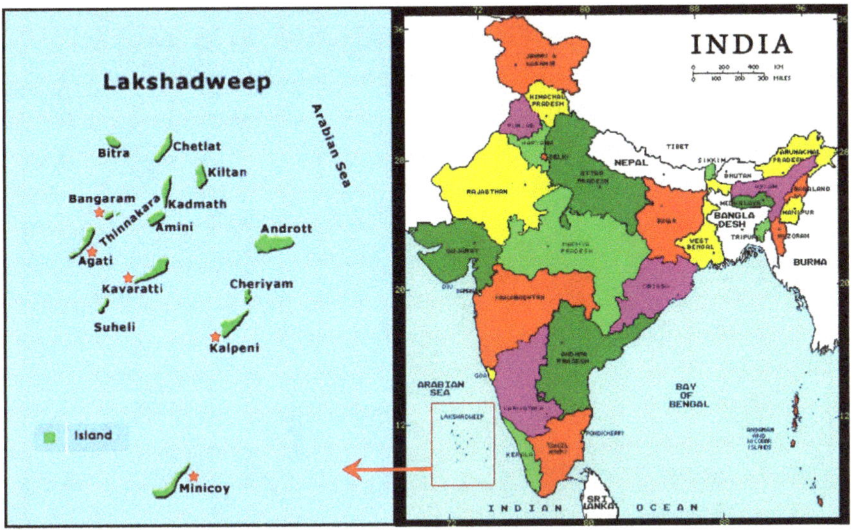

Fig. 1 Geographical location of the sampling stations in the Lakshadweep island. *Source* https://www.mapsofindia.com/maps/india/india-political-map.htm

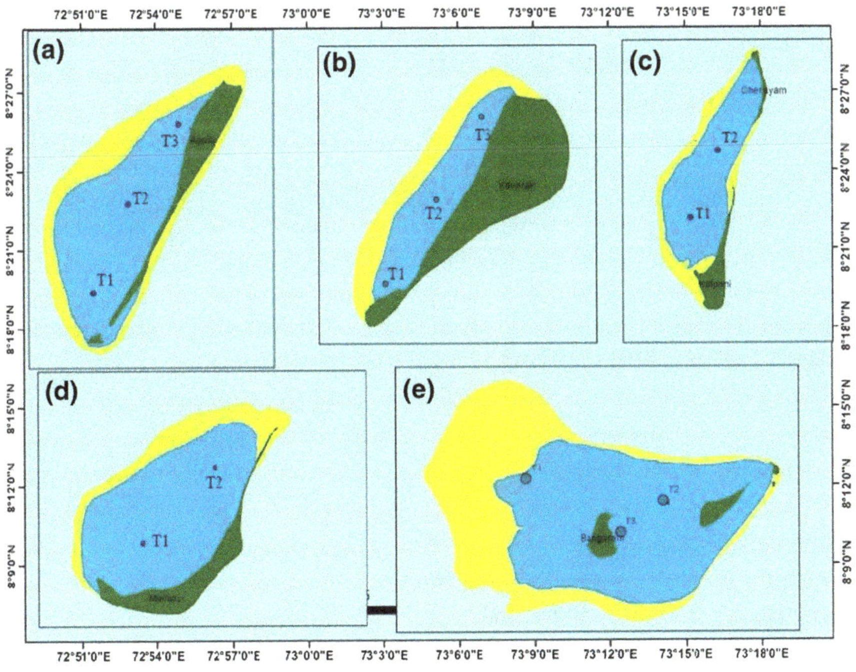

Fig. 2 Lagoon stations showing transects of **a** Agatti, **b** Kavaratti

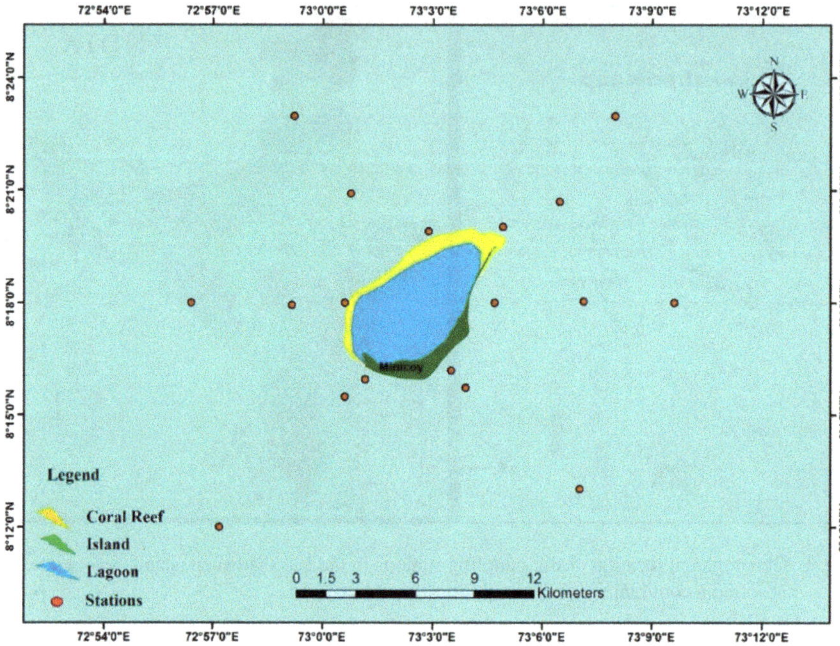

Fig. 3 Study area—cruise track of FORV Sagar Sampada Cruise #338 (Minicoy Island)

Wi and Soh 2013a, b; Wi et al. 2013). Morphological examination and dissection of specimens were done in glycerol distilled water mixture using fine needles on a cavity slide under Leica (Model DM 500) bright-field compound microscope. Drawings were made with the aid of the drawing tube. Specimens were measured using ScopePhoto (x64) software. The specimens were incorporated into the copepod collection of Department of Marine Biology, Microbiology and Biochemistry, School of Marine Sciences, Cochin University of Science and Technology (catalog number MBM/DBT/01/14, MBM/DBT/02/14, MBM/DBT/1-21/2017).

3 Results

Family: Corycaeidae
Genus: *Corycaeus* (*Corycaeus*) Dana, 1845
***Corycaeus crassiusculus* Dana, 1849**

Material examined: Sorted out the samples from plankton samples collected from Kavaratti, Minicoy (open ocean and lagoon) and Bangaram.

Female (Fig. 4): Total length 0.78–0.8 mm. Prosome (PR) four segmented, frontal margin arc-shaped, with two large separate cuticular lenses (Fig. 4a). Prosome about twice longer than urosome including caudal rami (2.6:1.3); about 1.8 times as long

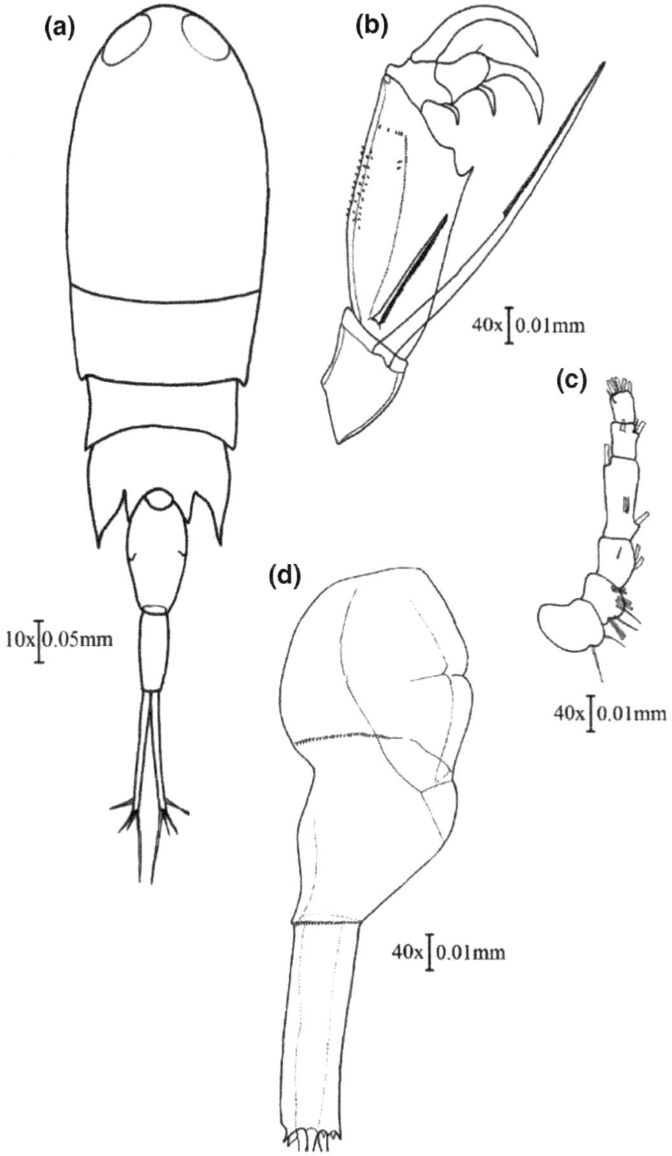

Fig. 4 *Corycaeus crassiusculus* female. **a** Habitus (dorsal); **b** antenna (A2); **c** antennule; **d** urosome (lateral view)

as wide (2.6:1.4). The genital segment is shorter than anal somite and caudal rami combined. Urosome (UR) two-segmented with very divergent caudal rami (Fig. 4d). Genital Somite (GS) overlaps anal somite at the dorsal margin. Anal somite is rectangular shaped with its distal margin ornamented with spinules ventrolaterally; 0.98 times as long as wide at base; slightly shorter than genital somite. Antenna (A2) four-segmented (4b). Coxobasis 1.6 times longer than wide; inner distal margin bears a long stout seta fringed with minute spinules along the inner distal and terminal margin. Endopod three-segmented and unequal; first endopodal segment robust, extremely longer than other two endopodal segments, about 2.3 times as long as wide at base; bears unipinnate setae, on inner proximal margin; 2.5 times shorter than coxobasal setae. Midventral surface vertically adorned with smooth denticles along the length of the first endopodal segment. Outer lateral margin randomly decorated with small denticles in which anterolateral margin bears a serial row of five denticles; marginal two are placed equidistant and other three serially. Adjacent to the midventral row of denticles, along the anteroventral margin two more denticles are also present in which the proximal one is the longest. Inner distal margin formed of two comparatively stout teeth like process. Second endopodal segment, shortest of the three bears three elements. Third endopodal segment cylindrical slightly as long as wide at the base, with a hump-like protrusion at the distal part bearing naked spiniform setae and is drawn into a curved terminal claw with a small blunt spine at inner base. Antennule (A1) short, six segmented (Fig. 4c).

Remarks: Females of *C. crassiusculus* are largely identified by the overlapping of the genital segment on the anal segment at the dorsal margin; described here is the specimen characterized by body length of 0.78 mm; two-segmented urosome with very divergent caudal rami; ventro-lateral ornamentation of the anal somite; six segmented antennule; ornamentation of the first endopodal segment of the antenna. Descriptions of *C. crassiusculus* by Dana's (1852–55) were based exclusively on male specimens. Yet, the female of *C. venustus* described in the same papers were later identified by Dahl (1912) as the female of *C. crassiusculus*. Therefore, the name *C. venustus* was dropped. The present specimen from the waters of the Lakshadweep Islands on comparison with all the known species of the genus show the closest similarity with *C. crassiusculus* in shape, size, features of the appendages, arrangements of spines and setae and other characteristics.

Distribution: Atlantic, Mediterranean, Indo-Pacific, Japan Sea, North Pacific (http://copepods.obs-banyuls.fr/en).

***Corycaeus speciosus* Dana, 1849**

Material examined: Sorted out the samples from plankton samples collected from Kavaratti, Minicoy (open ocean and lagoon) and Bangaram.

Female (Fig. 5): Total length 1.80–2.33 mm. Prosome four segmented, 2.05 times longer than wide and 1.4 times as long as urosome. Dorsally visible cuticular lenses present (Fig. 5a). Urosome two-segmented 3.7 times longer than wide. Almost rhomboid-shaped genital double somite 1.95 times longer than wide. AS (Anal Somite) 1.3 times as long as wide. CR (Caudal Rami) divergent at its base, 1.48 times as long as wide. The ratio of the length of urosomal somites and CR

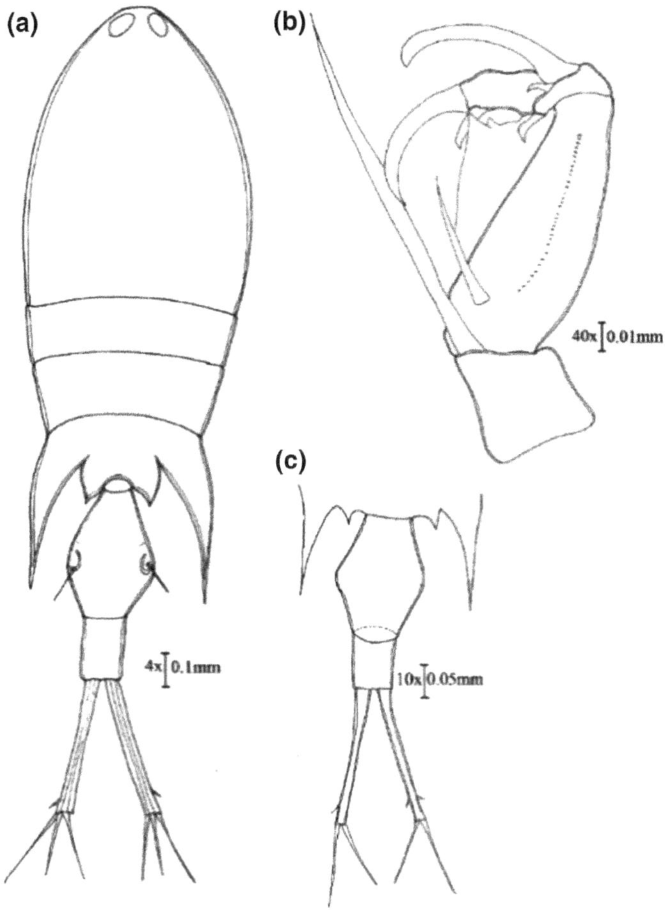

Fig. 5 *Corycaeus speciosus* female. **a** Habitus (dorsal); **b** A2; **c** urosome

10:4.7:1.7 (Fig. 5c). A2 four segmented. Coxobasis with a coxobasal setae which is 2.75 times longer than endopodal setae. 1st endopodal segment bears a short endopodal setae. 2nd endopodal segment bears 3 elements-one larger and two small. Lateral margin bears numerous spines. 3rd endopodal segment bears two elements (Fig. 5b).

Male (Fig. 6): Total length 1.03–1.15 mm. Prosome four segmented (Fig. 6a shows first and 2nd segment fused) 2.05 times longer than wide and 1.6 times as long as urosome. Urosome 2 segmented 2.8 times longer than wide. Genital double somite 1.41 times longer than wide. AS (Anal Segment) 1.81 times as long as wide. CR (Caudal Rami) parallel, 3.66 times as long as wide. Length ratio of urosomal somites and CR 2.4:1:1.1 (Fig. 6a). A2 four segmented. Coxobasis with a coxobasal setae which is 1.38 times longer than endopodal setae. 1st endopodal segment bears a short endopodal setae. 2nd endopodal segment bears two elements. Lateral margin bears

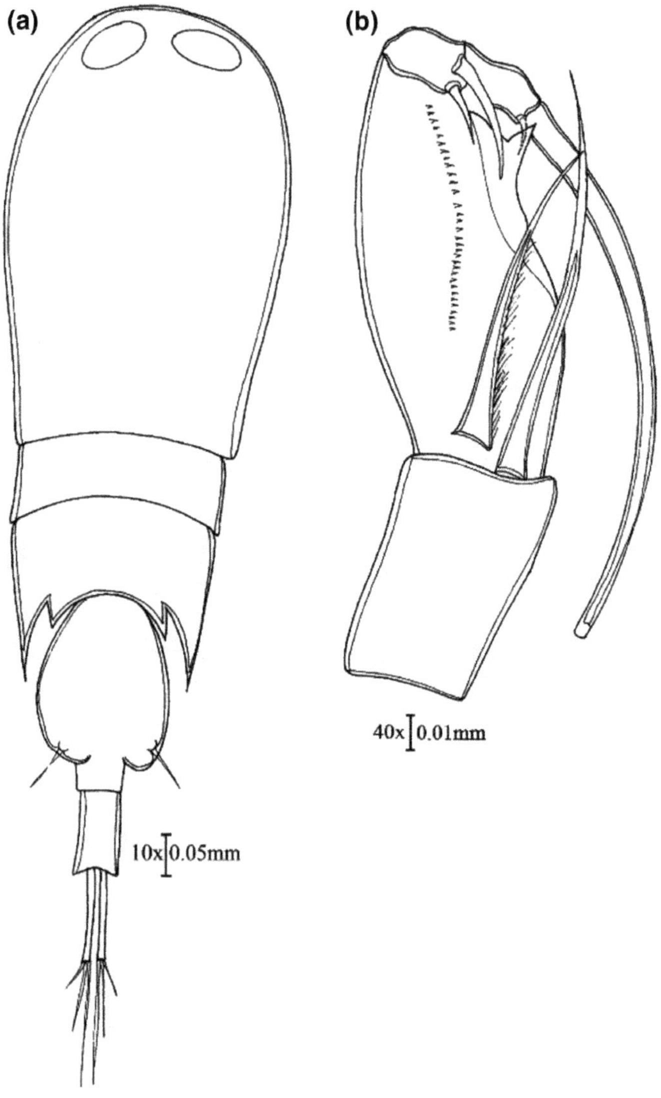

Fig. 6 *Corycaeus speciosus* male. **a** Habitus; **b** A2 (antenna)

numerous spines. 3rd endopodal segment bears two elements of which one is drawn out as a claw (Fig. 6b).

Remarks: In Arabian Sea, this species was first reported by Sewell in 1947. *C. speciosus* is distinguished by its divergent CR, rounded forehead in female and broad one in male. Presence of minute spines on the mid-lateral region of the first endopodal segment of the A2 in both females and males is also a distinguishing character. Females are comparatively larger than males. *C. speciosus* is an epipelagic or sometimes bathypelagic species. According to Motoda (1963), this species forms the most common *Corycaeus* in Hawaiian waters and Oahu Island. Kang et al. (1990) and Motoda (1963) noted that the length of the coxobasal setae of A2 was 2.8 times that of endopodal setae. Length of coxobasal to endopodal setae was 2.75 for ♀and 1.38 for ♂of Lakshadweep specimen. Dana (1849) opined on the length-width ratio of CR being 12 times for males. But this ratio was found be lesser in the males of Lakshadweep specimen (3.66). Björnberg (1963) considered this species as a typical indicator of warm tropical oceanic water.

Distribution: Tropical, Subtropical, and North Temperate Atlantic, Mediterranean, Indo-Pacific, East china Sea, Japan Sea, South and North Pacific. In the Arabian Sea, this species was first reported by Sewell in 1947 (http://copepods.obs-banyuls.fr/en).

Corycaeus clausi F. Dahl, 1894

Material examined: Sorted out the samples from plankton samples collected from Kavaratti, Kalpeni, Minicoy (open ocean and lagoon), and Agatti.

Female (Fig. 7): Length 1.50–1.6 mm. Prosome four segmented, broad, 2.06 times as long as wide and 1.57 times as long as urosome. Two segmented urosome. GS is

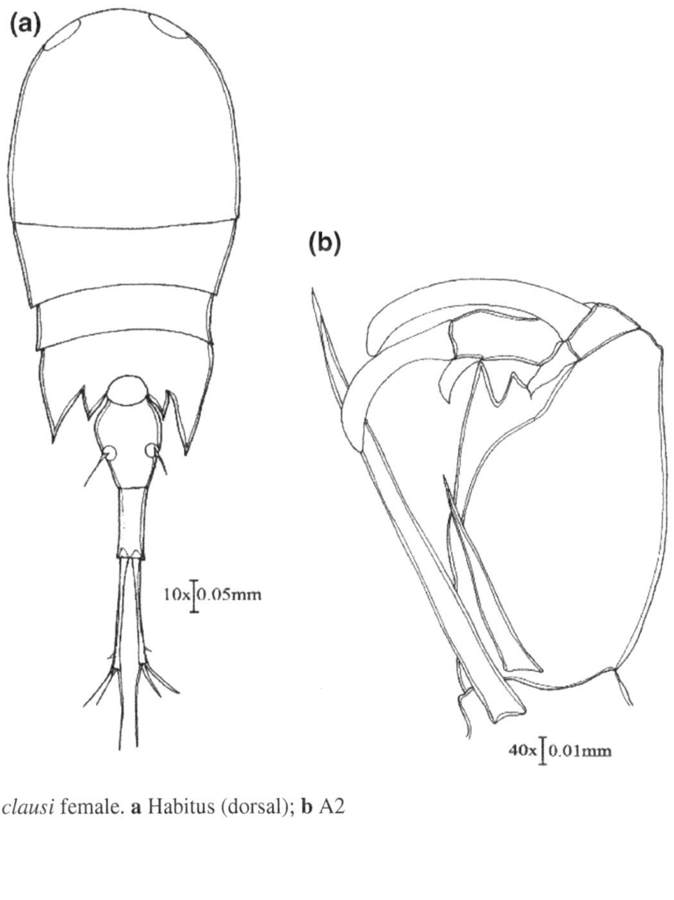

Fig. 7 *C. clausi* female. **a** Habitus (dorsal); **b** A2

as long as wide. CR more than thrice as long as wide at the proximal. Proportional lengths of the urosomal segments and CR are 2.5:2.2:3.2 (Fig. 7a). Antenna (A2) four segmented with coxobasal setae longer than endopodal setae. Lateral margin of the first endopodal segment adorned with a row of minute spines. 2nd endopodal segment with two spines, one larger and one smaller. 3rd endopodal segment also with two spines, one larger which is drawn out as a claw and one smaller (Fig. 7b).

Male (Fig. 8): Length 1.37–1.5 mm. Prosome four segmented where cephalosome shows incomplete fusion with the first prosomal segment; 1.80 times longer than wide and 1.69 times longer than urosome. Proportional lengths of the urosomal segments and CR is 3.6:1.4:2.5. GS 1.6 times as long as wide. CR about 2.5 times as long as wide at the proximal end (Fig. 8a). Antenna (A2) four segmented with coxobasal setae longer than endopodal setae. Lateral margin of the first endopodal segment adorned with a row of minute spines. 2nd endopodal segment with two spines, one larger and one smaller. 3rd endopodal segment also with two spines, one larger which is drawn out as a claw extending up to the coxobasal segment and one smaller (Fig. 8b).

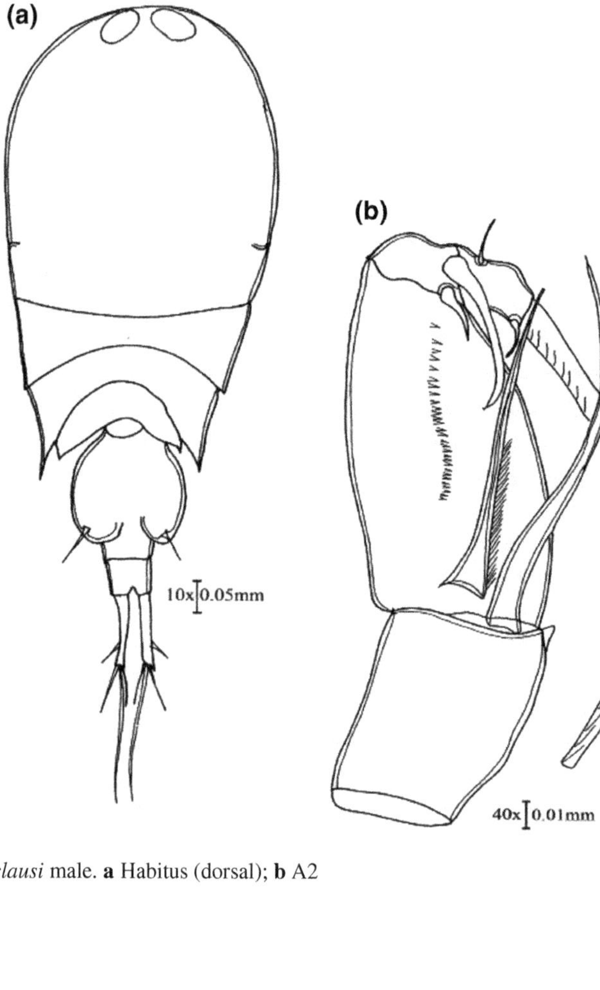

Fig. 8 *C. clausi* male. **a** Habitus (dorsal); **b** A2

Remarks: The diagnostic feature of *C. clausi* is the presence of a row of minute spines on the lateral portion of first endopodal segment of the A2. Comparing the measurements, armature and spine ornamentation of the specimens collected from various islands of Lakshadweep clearly established that the present specimens belong to *C. clausi*. Sewell (1947) considered this species as a variety of *C. crassiusculus*. Boxshall and Halsey considered this as full generic status pending phylogenetic revision of the family at the generic level.

Distribution: Epi-mesopelagic, recorded from temperate waters of the Atlantic and rarely from the Indo-pacific (http://copepods.obs-banyuls.fr/en).

Corycaeus vitreus Dana, 1849

Material examined: Sorted out the samples from plankton samples collected from Kavaratti, Kalpeni, Minicoy (open ocean and lagoon), Agatti and Bangaram.

Male (Fig. 9): Total length 1.06–1.15 mm. Prosome three-segmented; cephalosome fused with first prosomal segment; 1.75 times longer than wide and 1.46 times as long as urosome (Fig. 9a). Urosome 2 segmented (Fig. 9c); 3.9 times longer than wide. Genital double somite 1.33 times longer than wide. AS (Anal Somite) 2.97 times as long as wide. CR (Caudal Rami) slightly divergent at its base, 1.92 times as long as wide. The ratio of urosomal somites and CR 2.5:0.75:1.25. Antenna (A2) four segmented with coxobasal setae shorter than endopodal setae. 2nd endopodal segment with two elements, one long and one short at base. 3rd endopodal segment drawn out a long terminal claw extending up to coxobasis (Fig. 9b).

Remarks: Dahl (1912) defined the characters of the male specimens of this species. Only Farran (1936) reported female specimens till date. He opined that the male of *C. vitreus* differs from that of *C. crassiusculus* and *C. clausi* by its short head broad anteriorly and tapers posteriorly and short AS. He collected *C. vitreus* from New Zealand but reported it as the female of *C. crassiusculus*. Later he reported a specimen collected from Great Barrier Reef as the female specimen of *C. vitreus* for the first time. He opined that the female of *C. vitreus* to be more than that of *C. crassiusculus*. Only male specimens were observed. When Tanaka's male specimens were compared with the Lakshadweep specimen, some of the characters were comparable even with slight variations. They included (1) PR to UR ratio being lower in Lakshadweep specimen than that of Tanaka (1.46 vs. 1.5); (2) Length width proportion of CR being less in Lakshadweep specimen (1.92 vs. 5); (3) AS wider at the proximal end than at distal end; (4) ornamentation of A2.

Distribution: Ocean and Coastal, Pacific Ocean, Coast of Western Australia (http://copepods.obs-banyuls.fr/en).

Genus *Corycaeus (Urocorycaeus)* (Dahl 1912)
Urocorycaeus lautus Dana, 1848

Material examined: Sorted out the samples from plankton samples collected from Kavaratti, Agatti, and Minicoy (open ocean and lagoon).

Male (Fig. 10): Total Length 1.48–1.6 mm. Prosome three-segmented with frontal margin arc-shaped, 2.34 times as long as wide and 1.16 times as long as urosome. Urosome two segmented. Proportional lengths of GS:AS:CR is 2.8:2:5.2. GS bulgy

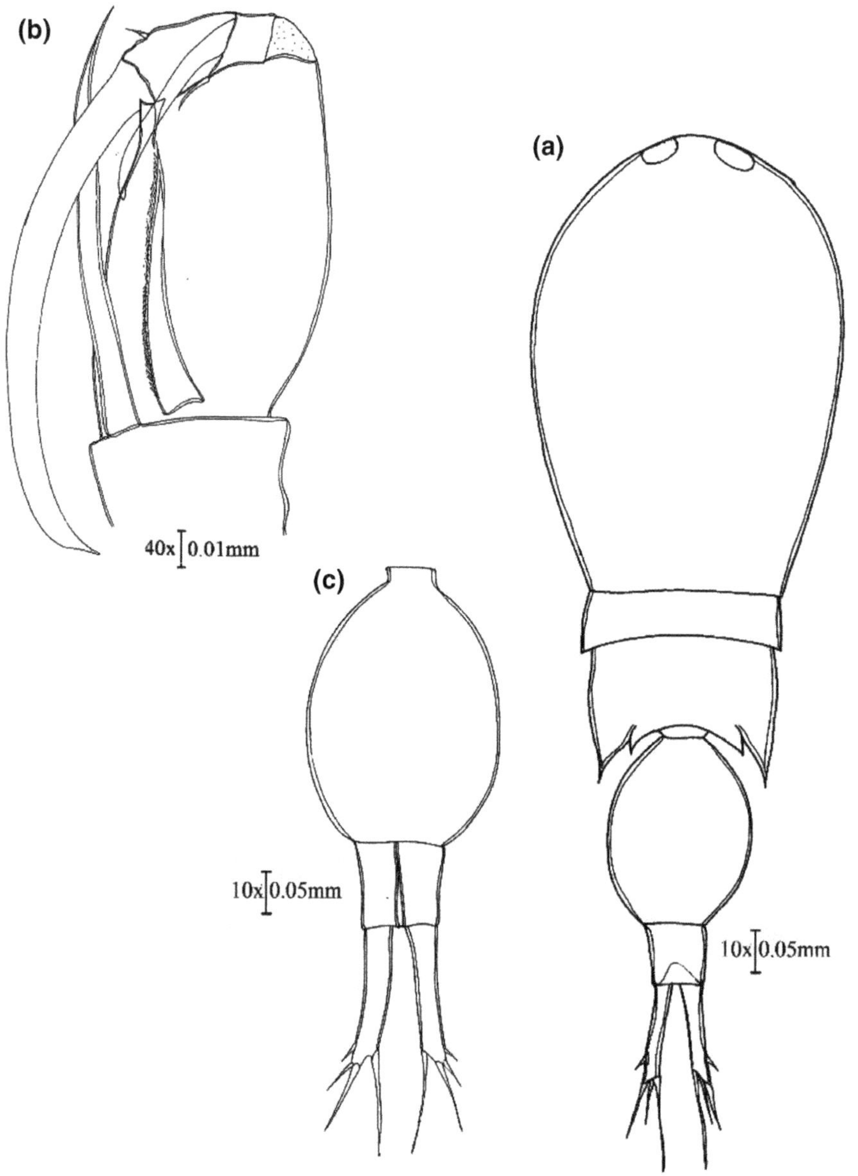

Fig. 9 *C. vitreus* male. **a** Habitus; **b** A2; **c** urosome

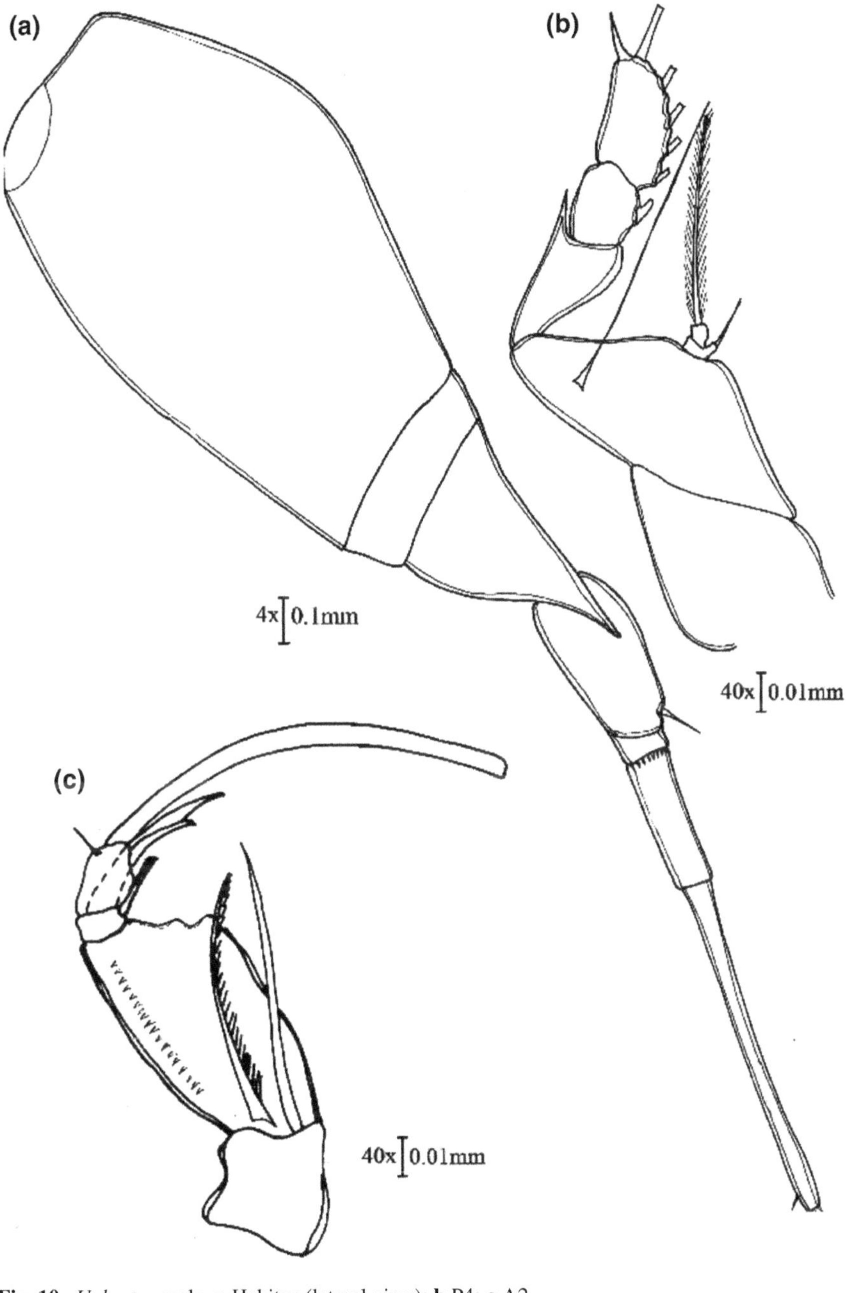

Fig. 10 *U. lautus* male. **a** Habitus (lateral view); **b** P4; **c** A2

double somite, 2.33 times as long as wide. Rectangular shaped AS, 3.33 times as long as wide which is not as long as GS; Slender, long and slightly divergent CR which is 9.89 times as long as wide. CR longer among the urosomal segments than and 1.08 times as long as the rest of the urosomal segments (Fig. 10a). P4 with two setae (one long and one rudimentary) in the endopodite (Fig. 10b). A2 (Fig. 10c) four segmented. Unipinnate endopodal setae shorter than coxobasal setae. Lateral margin of the first endopodal segment adorned with a row of minute spines. 2nd endopodal segment with two spines, one larger and one smaller. 3rd endopodal segment also with two spines, one larger drawn out as a claw and the other smaller at the base.

Remarks: *U. lautus* male specimens can be recognized by the presence of 2 setae (1 long and 1 rudimentary) in the P4 endopodite; UR differentiated into AS and GS; length ratio of CR to rest of the urosome. The present specimen agrees with the above said distinct characters. The length ratio of CR to rest of the abdomen (1.02) is also comparable with the previous descriptions like 1.2 (Farran 1911); 1.3 (Dahl 1912); 1.2 (Mori 1937); 1 (Dakin and Colefax 1940); 1.3 (Tanaka 1957); 1.49 (Motoda 1963). All these confirms the taxonomic integrity of *U. lautus* male in Lakshadweep sea.

Distribution: Tropical, Subtropical and North Temperate Atlantic, Mediterranean, Caribbean Sea, Indo-Pacific, East China Sea, Japan Sea, Pacific Ocean (http://copepods.obs-banyuls.fr/en).

Urocorycaeus furcifer Claus, 1863

Material examined: Sorted out the samples from plankton samples collected from Kavaratti, Agatti, Kalpeni, and Minicoy (open ocean and lagoon).

Male (Fig. 11): Total Length 1.28–1.35 mm. Prosome three-segmented with frontal margin arc-shaped, 2.44 times as long as wide. Prosome 1.38 times as long as urosome. Urosome two segmented. GS bulgy double somite, 1.65 times as long as wide. Rectangular shaped AS, 0.33 times as long as wide which is not as long as GS; Slender, long and slightly divergent CR which is 5.25 times as long as wide. CR longer among the urosomal segments. Proportional lengths of GS:AS:CR is 4:2.1:4.2 (Fig. 11a). Antenna (A2) four segmented with coxobasal setae only slightly longer than endopodal setae unipinnate. Proximal lateral margin of the first endopodal segment ornamented with minute spines. 2nd endopodal segment with two elements, one longer than the other. 3rd endopodal segment with two elements; one long drawn out as a terminal claw and other being too short at the base (Fig. 11b).

Remarks: The most distinct character of the genus *Urocorycaeus* is that the CR in both sexes are lengthy measuring at least twice the length of the rest of the UR and almost parallel, slightly divergent at the distal ends. Differentiation of UR into GS and AS is another distinguishing character of a male specimen of *U. furcifer* from that of its allied species *U. longistylis*. Here in the present specimen from Lakshadweep islands has this differentiation very well that confirms its taxonomic identity. In the present specimens, proportional lengths of Urosomal segments is 40:21:42. This is at variance with the description given by Cervigon (1964) (30:12:58). Tanaka (1957) reported only female specimens.

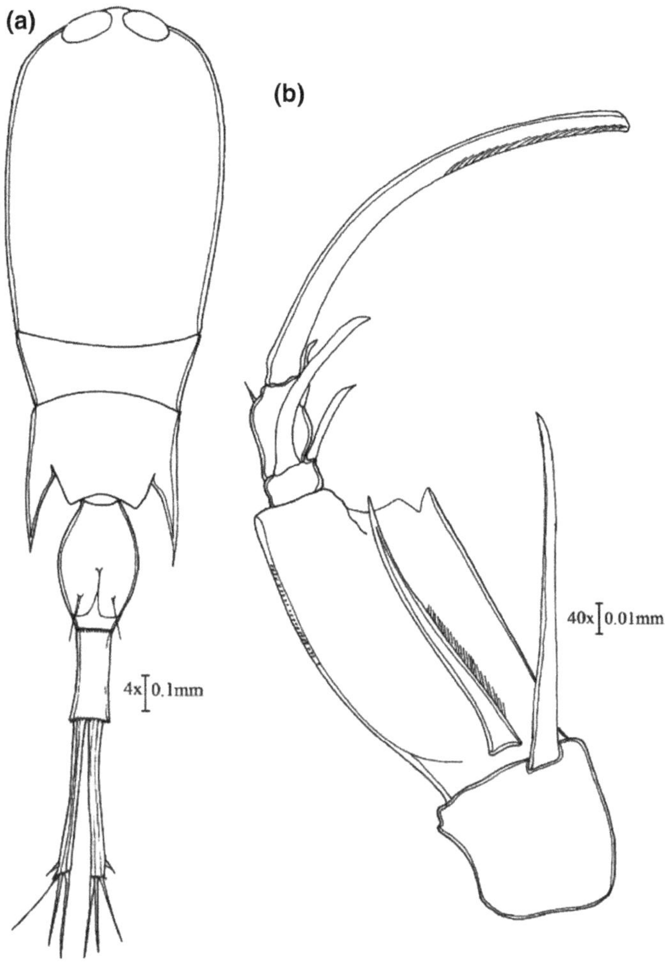

Fig. 11 *U. furcifer* male. **a** Habitus; **b** A2

Distribution: Recorded from Indo-Pacific, northern coast of South America, Mediterranean and Sargassum Seas (http://copepods.obs-banyuls.fr/en).

Genus *Corycaeus (Onychocorycaeus)* (Dahl 1912)
***Onychocorycaeus catus* Dahl F, 1894**

Material examined: Sorted out the samples from plankton samples collected from Kavaratti, Kalpeni, Minicoy (open ocean and lagoon), Bangaram, and Agatti.

Female (Fig. 12): Body cylindrical, tapering posteriorly. Total length 0.65–0.66 mm. Prosome four segmented, more than twice as long as urosome including caudal rami about 1.75 times as long as wide (Fig. 12a). Urosome two-segmented with divergent caudal rami. Genital somite oval, 1.5 times as long as greatest width at the anterior mid-region, longer than AS and CR combined; AS rectangular shaped,

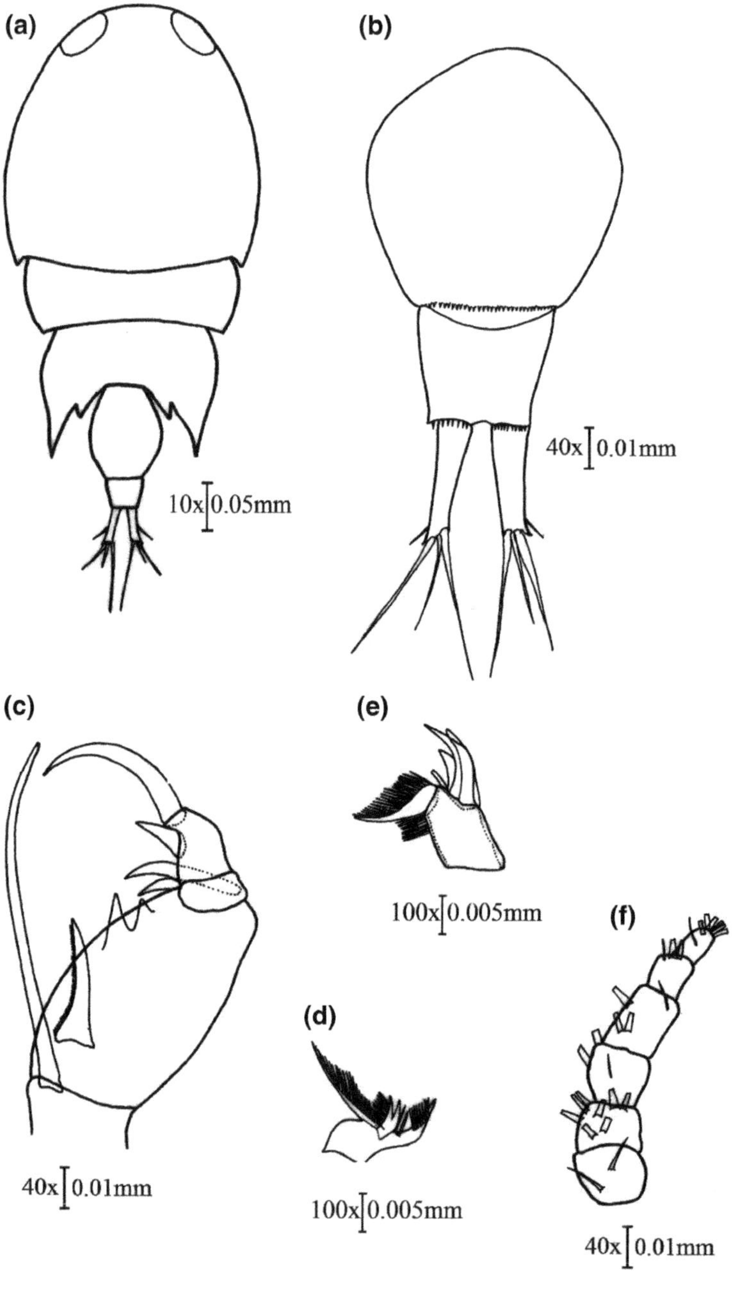

Fig. 12 *O. catus* female. **a** Habitus; **b** urosome; **c** A2; **d** maxilla; **e** mandible; **f** antennule (A1)

about 1.3 times as long as wide at base; distal margin bears spinules ventrolaterally; 3.9 times shorter than genital somite and 1.3 times shorter than caudal rami. Caudal rami divergent, 1.67 times longer than wide at the base, about 0.35 times shorter as long as genital somite and 1.38 times as long as anal somite (Fig. 12b). A2 four segmented with coxa and basis with strong unipinnate setae on inner distal margin. Endopod three-segmented and unequal in length; 1st segment bearing short unipinnate seta on ventral proximal margin shorter than coxobasal seta, inner distal margin formed into two stout teeth. 2nd segment short bearing two elements (a) curved stout short spine arising from outer distal margin and is longer than the other and (b) comparatively smaller spine arising from the inner margin; 3rd segment cylindrical, 1.2 times as long as wide at base, armed with a curved terminal claw and a short spine arising from the inner distal margin (Fig. 12c). Maxilla (Fig. 12d) with syncoxa unarmed and unornamented. Inner margin of the allobasis produced into spiniform process and bears two elements; one is broad and robust with the comb-like spine; the other is smaller than former with smaller combs but have many spinous processes next to it. Mandible (Fig. 12e) with gnathobase bearing two elements ie, spine and blade where the spine is slightly broad and robust. Blade forms spinous process surrounded by a patch of spinules around base. Antennule (Fig. 12f) short, six segmented.

Remarks: When compared with the descriptions of Tanaka (1957) from Japanese waters, females of *O. catus* from Kavaratti waters, Lakshadweep showed almost similar length ratio of PR:UR (including CR) where PR being more than two times the length of UR (4.9:1.9 vs. 9:4). However, a few morphological variations from the former description were also there on the total body length being smaller (0.65 vs. 0.93–1 mm), the length-width ratio of the anal somite being slightly different (4.5:3.5 vs. 7:8), the length-width ratio of the CR being smaller (1.67 vs. 4) and length the proportion of the GS:AS:CR being much large (177.5:45:62.5 vs. 58:20:22). On the contrary, females of *O. catus* described by Motoda (1963) from Hawaiian waters differed from those of the Kavaratti specimens in the proportional lengths of PR:UR (including CR) where PR about twice the length of UR in Hawaiian waters versus more than twice in Kavaratti specimen and the total body length being larger (1.14 vs. 0.65 mm) where the morphological characteristics such as GS longer than AS and CR combined; AS 1.3 times as long as wide (4.5:3.5) and slightly shorter than CR, were found to be similar. In contrast, from the descriptions of female *O. catus* by Karanovic (2003) from Australian waters, the body length seems larger being 1.06 mm when compared to 0.65 mm of Kavaratti specimen. Variations also appear in the length-width ratio of prosome which being larger in Kavaratti specimen from that of Australian specimen (1.75 vs. 1.0) as well the details like surface of the cephalic shield of the prosomites with many small sensilla and cuticular pores. While Karanovic (2003) explains that genital somite is only slightly longer than wide in Australian specimen, Kavaratti specimens varies from it by the genital somite being 1.5 times longer than the greatest width and anal somite about 0.8 times as long as wide in Australian specimen which is smaller to that of Kavaratti specimen (0.8 vs. 1.3). Anal somite 3.9 times shorter than genital somite in Kavaratti specimen while that of Australian specimen is only 0.4 times as long as genital somite, which explains another variation.

Distribution: Recorded from the Indo-Pacific Ocean, Lakshadweep Sea, Great Barrier Reef Sea, Arabian Sea and very common in the warm waters of Japan (http://copepods.obs-banyuls.fr/en).

Onychocorycaeus agilis Dana, 1849

Material examined: Sorted out the samples from plankton samples collected from Kavaratti, Kalpeni, Minicoy (open ocean and lagoon), Agatti, and Bangaram.

Female (Fig. 13): Total length 1.34–1.38 mm. Prosome robust and four segmented, twice as long as wide. Prosome 1.66 times as long as urosome. Slender CR. Urosome 2 segmented. Proportional lengths of GS:AS:CR is 0.9:0.85:1.1 (Fig. 13a). Broad 1st endopodal segment of A2 bears two spine-like processes at the inner distal margin whose level is slightly displaced from that of second endopodal segment. Coxobasal setae thrice as long as endopodal setae. 2nd endopodal segment bears two spines. 3rd

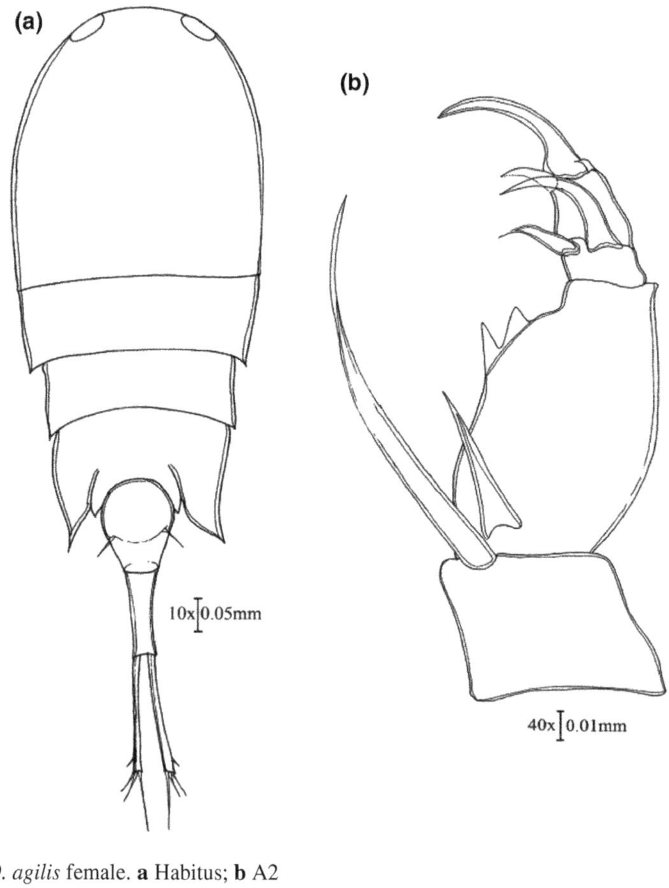

Fig. 13 *O. agilis* female. **a** Habitus; **b** A2

endopodal segment also bears two elements, one short and the other as a terminal claw (Fig. 13b).

Male (Fig. 14): Total length 0.7–0.8 mm. Prosome four segmented; 1.83 times as long as wide and 1.76 times as long as urosome. Slender CR. Urosome two segmented. Proportional lengths of GS:AS:CR is 2:1.3:2 (Fig. 14a). Broad 1st endopodal segment of A2 bears two spine-like processes at the inner distal margin whose level is slightly displaced from that of second endopodal segment. Coxobasal setae longer than endopodal setae. 2nd endopodal segment bears two slender spines. 3rd endopodal segment also bears two elements, one short and the other as a terminal long claw (Fig. 14b).

Remarks: Giesbrecht (1892) synonymized *C. gracilicaudatus* with male specimen of *O. agilis* of Dana, Dahl (1912). Even though Scott (1909) presented only a brief note on the female specimen characters, his *C. gracilicaudatus* was also considered synonymous to *O. agilis*. The diagnostic features like (1) swelling of the GS at the middle and almost as long as furca (Caudal Rami) (0.81:1); (2) AS slightly

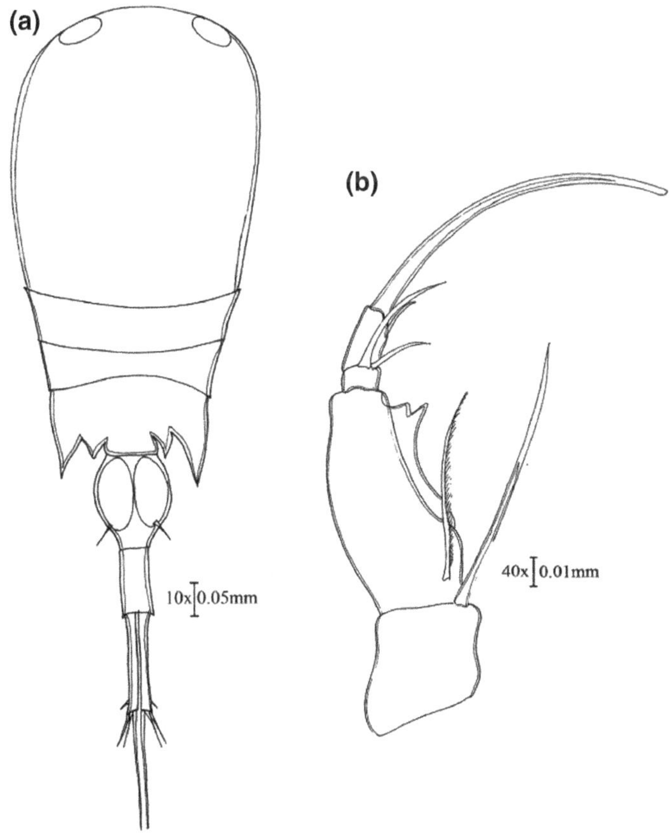

Fig. 14 *O. agilis* male. **a** Habitus; **b** A2

shorter than CR (0.85:1.1) and more than twice (3.4) as long as wide; (3) Coxobasal setae 2.6 times as long as endopodal setae; (4) first endopodal segment of A2 bears 2 spine like process at the inner distal margin, of the female specimens of *O. agilis* from Lakshadweep waters match with the depictions of earlier works. This undoubtedly proves the taxonomic status of *O. agilis*.

Distribution: Warm waters of Japan, tropical regions of Atlantic, South and North Pacific, Caribbean Sea and Indian Ocean (http://copepods.obs-banyuls.fr/en).

Onychocorycaeus giesbrechti Dahl, 1894

Material examined: Sorted out the samples from plankton samples collected from Kavaratti, Minicoy (open ocean and lagoon), Agatti, and Bangaram.

Female (Fig. 15): Total Length 1.05 mm. Large prosome 1.73 times as long as wide and 2.13 times as long as urosome. Trimmed CR. Urosome two segmented. GS ovate 1.33 times as long as wide. Ventral surface of the GS with a median hook. AS slightly divergent at the base, 1.6 times as long as wide. CR 1.16 times as long as wide.

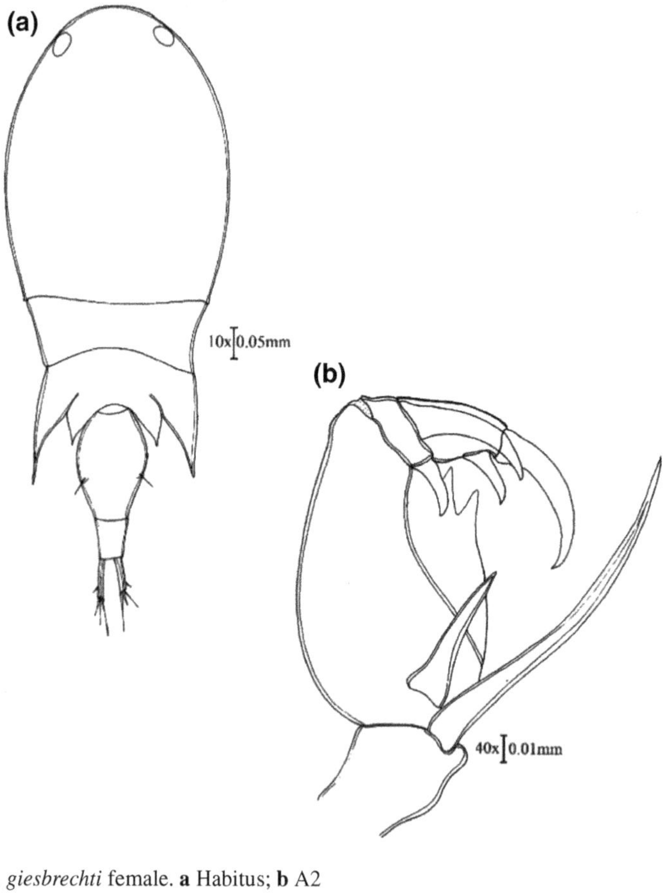

Fig. 15 *O. giesbrechti* female. **a** Habitus; **b** A2

Proportional lengths of GS:AS:CR is 1:0.4:0.35 (Fig. 15a). A2 four segmented. 1st endopodal segment about 1.7 times as long as wide at base bearing short endopodal seta on ventral proximal margin 2.58 times shorter than coxobasal setae; inner distal margin formed into two stout teeth. 2nd segment short bearing two elements; 3rd segment cylindrical, armed with a curved terminal claw and a short spine arising from the inner distal margin (Fig. 15b).

Male (Fig. 16): Total Length 0.967 mm. Prosome robust, 2.15 times as long as wide. Prosome 1.22 times as long as urosome. Trimmed CR. Urosome two segmented. GS ovate 1.25 times as long as wide. AS with parallel lateral margins, 2.57 times as long as wide. CR 4.2 times as long as wide. Proportional lengths of GS:AS:CR is 1.25:0.9:2.1 (Fig. 16a and c). A2 four segmented. 1st endopodal segment bears endopodal seta shorter than coxobasal setae; inner distal margin formed into stout teeth. 2nd segment short bearing two elements; 3rd segment cylindrical, armed with a curved terminal claw and a short spine arising from the inner distal margin (Fig. 16b).

Remarks: *C. giesbrechti* female described by Tanaka differs from those described by Dahl in the proportional lengths of the abdominal segments and CR (46:28:28 for latter and 41.5:17.25:19.5 for former) and in the more robust GS. The male specimens of Tanaka closely resemble *C. agilis*. But the shape of the 4th thoracic segment that is more tapered, slender GS with a small ventral hook and the large cylindrical part are the most distinguishing characters when compared with from those of *C. agilis*, *O. giesbrechti* described during the present study and proportions of the relative segments and the body size was almost equal to those of earlier descriptions. The diagnostic features like (1) Ovoid PR in female and robust PR in male; (2) GS ovate in both male and female; (3) A median hook in the ventral surface of both male and female; (4) Length width proportions of the PR, GS and AS of the specimens of *O. giesbrechti* from Lakshadweep waters match with the depictions of previous works. This proves the taxonomic status of *O. giesbrechti*.

Distribution: South Africa, Tyrrhenian Sea, Adriatic Sea, Ionian Sea, Aegean Sea, Thracian Sea, Lebanon Basin, Suez Canal, Red Sea, Madagascar, Indian ocean, China Seas (http://copepods.obs-banyuls.fr/en).

Onychocorycaeus latus Dana, 1849

Material examined: Sorted out the samples from plankton samples collected from Kavaratti, Kalpeni, Minicoy (open ocean and lagoon), Agatti, and Bangaram.

Male (Fig. 17): Total length 0.813 mm. Prosome four segmented and robust with two unequal lateral flaps at the distal margin but less widely than proximal margin, 1.97 times as long as wide (Fig. 17a). Prosome 1.59 times as long as urosome. UR two-segmented (Fig. 17c). GS elongate ovate and extended anteriorly 1.67 times as long as wide. Ventral surface of the GS with no median hook. AS with parallel lateral margins, 2.5 times as long as wide. CR 3.53 times as long as wide. GS 1.45 times shorter than AS and CR combined. Proportional lengths of GS:AS:CR is 0.72:0.45:0.6. A2 four segmented. 1st endopodal segment bears a uniramous endopodal setae which are shorter than coxobasal setae. 2nd endopodal segment bears two unequal setae at its

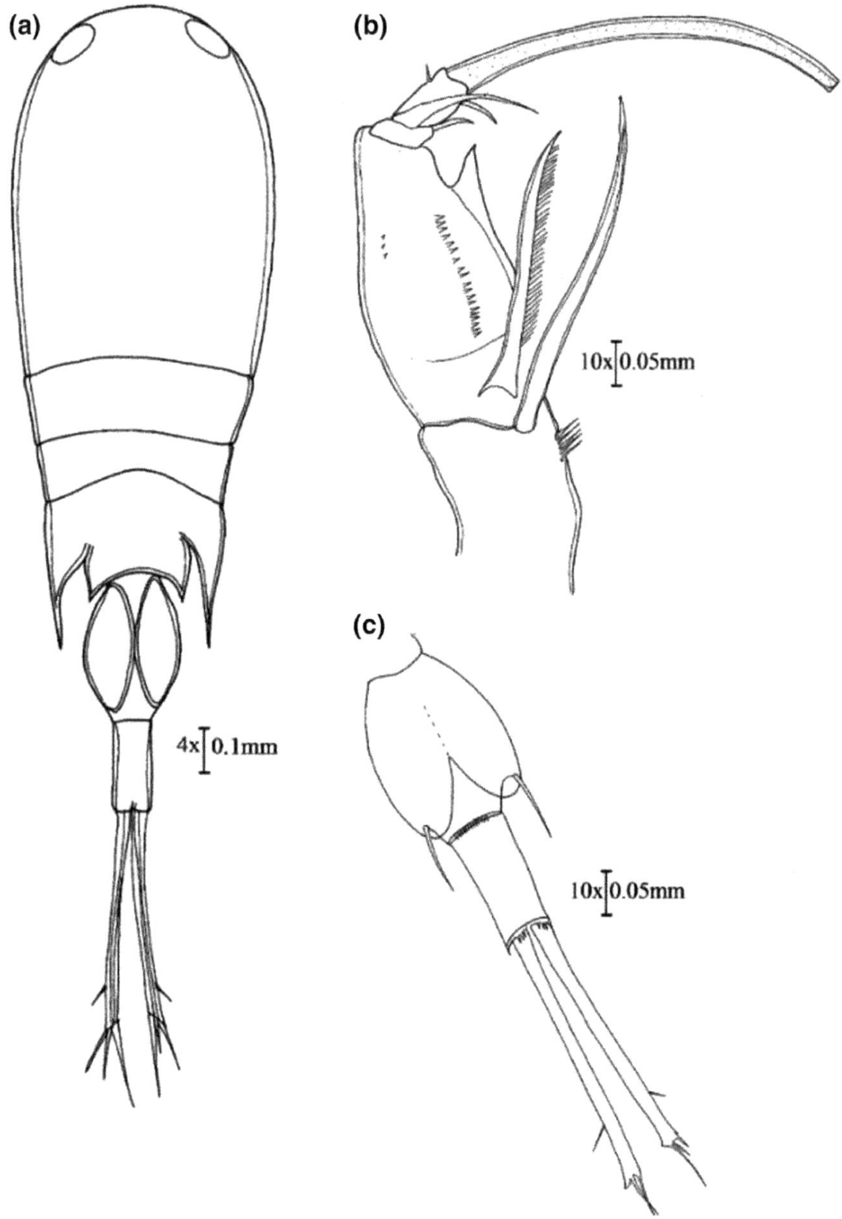

Fig. 16 *O. giesbrechti* male. **a** Habitus; **b** A2; **c** urosome

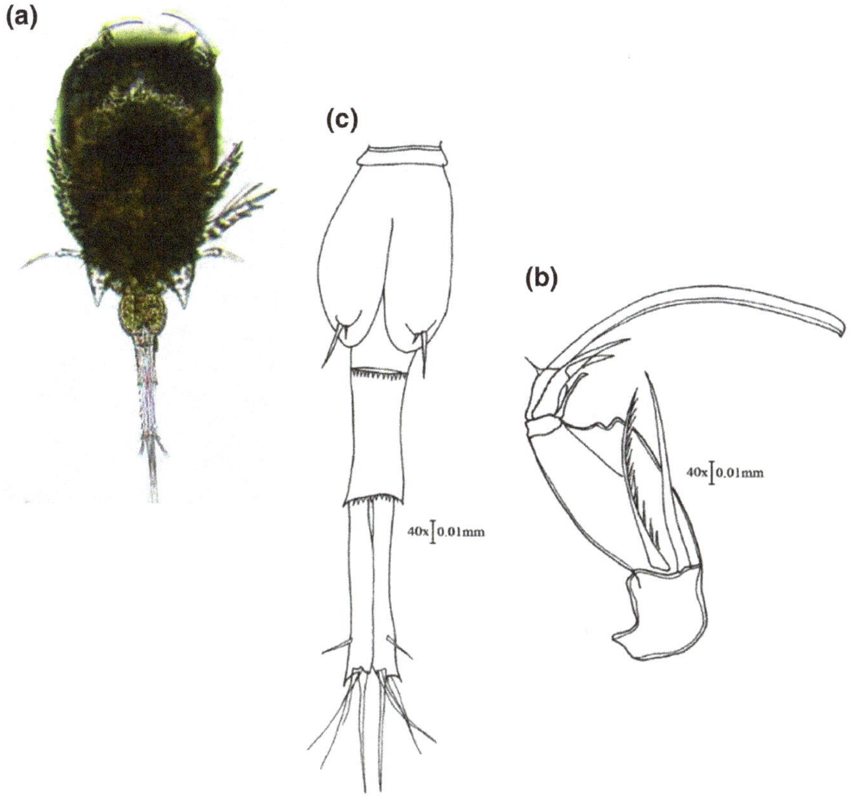

Fig. 17 *O. latus* male. **a** Habitus; **b** A2; **c** urosome

inner distal margin. 3rd segment armed with a long terminal claw and a short spine arising from the inner distal margin (Fig. 17b).

Remarks: Dana's *C. laticeps* was synonymized with this species. *O. latus* species is definitely identified by the ornamentation of A2, number of spines in P4 endopod and anteriorly extended GS. Male specimens differ from female in the appearance of lateral flaps and its width at the distal margin of PR as well as overall shape of PR. A2 in both sexes is almost similar. Motoda (1963) got only male specimens from Oahu Island. When compared, Lakshadweep specimens agreed with the features like GS shorter than AS and CR combined; body length; A2 ornamentation; A2 terminal claw exceeding the combined length of 1st and 2nd segments; but varied in AS length-width ratio being slightly more (2.5) than Oahu specimen which was only 2.3 times. In contrast, Dahl (1912) mentioned in the GS length being about as long as CR and AS combined but is Lakshadweep specimen the same was shorter about 1.45 times. However, GS being almost twice as long as wide in Dahl's specimen was comparable with the present specimen (1.67).

Distribution: Mediterranean, Japan Sea, South and North Pacific Ocean (http://copepods.obs-banyuls.fr/en).

Onychocorycaeus pumilus **Dahl M, 1912**

Material examined: Sorted out the samples from plankton samples collected from Kavaratti, Minicoy (open ocean and lagoon), Agatti, and Bangaram.

Female (Fig. 18): Total Length 0.671–0.68 mm. Prosome three-segmented; 1.56 times as long as wide. Urosome two-segmented; 2.8 times as long as wide. Presence of wing-like expansion of the 3rd thoracic segment up to GS and small posterior protuberance of 4th thoracic segment (Fig. 18a). A2 four segmented. Coxobasal setae longer than endopodal setae. 1st endopodal segment bears a short spine at the inner distal margin. 2nd segment is short bearing two spines like process. 3rd segment bears two spines, one at the inner distal margin and the other drawn out as a terminal claw (Fig. 18b).

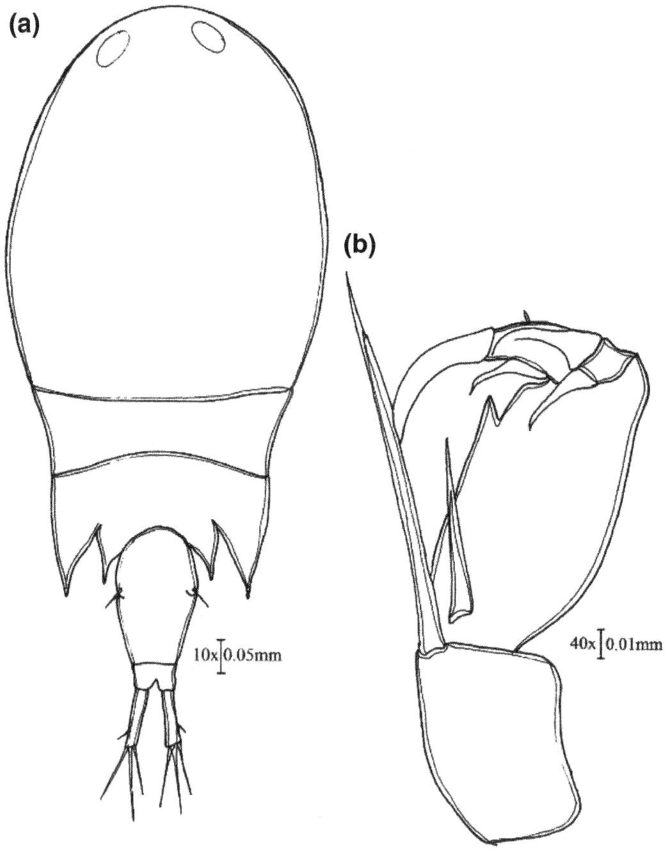

Fig. 18 *O. pumilus* female. **a** habitus; **b** A2

Remarks: Dahl (1912) and Tanaka (1957) could only report male species whereas female species were reported by Chen et al. (1974) and Sewell (1947). However, Sewell had already reported *O. pumilus* from Arabian Sea which is a part of the Indian Ocean. Wing like expansion of the 3rd thoracic segment up to GS and small posterior protuberance of 4th thoracic segment is mainly a characteristic of Onychocorycaeus genus and obviously of *O. pumilus* too. Tanaka regarded both *O. medius* and *O. pumilus* as identical species. After Tanaka, Farran's specimen of *O. pumilus* from the Great Barrier Reef were more like *O. pumilus* itself than to *O. medius*. Anyhow, its presence in extreme North Pacific is said to be surprising and Boxshall and Halsey considers this subgenus as full generic status pending phylogenetic revision.

Distribution: South Africa, Ibero-moroccan Bay, Suez Canal, Red Sea, Arabian Sea, Madagascar, Indonesia-Malaysia, Sarawak-Bintulu coast, Bismarck Archipelago, Philippines, China Seas, Taiwan, Japan (Kuchinoerabu Island), Japan, Australia (North West Cape, Great Barrier), New Caledonia, off Hawaii, Pacific Ocean (http://copepods.obs-banyuls.fr/en).

***Onychocorycaeus pacificus* Dahl F, 1894**

Material examined: Sorted out the samples from plankton samples collected from Kavaratti, Minicoy (open ocean and lagoon), Agatti, and Bangaram.

Female (Fig. 19): Total Length 1.12–1.15 mm. Large three-segmented prosome 1.96 times as long as wide and 1.52 times as long as urosome. 3rd pedigerous somite with lateral pointed lateral flaps but 4th pedigerous somite with short blunt flaps. Urosome two segmented. GS ovate 1.14 times as long as wide. AS slightly divergent at the base, 3 times as long as wide. CR 6.1 times as long as wide. Proportional lengths of GS: AS CR is 1.25:1.05:1.85 (Fig. 19a). A2 four segmented. Coxobasal setae longer than endopodal setae. 1st endopodal segment bears a short spine at the inner distal margin. 2nd segment is short bearing two spines like process. 3rd segment bears two spines, one at the inner distal margin and the other drawn out as a claw (Fig. 19b).

Remarks: Sewell considers *O. pacificus* and *O. ovalis* as local races of the same species. In Sewell's specimen, the ratio of PR:UR was 1.94 and that of Lakshadweep specimen being 1.52. Even with slight variation, this was comparable. The ornamentation of A2; number of spines in P4 endopod; the shape of urosomal segments all these confirmed the morphological identity of *O. pacificus* female specimen.

Distribution: Indian Ocean, Pacific Ocean, Japanese waters (http://copepods.obs-banyuls.fr/en).

Genus *Corycaeus (Ditrichocorycaeus)* (Dahl 1912)
***Ditrichocoryceus andrewsi* Farran, 1911**

Material examined: Sorted out the samples from plankton samples collected from Kavaratti, Minicoy (open ocean and lagoon), Agatti, and Bangaram.

Female (Fig. 20): Total body length 1–1.07 mm. Prosome four segmented, frontal margin arc-shaped, with two large separate cuticular lenses; about 1.97 times longer than urosome and 2.15 times as long as wide. Urosome two-segmented with slightly divergent caudal rami. GS somite broadish oval and widest at the middle and 1.34

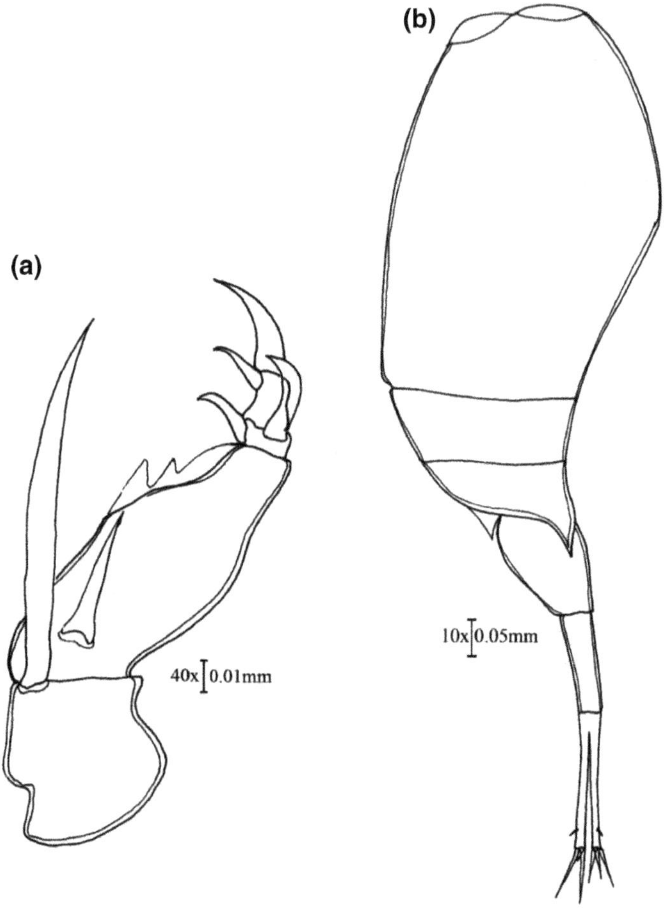

Fig. 19 *O. pacificus* female. **a** Habitus (lateral); **b** A2

times longer than wide. AS 1.66 times long as wide. CR about 1.38 times longer than wide. Proportional lengths of the GS:AS:CR is 3.5:2:1.8. GS is as long as anal somite and caudal rami combined (Fig. 20a). Antenna four-segmented bearing three endopodal segments. 1st endopodal segment robust, bears unipinnate setae shorter than coxobasal setae. Outer distal margin produced into a spine. 2nd endopodal segment, shortest of the three, bears two elements. 3rd endopodal segment armed with a long robust terminal claw and a short spine at its base (Fig. 20b).

Male (Fig. 21): Total body length 0.712–0.8 mm (Fig. 21a). Prosome four segmented, frontal margin arc-shaped, with two large separate cuticular lenses; about 1.5 times longer than urosome and 1.46 times as long as wide. The genital segment is as long as anal somite and caudal rami combined (205:102.5:105). Urosome two-segmented with slightly divergent caudal rami. Genital double somite (GDS) or Genital somite (GS) broadish oval and widest at the middle and 1.52 times longer

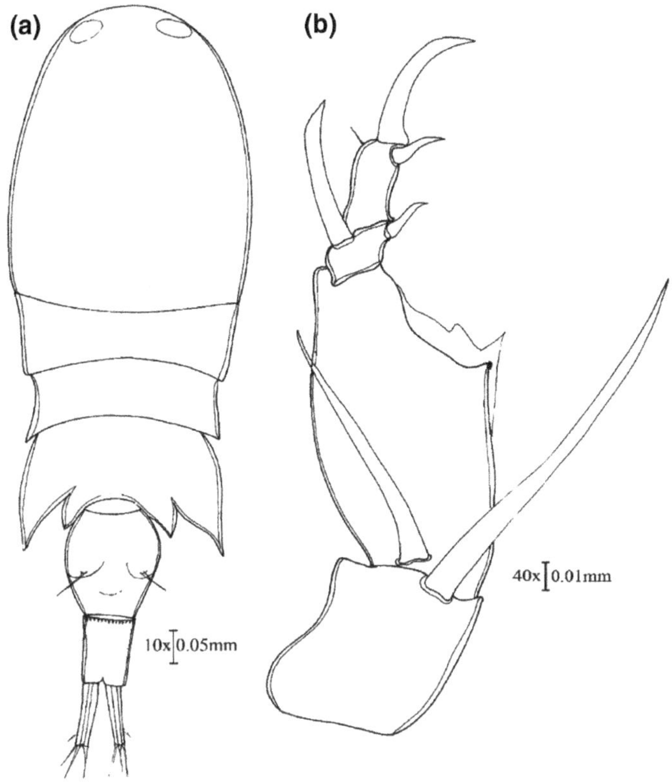

Fig. 20 *D. andrewsi* female. **a** Habitus; **b** A2

than wide. Anterior margin bears a ventral hook which is evident in the lateral view. AS 1.5 times long as wide. CR long and divergent, about 5 times longer than wide (Fig. 21a). Proportional lengths of GS:AS:CR is 7.6:3.6:4. Antenna four-segmented bearing three endopodal segments. 1st endopodal segment robust and barrel-shaped bears unipinnate setae shorter than coxobasal setae. Outer distal margin produced into a spine. Midventral surface vertically adorned with denticles gradually increasing in length at distal end. 2nd endopodal segment, shortest of the three, bears two elements. 3rd endopodal segment armed with a long robust terminal claw extending up to coxobasis with a blunt tip and a short slender spine at its base (Fig. 21b). Maxilliped (Fig. 21c) three-segmented. Syncoxa unarmed 1.5 times long as wide at the base; oval, solid and expanded basis, 2.17 times as long as wide at base, with inner distal and terminal margin tassled with two rows of spinules longitudinally arranged. Two segmented subchela which consists of unarmed proximal endopodal segment and distal endopodal segment, about 35 times as long as wide at the base, drawn into long curved claw extending up to syncoxa, armed with two bipinnate setae at the proximal and lateral margin. Maxilla (Fig. 21d) three-segmented with syncoxa unarmed. Allobasis produced distally into curved stout spine fringed with setae. Inner margin of the allobasis bears processes in which one with a comb like spine and the other one pronged into two in which one is unipinnate small spine adjacent to comb-like spine and the other one long, extending with numerous spines in it. Maxillule (Fig. 21e) with precoxal arthrite bearing four articulated spine-like process: innermost one ‘a’ with distal area pointed; element ‘b’ almost equal in length to that of ‘a’, solid and distal margin has spine-like process; element ‘c’, 2.73 times shorter than ‘a’ and ‘b’ and short; element ‘d’ shortest and naked. Length ratio of the elements 22.5:22.5:7.5:5. Legs 1–3 (Fig. 21f–h) with coxa, basis and three-segmented rami. Intercoxal sclerite well-developed, P1 and P2 with plumose inner seta; basis of P1 and P3 with outer seta; exopods distinctly longer than endopods.

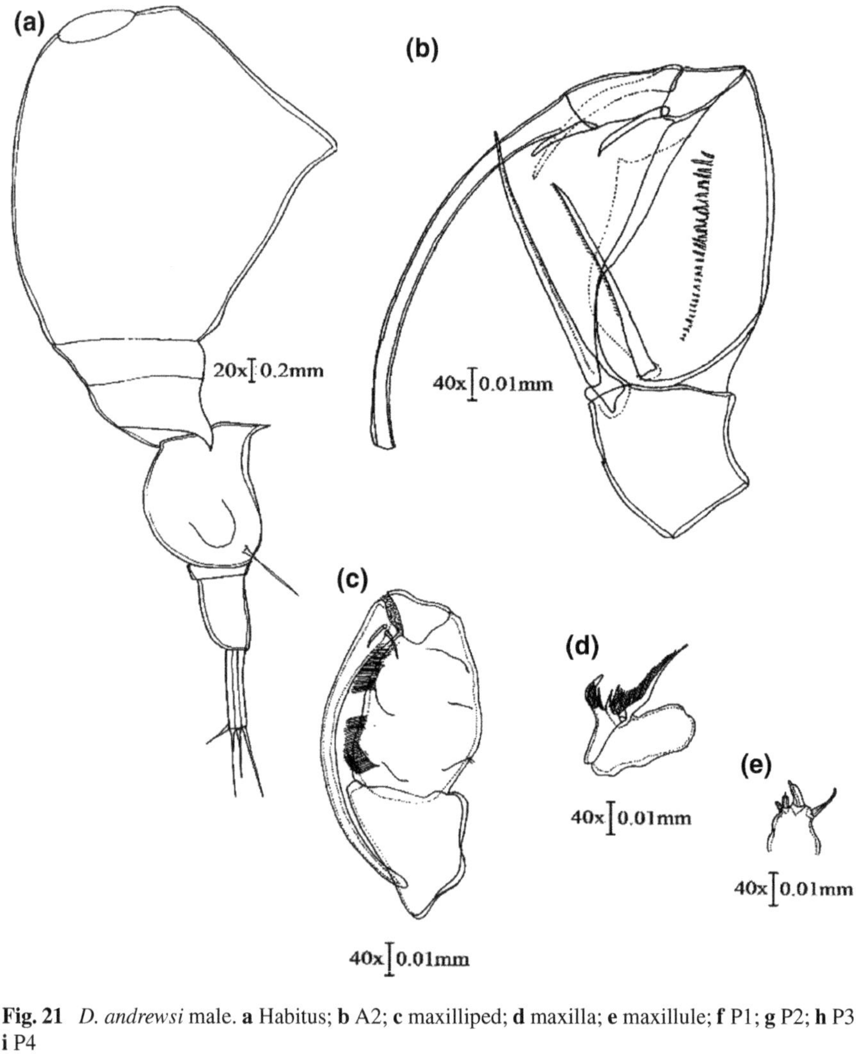

Fig. 21 *D. andrewsi* male. **a** Habitus; **b** A2; **c** maxilliped; **d** maxilla; **e** maxillule; **f** P1; **g** P2; **h** P3; **i** P4

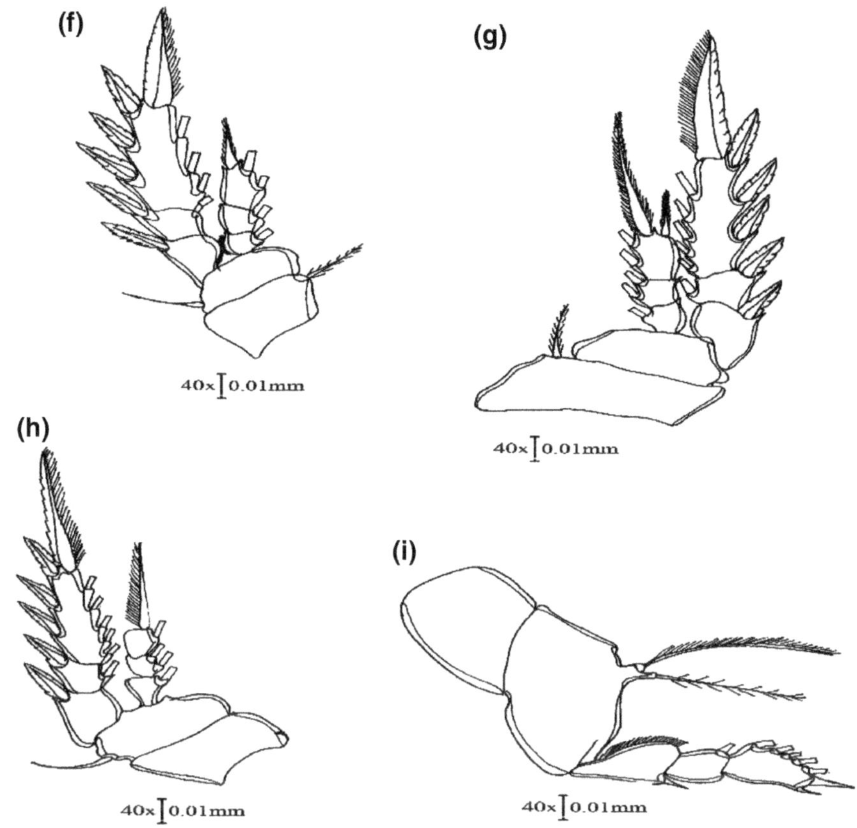

Fig. 21 (continued)

P4 endopod reduced into a knob-like segment with two plumose setae arising from it, one slightly longer than the other (Fig. 21i).

Remarks: Genus *Ditrichocorycaeus* M. Dahl (1912) is characterized by the procession two setae on endopod of P4 and divided into two groups based on relative lengths of the caudal ramus of which the first group has short caudal rami almost equal to anal somite but shorter than genital somite and the second group is characterized by the possession of caudal ramus longer than both anal somite and genital somite (Dahl 1912; Tanaka 1957). The male specimen of *D. andrewsi* belongs to the first group. The combination of general morphological features such as the shape of the genital double somite, the presence of ventral hook; slight swelling about the middle lateral margins of the second thoracic segment: width length ratio of the genital segment; presence of median longitudinal row of spinules/denticles in the first endopodal segment; total body length; PR:UR ratio is consistent with the description of *D. andrewsi* male by Tanaka (1957). However, variations were also noticed, from the illustrations given by Tanaka (1957) of Japanese waters in proportional lengths of the urosomal somites and CR greater (58:21:21) as compared to Kavaratti specimens (7.6:3.6:4); length-width ratio of the CR being less than that of Kavaratti specimen (4 vs. 5). *Corycaeus trukicus* Mori is the synonym of *D. andrewsi* in the short caudal rami whose length is almost equal to that of the anal somite, short wing-like a process on posterior ends of metasomal somite 3 in the male. Although P4 and caudal setae are considered as important identification keys to the genus *Ditrichocorycaeus,* the same has not been described by Tanaka (1957).

Distribution: Pacific coast, Japanese waters, Indo-Pacific Ocean, Truk Island (http://copepods.obs-banyuls.fr/en).

***Ditrichocoryceus affinis* McMurich, 1916**

Material examined: Sorted out the samples from plankton samples collected from Kavaratti, Kalpeni, Minicoy (open ocean and lagoon), Agatti, and Bangaram.

Female (Fig. 22): Total body length 0.95–1 mm (Fig. 22a). Prosome five segmented with 4th metasomal segment produced into lateral flaps. Prosome 2.27 times as long as wide and 1.8 times as long as urosome. Genital segment double somite 1.06 times as long as wide. AS 1.41 times as long as wide. CR 1.73 times as long as wide. Proportional lengths of GS:AS:CR is 5.8:2.6:5.2 (Fig. 22a and b). A2 four segmented with three endopodal segments. 1st endopodal segment bears short setae which is shorter than coxobasal setae. 2nd endopodal segment, short, bears two spines and 3rd endopodal segment adorned with a claw like process and a short spine (Fig. 22c).

Remarks: Davis (1949) considers *Corycaeus japonicus* to be the synonym of *D. affinis*. *D. affinis* can be identified by CR shorter than urosome, two-segmented urosome, the ventral keel is not protruded, P4 endopod with two setae and divergent CR. Tanaka (1957) reported length-width ratio of CR for a female to be 9 and 1.7 times its AS. But Lakshadweep specimen exhibited slight variation but comparable with CR length-width ratio as 1.73 and CR:AS ratio as 2. As per Kang et al. (1990) from Korean waters, CR was 7–8 times its width which varied with Lakshadweep specimen (1.73). He also mentioned on the GS being as long as CR and bears a ventral process was consistent with the present study female specimen. Male specimens were not observed during the study period.

Distribution: Distributed in the inland waters of Japan, Korean waters Namibia, Red Sea, The Gulf of Oman; Straits of Malacca, Gulf of Thailand, Hong Kong, China Seas, Korea Strait, Japan Sea, Japan, Kuchinoerabu Island, Ariake Bay, Seto Inland Sea, Tanabe Bay, Tokyo Bay, Charlotte Queen Is., British Columbia, Vancouver Is., Nitinat Lake, Strait of Georgia, San Francisco Estuary, W Baja California (Magdalena Bay), Gulf of California (http://copepods.obs-banyuls.fr/en).

***Ditrichocoryceus dahli* Tanaka, 1957**

Material examined: Sorted out the samples from plankton samples collected from Kavaratti, Minicoy (open ocean and lagoon), Agatti, and Bangaram.

Male (Fig. 23): Total body length 0.687–1.09 mm. Prosome four segmented; frontal margin arc-shaped, with two large separate cuticular lenses; prosome about 1.31 times longer than urosome; 1.96 times as long as wide (Fig. 23a). GDS broadish oval and widest at the middle 2.2 times as long as wide; AS 3.28 times as long as wide.

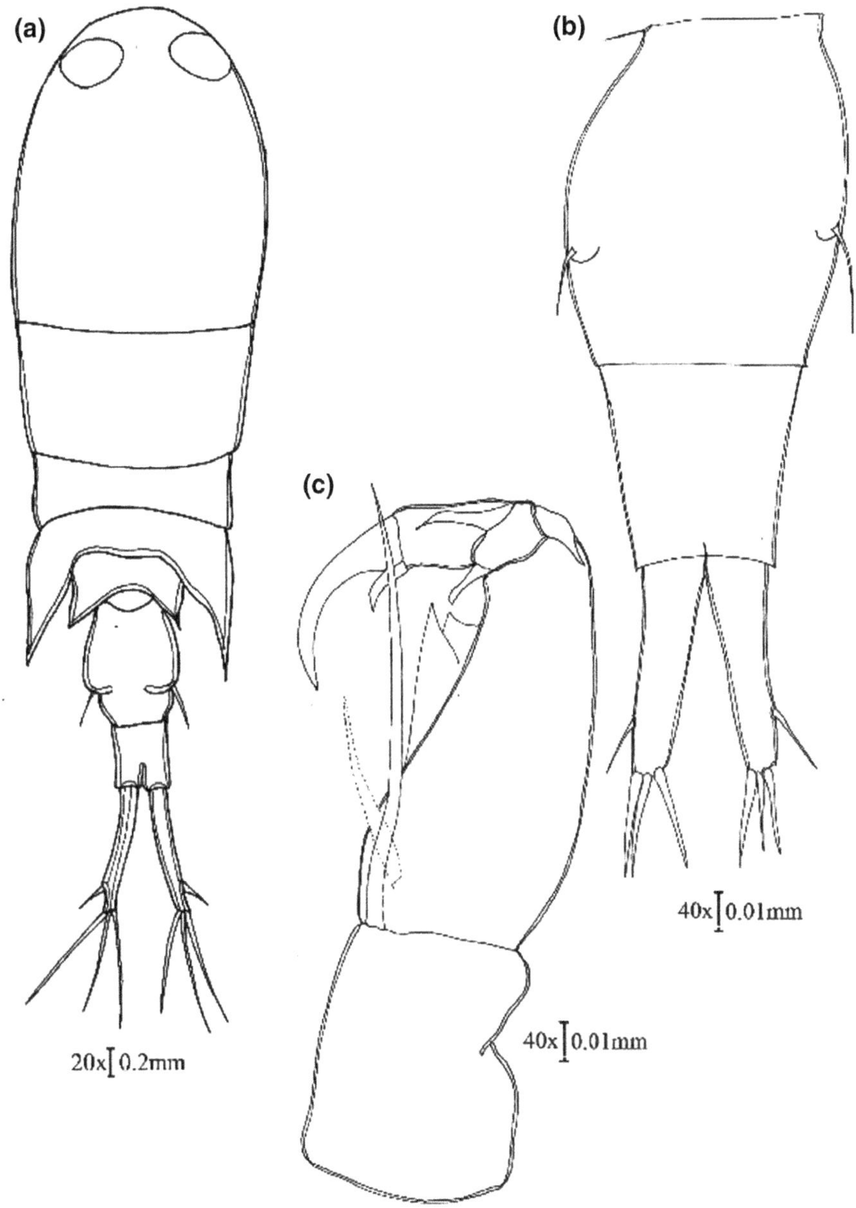

Fig. 22 *D. affinis* female. **a** Habitus; **b** urosome; **c** A2

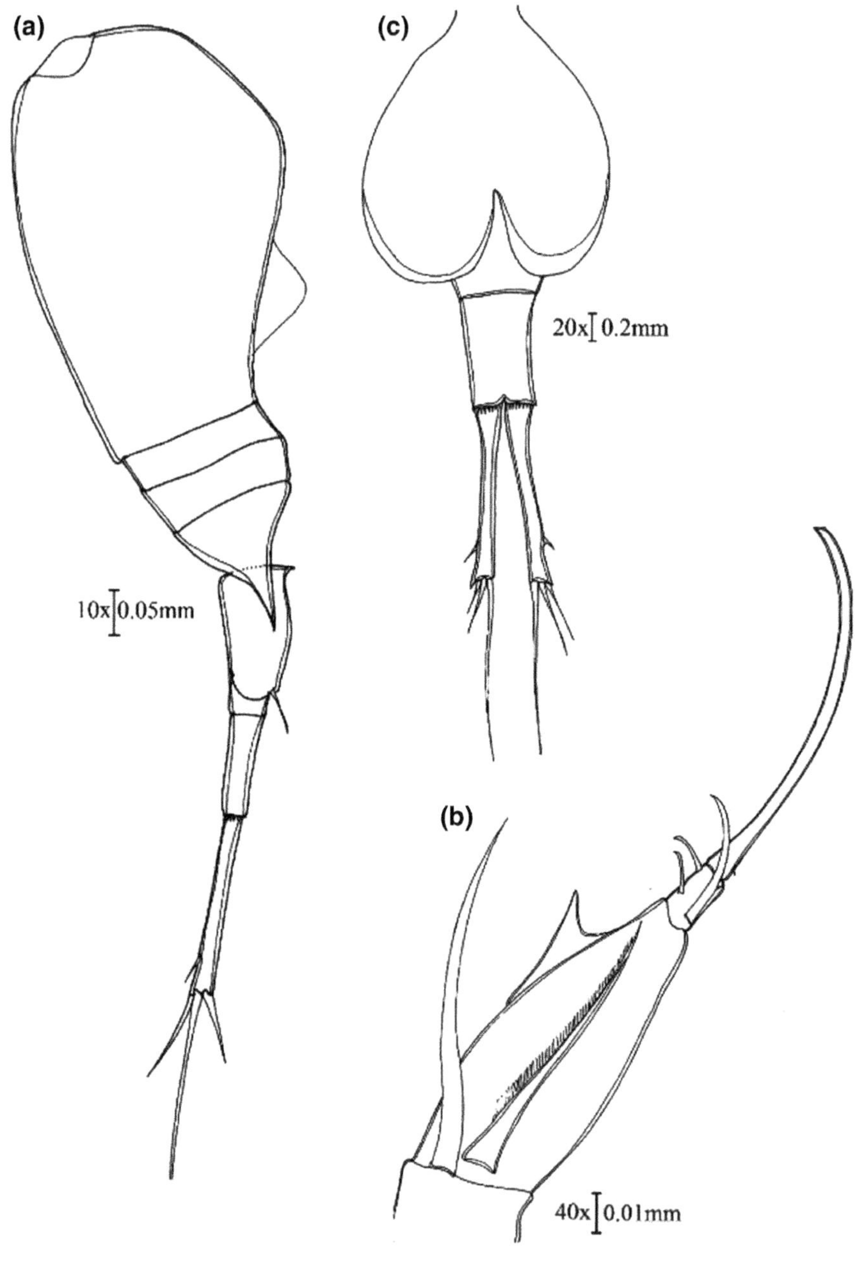

Fig. 23 *D. dahli* male. **a** Habitus (lateral); **b** A2; **c** urosome

Urosome two-segmented with slightly divergent caudal rami at the base Proportional lengths of the urosomal somites and CR is 1.65:1.15:2 Caudal rami long and divergent at the base, about 8 times longer than wide (Fig. 23c). The shape of A2 is very peculiar. A 2 four segmented with three endopodal segments. 1st endopodal segment bears a unipinnate setae shorter than coxobasal setae. 2nd endopodal segment 3 slender spines of which one is longer. 3rd endopodal segment is drawn out as a long slender claw which extends upwards (Fig. 23b).

Remarks and Discussion: As per Tanaka (1957), this species is synonymous to *D. lubbocki*, *D. tenius* and *D. africanus*. *D. dahli* is distinguished by the length of the 3rd endopodal segment being longer, the oval genital segment with a small ventral hook and long CR.

When compared with previous descriptions, the present specimens showed variations as well as similarities with the original description of Tanaka (1957). Proportional lengths of urosomal segments and CR slightly varied (1.65:1.15:2) with that of Tanaka being 5.0:1.8:3.2. However, CR length-width proportion was similar being (8 vs. 8) for Lakshadweep and Tanaka's specimen respectively. However, no female specimen could be met in the present study. Recently, Al-Yamani et al. (2011) reported both sexes of *D. dahli* from Northwestern Arabian Gulf of which for the male specimens, prosome being less than twice as urosome, was very much like Lakshadweep specimen (1.31); length of CR being more than GS was found be almost similar (3.0:1.8:3.8 vs. 1.65:1.15:2) and CR being longer than AS.

Distribution: Epipelagic, Coastal and oceanic, Subtropical and tropical, Indian and Pacific Oceans (http://copepods.obs-banyuls.fr/en).

Ditrichocoryceus tenius Giesbrecht, 1891

Material examined: Sorted out the samples from plankton samples collected from Kavaratti, Kalpeni, Minicoy (open ocean and lagoon), Agatti, and Bangaram.

Female (Fig. 24): Total body length 0.94–0.969 mm. Prosome four segmented, arc-shaped with two cuticular lenses; about 1.4 times longer than urosome; 1.79 times as long as wide. Urosome conspicuously narrower than prosome. GDS broadish oval and widest at the middle 1.68 times as long as wide; AS thrice as long as wide (3.1). Urosome two-segmented with slightly divergent caudal rami at the base. Proportional lengths of the urosomal somites and CR is 1.85:1.4:1.75. Caudal rami long, about 3.5 times longer than greatest width at base (Fig. 24a). A2 four segmented with three endopodal segments. 1st endopodal segment bears a setae shorter than coxobasal setae. 2nd endopodal segment two spines of which one is longer. 3rd endopodal segment bears two elements of which one is drawn out as a terminal claw and the other short at the base (Fig. 24b).

Remarks: *D. tenius* in general manifestation it resembles *D. lubbocki* according to Scott (1909) who reported only female specimen. Two segmented urosome is a characteristic feature with long CR which is slightly divergent at the base. In Scott's specimen, CR is nearly twice as long as AS which is comparable to that of Lakshadweep specimen, being 1.25 times as long as AS. Dahl (1912) also had the same opinion on CR:AS ratio which was twice longer than AS. The feature in which GS and AS combined longer than CR, of Dahl's female specimen were

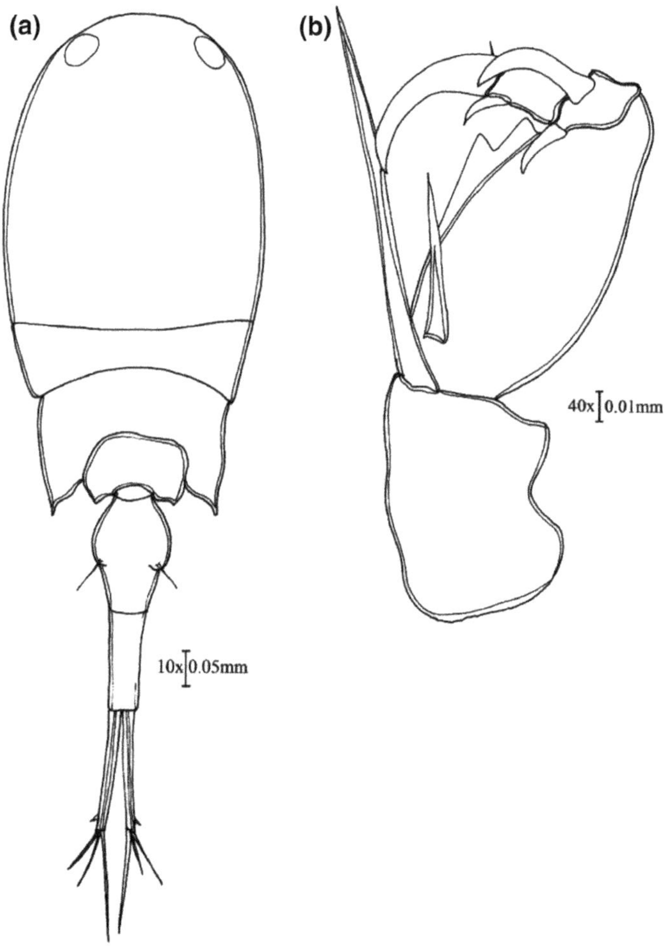

Fig. 24 *D. tenius* female. **a** Habitus; **b** A2

consistent with the Lakshadweep specimen. Giesbrecht (1892) also reported only female specimens. The present study also encountered only female specimens.

Distribution: Indian and Pacific Ocean (http://copepods.obs-banyuls.fr/en).

Ditrichocorycaeus subulatus Herrick, 1887

Material examined: Sorted out the samples from plankton samples collected from Kavaratti, Minicoy (open ocean and lagoon), Agatti, and Bangaram.

Female (Fig. 25): Total length 0.89–0.9 mm. Prosome four segmented; 1.76 times as long as wide and 1.96 times as long as urosome. 4th prosomal segment produced into lateral flaps; two-segmented urosome 2.90 times as long as wide. Oval shaped GS, 1.24 times as long as wide. AS rectangular shaped, 0.69 times shorter than GS and 0.94 times shorter than CR. CR is 4.75 times longer than wide. Proportional lengths

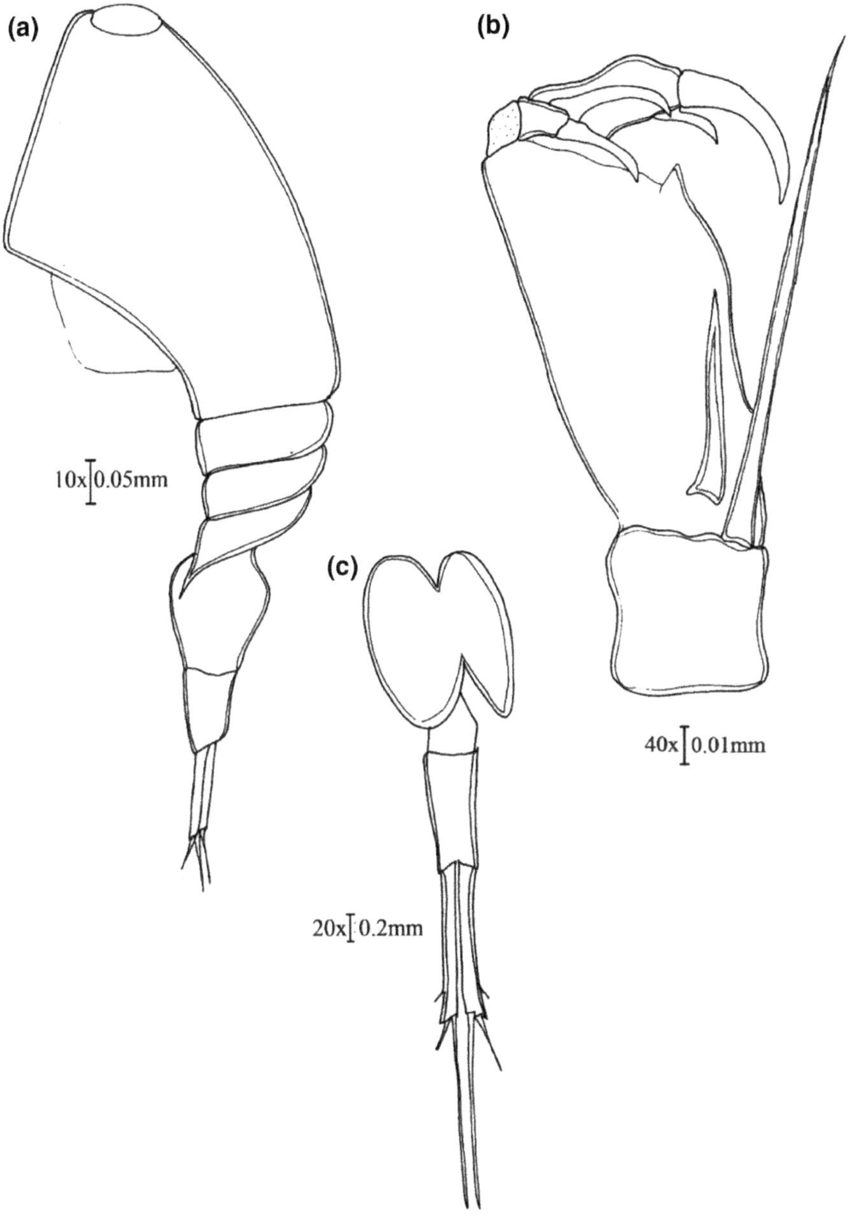

Fig. 25 *D. subulatus* female. **a** Habitus; **b** A2; **c** urosome

of GS:AS:CR 1.3:0.9:0.95 (Fig. 25a and c). A2 four segmented with three endopodal segments. 1st endopodal segment with endopodal setae shorter than coxobasal setae. 2nd endopodal segment with two spines. 3rd endopodal segment, cylindrical, bears two elements of which one is drawn out as a terminal claw and the other short at the base (Fig. 25b).

Remarks: Wilson (1949) synonymized it as *Corycaeus americanus*. According to him, characters like 3rd metasomal segment produced into lateral flaps agreed with the present description. However, variations also existed in the length ratio GS:AS being lower (1.44) for Lakshadweep specimen than that of former (2). He also mentioned on CR being longer than the other two segments combined also varied in the present study specimen. Nevertheless, the ornamentation of A2 and P4 endopod with two setae also confirmed the identity of *D. subulatus* female specimens as we could not come across any male specimens.

Distribution: Epipelagic, Coastal and oceanic, Subtropical and tropical, Indian and Pacific Oceans (http://copepods.obs-banyuls.fr/en).

***Ditrichocorycaeus lubbocki* Giesbrecht, 1891**

Material examined: Sorted out the samples from plankton samples collected from Kavaratti, Kalpeni, Minicoy (open ocean and lagoon), Agatti, and Bangaram.

Female (Fig. 26): Total length 0.94–0.99 mm. Prosome five segmented; broader than urosome 2.23 times as long as wide and 1.34 times as long as urosome. Lateral flaps of the 4th prosomal somite extend to the middle of the genital somite. A stout process arising from the lateral margin of the 5th prosomal somite Urosome two-segmented, 4.72 times as long as wide. Urosomal somite and CR in the proportional lengths 1.15:1.25:1.7 (Fig. 26a and c). Ventral surface of GS is flat without a ventral hook. A2 four segmented with three endopodal segments. Inner margin of coxobasis partly ornamented with spines. First endopodal segment with unipinnate endopodal setae too shorter than coxobasal setae and outer lateral margin adorned with numerous denticles or spine. 2nd endopodal segment with three spines. 3rd endopodal segment bears two elements of which one is drawn out as a small terminal claw and the other short at the base (Fig. 26b).

Remarks: *D. lubbocki* females agree well with the original description by Giesbrecht (1891) in its distinctive shape of GS/GDS which is distally swollen in dorsal view and absence of ventral hook. The absence of ventral hook was recorded as rounded eminence in figure by Sewell (1947), slightly round process by Chen et al. (1974), Zheng et al. (1982) and Wi et al. (2013). Presence of two well-developed teeth on the distal segment of first endopodal segment of A2 with outer lateral margin adorned with numerous denticles and presence of spines at the inner margin of coxobasis very well agreed with previous description of Wi et al. (2013).

Distribution: Epipelagic, coastal and oceanic, subtropical and tropical, Indian and Pacific Oceans (http://copepods.obs-banyuls.fr/en).

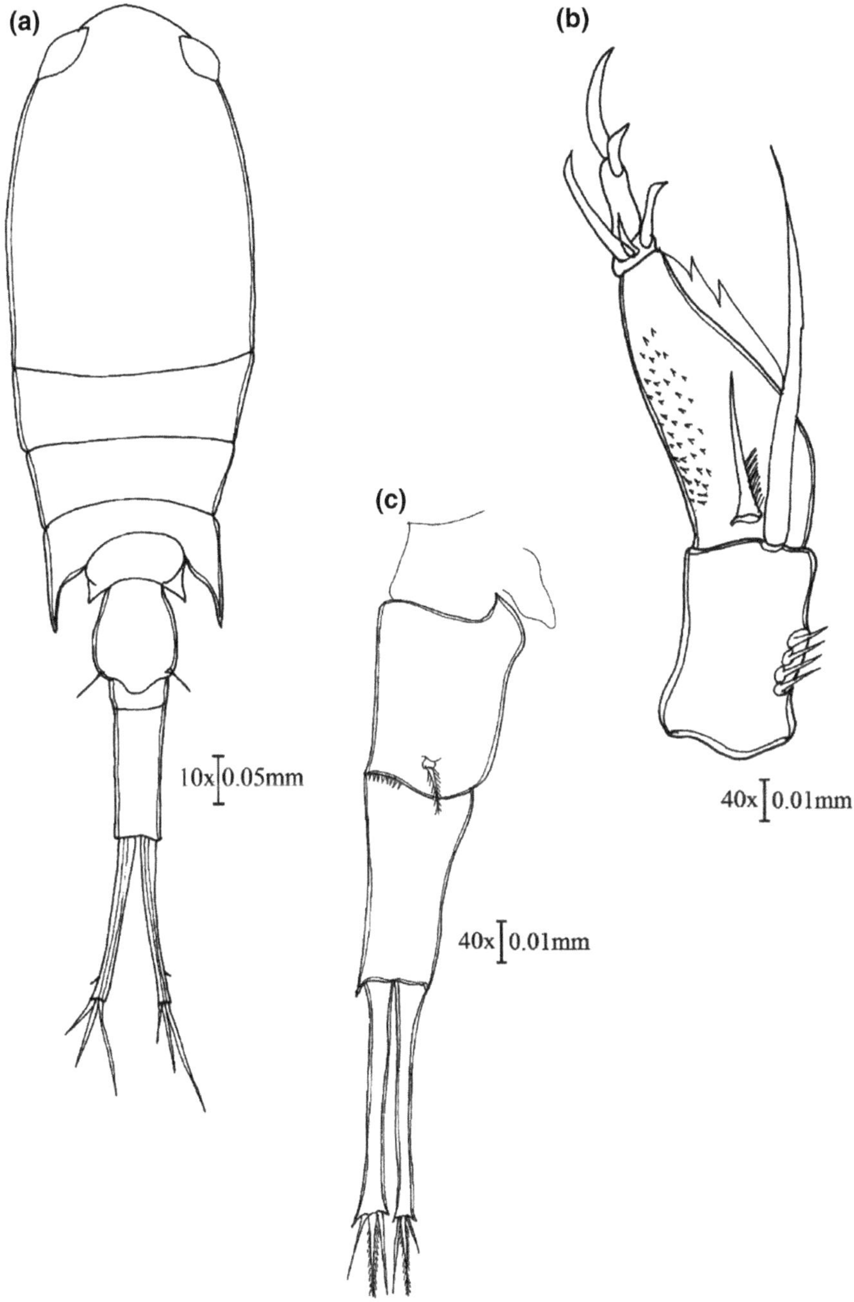

Fig. 26 *D. lubbocki* female. **a** Habitus; **b** A2; **c** urosome

Genus *Farranula* Wilson, 1932
Farranula gibbula Giesbrecht, 1891

Material examined: Sorted out the samples from plankton samples collected from Kavaratti, Kalpeni, Minicoy (open ocean and lagoon), Agatti, and Bangaram.

Female (Fig. 27): Total Length 0.866–0.88 mm. Ventral surface bears beak-like keel. Dorsal bump present on the second PR (Prosome) segment. Prosome 2.04 times as long as urosome. Urosome one segmented. The mid-part of GS swollen, 2.05 times as long as wide. Non-divergent CR, 7 times as long as wide. Proportional lengths of urosomal somites 0.41:0.21 (Fig. 27a). A2 four segmented with coxobasis and three endopodal segment. Bipinnate coxobasal setae at inner distal margin, slightly longer than endopodal setae which is also bipinnate. Inner distal margin of the first endopodal segment roughly serrated and outer lateral margin ornamented with a row

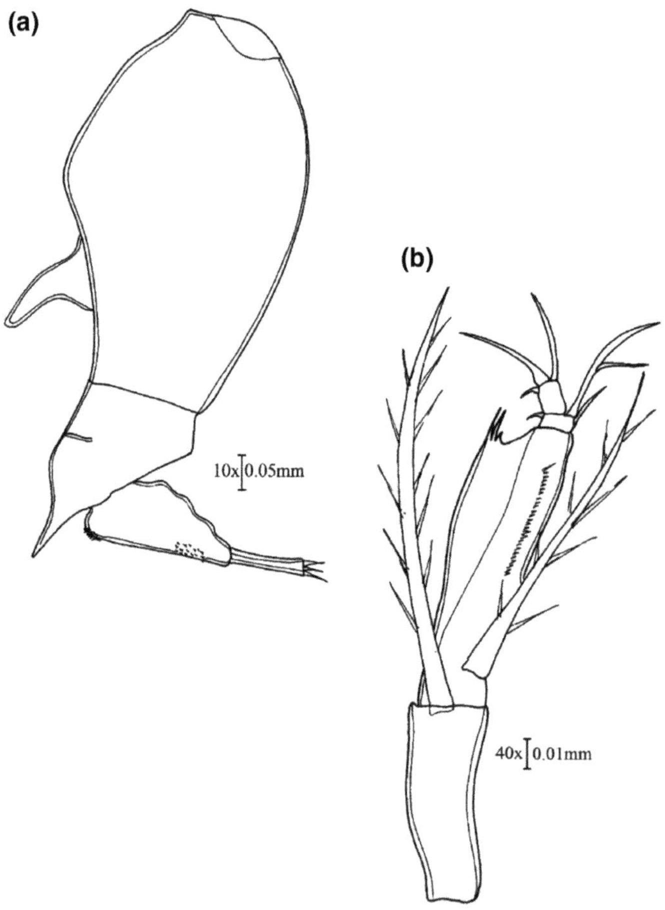

Fig. 27 *F. gibbula* female. **a** Habitus (lateral); **b** A2

of denticles. 2nd endopodal segment shortest, bearing three elements of which the longer one with a lateral branch. 3rd endopodal segment also bears three elements (Fig. 27b).

Remarks: The genus *Farranula* is distinct in the characters like two segmented prosome, combined GDS and AS, uniramous P4 lacking endopod in both sexes and all these features very well matched with Giesbrecht's (1893) descriptions. Females of *F. gibbula* are characterized by the exclusive shaped GDS as well as the presence of dorsoposterior bump of the second PR segment. This species has been previously reported by many authors in which descriptions were inaccurate. But Wi and Soh (2013b) from Korean waters gave a meticulous description on this species. The present study specimens were similar to the Korean specimen in the presence of above said distinguished characters which itself confirms the species identity. The morphometry of the Lakshadweep specimens was also consistent with that of Korean specimens. Because of the similar appearance in males of different species and lack of specific criteria for identification, it's very difficult to differentiate males. Farran (1911) opined on the difficulty in matching males to their respective females.

Distribution: Atlantic Ocean, Red Sea, Arabian Sea, Indo-Pacific, East China Sea, Japan Sea, South Pacific, North Pacific (http://copepods.obs-banyuls.fr/en).

***Farranula concinna* Dana, 1853**

Material examined: Sorted out the samples from plankton samples collected from Kavaratti, Kalpeni, Minicoy (open ocean and lagoon), Agatti and Bangaram.

Female (Fig. 28): Total Length 0.812–82 mm. Ventral surface bears beak-like keel. The second PR segment without a dorsal bump. Prosome 1.76 times as long as wide. The posterolateral margin of the second PR segment projected up to the middle of GS. Prosome 2.30 times as long as urosome. Urosome one segmented. GS irregularly humpbacked, 2.14 times as long as wide, greatest width at the proximal region. CR, 2.66 times as long as wide. Proportional lengths of GS:CR is 2.25:0.8. A2 four segmented with Coxobasis and three endopodal segments (Fig. 28b). Bipinnate coxobasal setae at the inner distal margin, slightly longer than endopodal setae which is also bipinnate. Inner distal margin of the first endopodal segment roughly serrated and outer lateral margin ornamented with a row of denticles. 2nd endopodal segment shortest, bearing three elements of which the longer one with a lateral branch. 3rd endopodal segment also bears three elements (Fig. 28b).

Male (Fig. 29): Total Length 0.766–0.77 mm. Ventral surface bears beak-like keel. Prosome two segmented. Absence of dorsal bump on the second PR segment. Prosome 1.08 times as long as urosome and 1.61 times as long as wide. Urosome one segmented. The mid-part of GS swollen and the fattest part below the mid-portion, 2.95 times as long as wide. Non-divergent CR, 9.3 times as long as wide. Proportional lengths of urosomal somites 6.5:2.8 (Fig. 29a). A2 four segmented with coxobasis and three endopodal segment. Bipinnate coxobasal setae at the inner distal margin, slightly longer than endopodal seta which is also bipinnate. Inner distal margin of the first endopodal segment roughly serrated and outer lateral margin ornamented with a row of denticles. 2nd endopodal segment shortest, bearing three elements of which the longer one with a lateral branch and the proximal small one serrated. 3rd

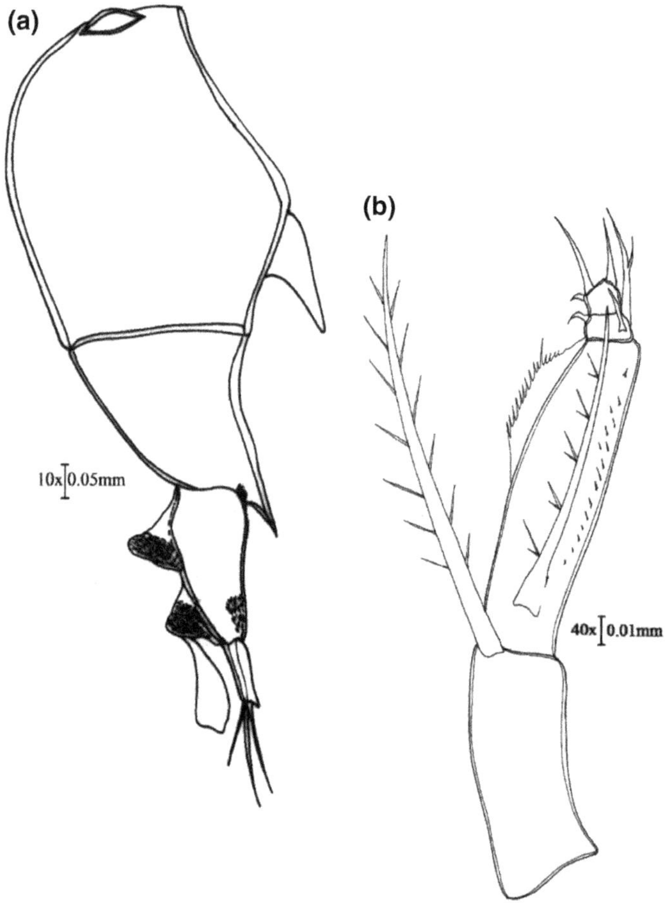

Fig. 28 *F. concinna* female. **a** Habitus; **b** A2

endopodal segment bears five elements of which longer one drawn out as a terminal claw extending up to coxobasis (Fig. 29b).

Remarks: *F. concinna* is similar to *F. gibbula* but smaller in size. Another difference from *F. gibbula* is the absence of humpback in both sexes in the second prosomal somite. When PR: UR ratio of Lakshadweep specimens for females and males (2.3 and 1.08) when compared with Tanaka's specimens (2.12 and 1.38), were comparable. CR 9.3 times as long as broad for Lakshadweep male specimens varied with that of Tanaka's (5). GS being slender in lateral view and 2.95 times as long as high for males were also consistent with that of Tanaka (3). These species have been previously reported by many authors in which descriptions were inaccurate. But Wi and Soh (2013b) from Korean waters gave a meticulous description of this species. However, the diagnostic characters which confirm the morphological identity of *F.*

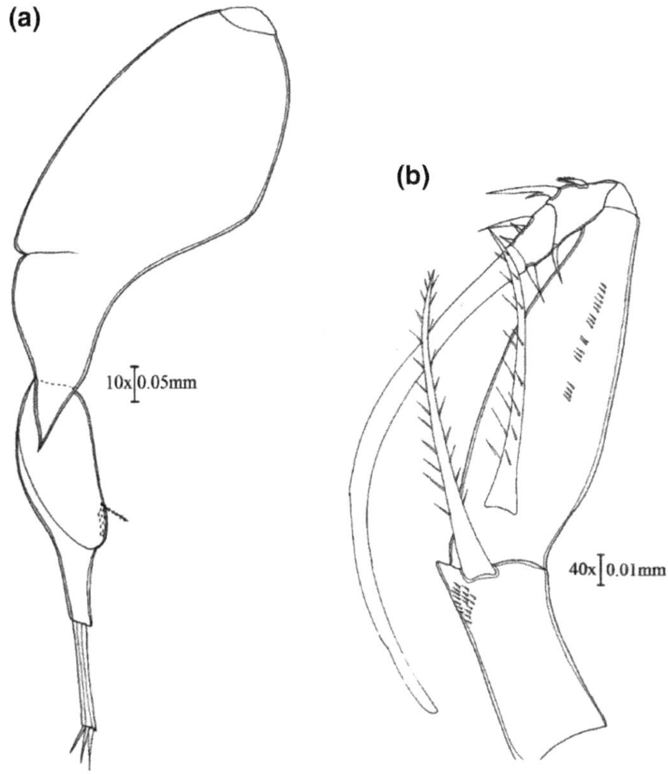

Fig. 29 *F. concinna* male. **a** Habitus (lateral); **b** A2

concinna were consistent with that of previous descriptions including the recent ones of Korean waters.

Distribution: North Atlantic, Indo-Pacific, East China Sea, South Pacific, North Pacific (http://copepods.obs-banyuls.fr/en).

***Farranula gracilis* Dana, 1849**

Material examined: Sorted out the samples from plankton samples collected from Kavaratti, Kalpeni, Minicoy (open ocean and lagoon), Agatti and Bangaram.

Female (Fig. 30): Total Length 0.997–1 mm. PR two-segmented with no mid-dorsal hump on the second PR segment as in *F. gibbula*; more than twice (2.4) times as long as wide; 2.29 times as long as urosome; ventral surface bears beak-like keel. Lateral view section appears more or less humpbacked. The postero-lateral margin of the second PR segment projected up to the middle of GS (Fig. 30a). Urosome one segmented, 2.83 times longer than wide. GS and AS combined. UR 3.09 times as long as CR. CR approximately 4.7 times longer than wide. Proportional lengths of GS:CR is 2.4:1.1. GS slender as long as wide, maximum width at the proximal region (Fig. 30b). A2 bears one plumose setae each on the basopodal and endopodal

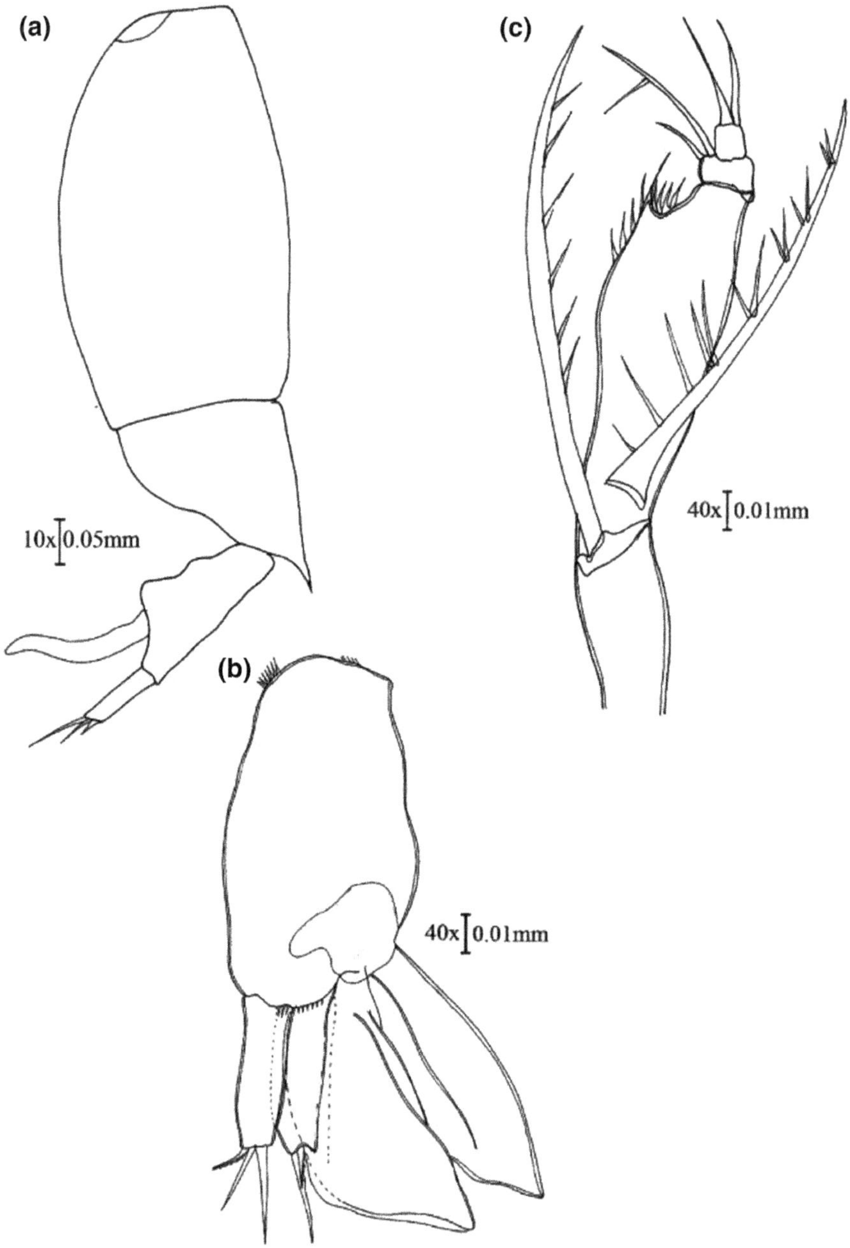

Fig. 30 *F. gracilis* female. **a** Habitus (lateral); **b** urosome; **c** A2

margin (Fig. 30c). A2 four segmented with coxobasis and three endopodal segments. Bipinnate coxobasal setae at inner distal margin, slightly longer than endopodal setae which is also bipinnate. Inner distal margin of the first endopodal segment roughly serrated. 2nd endopodal segment shortest, bearing two elements of which the longer one with a lateral branch. 3rd endopodal segment bears two elements (Fig. 30c).

Male (Fig. 31): Total Length 0.90–0.92 mm. PR two segmented with no mid dorsal hump on the second PR segment as in *F. gibbula*; more than twice (2.5) times as long as wide; 1.61 times as long as urosome. The postero-lateral margin of the second PR segment projected up to the middle of GS. Urosome one segmented, 3.03 times longer than wide. GS and AS combined. UR 3.38 times as long as CR. CR 2.88 times longer than wide. Proportional lengths of GS:CR is 3.1:1.3. GS slender 2.13 times as long as wide, maximum width at the proximal region (Fig. 31a, c). A2 four segmented with coxobasis and three endopodal segment. Bipinnate coxobasal setae at inner distal margin, slightly longer than endopodal setae which is also bipinnate. 2nd endopodal segment shortest, bearing three elements. 3rd endopodal segment bears three elements of which longer one drawn out as a terminal claw extending up to Coxobasis (Fig. 31b).

Remarks: After Björnberg (1963) this species when in large numbers is a good indicator of high saline (above 35.5) and warm waters (above 21 °C). According to Cervigon (1964) the length ratio of abdomen to caudal rami of *F. gracilis* female is 68:32 and 71:29. Specimens of *F. gracilis* is characterized by the absence of bump on second prosomal segment in females; one segmented urosome which is more than twice (female) and thrice (male) as long as broad; slender GS in males, shape and length width ratio of CR in both sexes. The combination of general morphological features such as shape of GS, narrow PR, presence of ventral hook in females; projection of posterolateral margin of the second thoracic segment: width length ratio of the UR more than twice in females and length width ratio more than thrice in males were consistent with the description of *F. gracilis* female by Cervigon (1964). However, characters were also comparable, from the illustrations given by Wi and Soh (2013b) of Korean waters in the length ratio of the PR:UR (2.7) slightly greater than those of Lakshadweep specimen (1.61); proportional lengths of GS:CR greater (2.5) as compared to Lakshadweep specimens (2.38) and absence of dorsal bump in both specimens.

Distribution: Madagascar, Indian and Pacific Ocean (http://copepods.obs-banyuls.fr/en).

Farranula rostrata Claus, 1863

Material examined: Sorted out the samples from plankton samples collected from Kavaratti, Kalpeni, Minicoy (open ocean and lagoon), Agatti, and Bangaram.

Female (Fig. 32): Total Length 0.710–0.85 mm. Ventral surface bears beak-like keel. Two segmented prosome, 2.28 times longer than wide. Cuticular lenses present. The second PR segment without a dorsal bump. The postero-lateral margin of the second PR segment projected up to the middle of GS. Prosome 1.23 times as long as urosome. Urosome one segmented, 4.2 times as long as wide. Length ratio of GS:CR 1.39, GS slender 2.48 times as long as wide, maximum width at the proximal region.

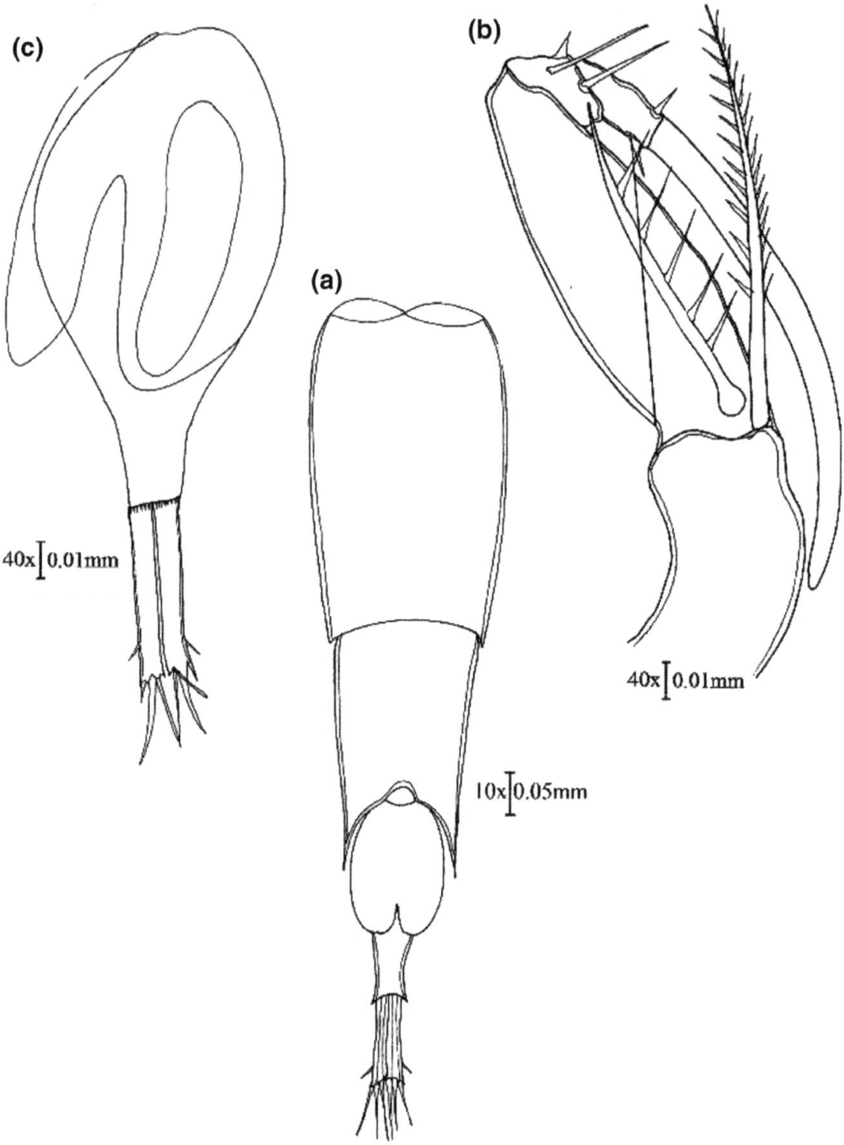

Fig. 31 *F. gracilis* male. **a** Habitus (dorsal); **b** A2; **c** urosome

CR, thrice as long as wide, slightly divergent at the distal end (Fig. 32a). A2 four segmented with coxobasis and three endopodal segments bipinnate coxobasal setae at inner distal margin, slightly longer than endopodal setae which is also bipinnate. Inner distal margin of the first endopodal segment roughly serrated. Second endopodal

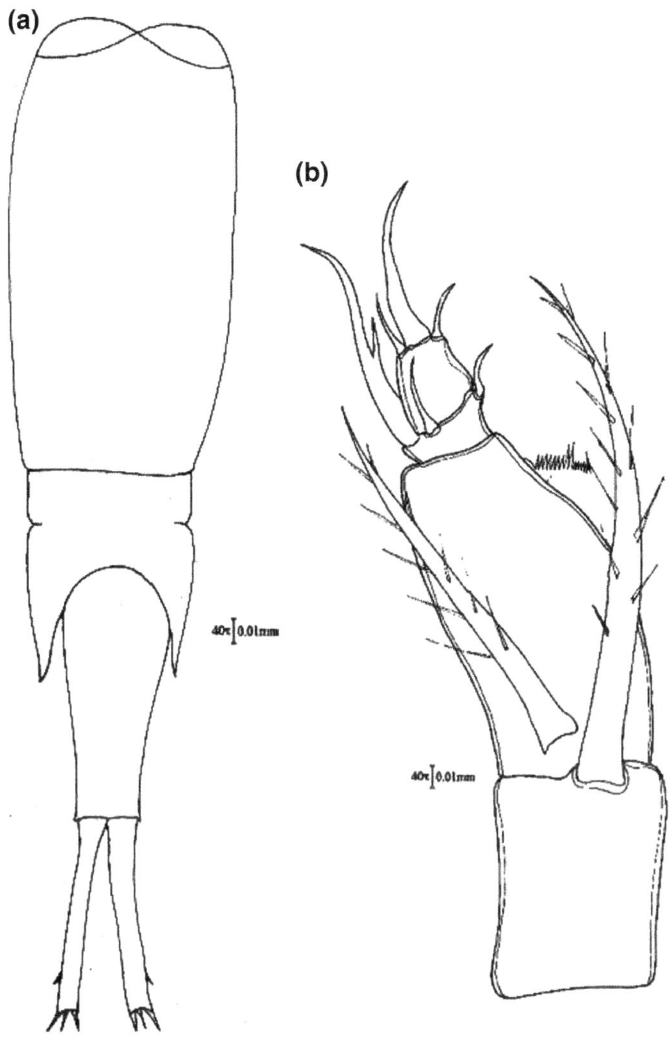

Fig. 32 *F. rostrata* female. **a** Habitus; **b** A2

segment shortest, bearing three elements of which the longer one with a lateral branch. 3rd endopodal segment bears three elements (Fig. 32b).

Remarks: *F. rostrata* was characterized by the ornamentation of A2 and GS shape. This species is considered to be a good indicator of surface waters of temperatures between 16 and 18 °C. When compared with previous literatures, length width ratio of prosome of Lakshadweep specimens were found to be comparable with Dahl (1912) and Itoh (1997) (2.28 vs. 2.4 and 2.2) respectively. Length-width ratio of GDS also were consistent with Giesbrecht (1892), Dahl (1912) and Itoh (1997) (2.1, 3 and 2.8) respectively but with slight variations in which present ratio being 2.4.

Distribution: Tropical Atlantic and the Mediterranean Sea, Indian Ocean, Pacific concern SE Australia, New Zealand, Indo-Pacific (http://copepods.obs-banyuls.fr/en).

Farranula curta **Farran, 1911**

Material examined: Sorted out the samples from plankton samples collected from Kavaratti, Agatti, Bangaram, and Minicoy (open ocean and lagoon).

Female (Fig. 33): Total Length 0.89–0.908 mm. Ventral surface bears beak-like keel. The second PR segment without a dorsal bump. The posterolateral margin of the second PR segment projected up to the middle of GS. Proportional lengths of GS:CR is 1.6:1. Prosome 1.75 times as long as urosome and 1.78 times as long as wide. Urosome one segmented. GS slender 2.66 times as long as wide. CR 2.85 times as long as wide (Fig. 33a). A2 four segmented with coxobasis and three endopodal segments

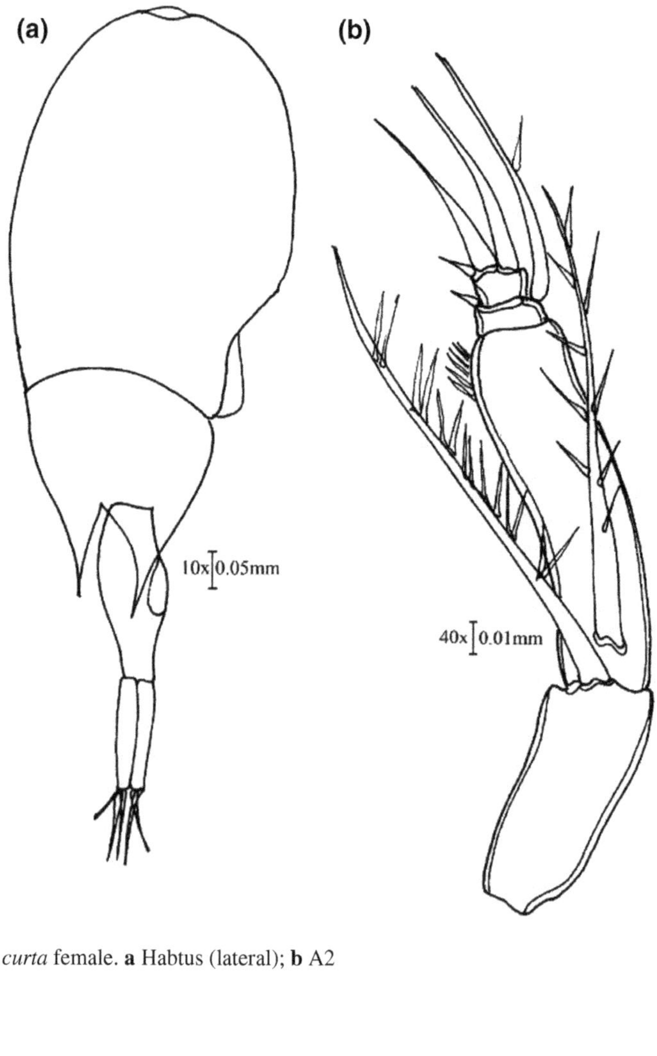

Fig. 33 *F. curta* female. **a** Habtus (lateral); **b** A2

bipinnate coxobasal setae at inner distal margin, slightly longer than endopodal setae which is also bipinnate. Inner distal margin of the first endopodal segment roughly serrated. 2nd endopodal segment shortest, bearing two elements of which the longer one with a lateral branch. 3rd endopodal segment bears two elements (Fig. 33b).

Remarks: The genus *Farranula* is distinct in the characters like 2 segmented prosome, combined GDS and AS, uniramous P4 lacking endopod in both sexes and all these features very well matched with Giesbrecht's (1893) descriptions. Females of *F. curta* is characterized by the absence of bump on second prosomal segment, slender GS and the shape of CR. The combination of general morphological features such as shape of GS, narrow PR, presence of ventral hook; projection of posterolateral margin of the second thoracic segment; length-width ratio of the GS (2.66 vs. 2.5) is consistent with the description of *F. curta* female by Tanaka (1957).

Distribution: Tropical Atlantic and the Mediterranean Sea, Indian Ocean, Pacific concern SE Australia, New Zealand, Indo-Pacific (http://copepods.obs-banyuls.fr/en).

4 Discussion and Conclusion

Twenty-three species have been identified based on morphological characteristics under five genera (*Corycaeus*, *Urocorycaeus*, *Onychocorycaeus*, *Ditrichocorycaeus*, and *Farranula*) belonging to the family Corycaeidae. Four species during the present study represented the genus *Corycaeus* which are *C. crassiusculus*, *C. speciosus*, *C. clausi* and *C. vitreus*. While females of *C. crassiusculus* could be identified by the overlapping of the genital segment on the anal segment at the dorsal margin, *C. clausi* males exhibited a row of minute spines on the lateral portion of first endopodal segment of the A2 which was absent is *C. vitreus* males. *C. speciosus* species was first reported in the Arabian Sea by Sewell in 1947 which is distinguished by its divergent CR, rounded forehead in female and broad one in male. Farran opined that the males of *C. vitreus* differs from that of *C. crassiusculus* and *C. clausi* by its short head which is broad anteriorly and tapers posteriorly short anal segment and fine transparent edge on the longest furcal setae.

Genus *Ditrichocorycaeus* M. Dahl (1912) is characterized by the procession two setae on endopod of P4. Due to deficient identification keys and more analogous morphological features, taxonomic ambiguity still continued until Wi et al. (2013) and Vidjak and Bojaniæ (2008) provided a detailed description of *D. dahli*, *D. lubocki*, *D. subtilis* and redescription of *D. minimus*. Genus *Ditrichocorycaeus* includes fourteen valid species that are widely distributed and so far been described from world oceans. From this, six species were recorded during the present study from Lakshadweep Sea, a part of the Indian Ocean. They are *D. andrewsi*, *D. affinis*, *D. dahli*, *D. tenius*, *D. subulatus* and *D. lubbocki*. Genus *Ditrichocorycaeus* is divided into two groups based on relative lengths of the caudal ramus. The first group has short caudal rami almost equal to anal somite but shorter than genital somite and the second group is characterized by the possession of caudal ramus longer than both anal somite and genital somite (Dahl 1912; Tanaka 1957). Based on this classification, out of the six species identified from Lakshadweep waters, *D. andrewsi*, *D. subulatus*, *D. affinis* can be categorized into group 1 and *D. dahli, D. tenius* and *D. lubbocki* into the second group. Nonetheless, the ecological and taxonomical record for *Ditrichocorycaeus* sp. of Indian Ocean barely exists. In this perspective, taxonomic identification of six species including scientific drawings during the present study seem significant in all aspects. Nevertheless, recently Johan et al. (2013) have recorded four species of *Ditrichocorycaeus* sp. from Malaysian waters and Veronica and Nagappa (2013) recorded one species of *Ditrichocorycaeus* sp. from Bay of Bengal, both of which indeed are the parts of Indian Ocean. Al-Yamani et al. (2011) also recorded three species of *Ditrichocorycaeus* from North western Arabian Gulf waters.

Among the seven species comprising the genus *Onychocorycaeus* Dahl 1912 which are widely distributed in world oceans, the present study recorded six species. They are *Onychocorycaeus giesbrechti* Dahl 1894, *Onychocorycaeus agilis* Dana 1849, *Onychocorycaeus catus* Dahl 1894, *Onychocorycaeus latus* Dana 1849, *Onychocorycaeus pacificus* Dahl 1894, and *Onychocorycaeus pumilus* Dahl 1912. *O. catus* and *O. pumilus* had already been reported by Karanovic (2003) from Australian waters. *O. catus* from Kavaratti island of Lakshadweep archipelago is the smallest ever reported (0.65 mm) (Radhika et al. 2014b). Most of the *Onychocorycaeus* species are distinguished mainly by the ornamentation of A2 and the number of setae on P4 endopod being one. Most of male specimens look alike in shape as well A1 ornamentation. However, females differ in the above-said character. The inner distal margin of the first endopodal segment of A2 produced into two stout teeth is yet another distinguishing character, especially in females. Even though secondary, the shape and length-width ratio of prosome also differs in females but in males, the prosome shape is broad in almost all species.

Genus *Urocorycaeus* is characterized by the presence of long and nearly parallel but ends slightly divergent CR in both sexes measuring at least twice the length of the rest of the urosome. From this genus, only two species could be taxonomically identified which is distributed in Indian, Pacific, and Atlantic oceans. They are *U. furcifer* and *U. lautus*. Species under this genus is primarily identified by the lengthy CR measuring at least twice its length. Another distinguishing character of male specimen of *U. furcifer* from that of its allied species *U. longistylis* is the differentiation of UR into GS and AS. *U. lautus* can be differentiated by the presence of 2 setae (1 long and 1 rudimentary) in the P4 endopodite where *U. furcifer* has only one. Size of the former is too large when compared to the latter one. Here in the present specimens from Lakshadweep Islands has this differentiation very well that confirms their taxonomic identity.

The most conspicuous characters that make the *Farranula* sp. distinct in both sexes are (1) two segmented prosome; (2) presence of ventral keel like knob in females; (3) combined GS and AS; (4) uniramous P4 lacking the endopod; (5) exopodal spines lacking for P1 to P3 except for the distal and terminal spines on the distal exopodal segment. Additional morphological features that was included by Wi and Soh (2013a, b) with the salient features of *Farranula* species from Korean waters are (1) P1 and P3 of *Farranula* species lack coxal setae, which are present on those legs of other

Corycaeidae; (2) a vestigial coxal seta on P2 coxa, and a minute process on the inner margin; and (3) the basis of P4 is fringed with spinules along inner median margin.

The genus *Farranula* includes seven species of which five were found in Lakshadweep waters during the present study. Due to the small size (more or less 1 mm), very similar morphology, as well as non-detailed previous descriptions of *Farranula* species lead to difficulty in identification. Wi and Soh (2013a, b) mentioned about the incompleteness and mistakes of three *Farranula* species made in previously published descriptions by Motoda (1963). Those included the lack of ornamentation on the surface of the first endopodal segment and coxobasis of the antenna; dorsal habitus of *F. carinata* (Motoda 1963: Fig. 21a) probably represents *F. curta*, judging by figure of genital double-somite and degree of expansion of the pleural areas of second prosomal somite; the incorrect setal formula for P4 in all three species. The dorsal habitus of female *F. rostrata* described by Chen et al. (1974) is also considered as *F. curta*, because of the shape of the genital double-somite and the relatively longer caudal ramus as compared with *F. rostrata*. Besides the lack of specific criteria for the identification of males of different species, their similar appearance also proved obscured in species differentiation. *F. gibbula*, *F. concinna* and *F. carinata* could not be matched by Farran (1911) to their respective females. Moreover, neither any specific difference of *F. gracilis* males could be displayed by both Dana (1852) and Dahl (1912) from the Atlantic Ocean. Without a proper examination of the dissected appendages and also because of similar length-width proportions of PR and CR, identification of *F. concinna* and *F. carinata* seems very difficult.

Acknowledgements The authors owe their gratitude to Head of the Department of Marine Biology, Microbiology and Biochemistry, School of Marine Sciences, Cochin University of Science and Technology for the facilities and the Department of Science and Technology and Union Territory of Lakshadweep for research permission and other logistic support and necessary facilities for undertaking the study. This study forms a part of a major project "Taxonomy **and genetic characterization of pelagic copepods (Crustacea) from the marine habitats of the South West Coast of India**" funded by Department of Biotechnology, Government of India (BT/PR4258/AAQ/3/575/2011).

References

Al-Yamani FY, Skyrabin VA, Al Husaini M, Khvorov AS, Prusova IY (2011) Marine zooplankton practical guide for the northwestern Arabian Gulf, vol 2. Lucky Press, Kuwait. ISBN 978-99966-95-07-0

Anon (1968) Zooplankton sampling. Monographs on oceanographic methodology. UNESCO, Paris

Björnberg TKS (1963) On the marine free-living copepods off Brazil. Bol Inst Oceanogr São Paulo 13(1):3–142

Boxshall GA, Halsey SH (2004) An introduction to copepod diversity. The Ray Society of London, 966 pp

Cervigon F (1964) Los Corycaeidae del Caribe sur-oriental (Copepoda, cyclopoida). Mem Soc Cienc Nat Le Salle 24:163–201

Chen QC, Zhang SZ, Zhu CS (1974) On planktonic copepods of the Yellow Sea and the East China Sea. 2. Cyclopoida and Harpacticoida. Stud Mar Sin 9:64–66

Dahl M (1912) Die Copepoden der Plankton-Expedition I. Die Corycaeinen. Mit Beru¨cksichtigung alle bekannten Arten. Ergebn. Plankton Expedition. Humboldt-Stiftung 2:1–134
Dakin WJ, Colefax AN (1940) The plankton of the Australian coastal waters off New South Wales. Monograph 1, Publications of the University of Sydney Department of Zoology, 215 pp
Dana JD (1849) United States exploring expedition during the years 1838, 1839, 1840, 1841, 1842 under the command of Charles Wilkes, vol 8
Dana JD (1852–55) Crustacea. U. S. exploring expedition during the years 1838-42, under the command of Charles Wilkes, vol 8
Davis CC (1949) Notes on the plankton of Long Lake, Dade County, Florida, with descriptions of two new copepods. Q J Fla Acad Sci 10(2 & 3):79–88
Farran GP (1911) Plankton from Christmas Island, Indian Ocean. I. On copepoda of the family Corycaeidae. Proc Zool Soc Lond 282–296
Farran GP (1936) Copepoda. Great Barrier Reef Expedition, 1928–29. Sci Rep 5(3):11–141
Giesbrecht W (1892–93) Systematik und Faunistik der pelagicschen Copepoden des Golfes von Neapel und der angrenzenden Meeres- Abschnitte. Fauna u Flora Golf Neapel, vol 19, pp 1–831, pls 1–54
Harris RP, Wiebe PH, Lenz J, Skjoidal HR, Huntley M (2000) ICES zooplankton methodology manual. Academic Press, London, CA 669 pp
Itoh H (1997) Family Corycaedae. In: Chihara M, Murano M (eds) An illustrated guide to marine plankton in Japan. Tokai University Press, Tokyo, pp 967–977
Jagadeesan L, Jyothibabu R, Anjusha A, Arya PM, Madhu KR, Muraleedharan K, Sudheesh (2013) Ocean currents structuring the mesozooplankton in the Gulf of Mannar and the Palk Bay, southeast coast of India. Prog Oceanogr 110:27–48. http://dx.doi.org/10.1016/j.pocean.2012.12.002
Johan I, AbuHena MK, Idris MH, Arshad A (2013) Taxonomic composition and abundance of zooplankton Copepoda in the coastal waters of Bintulu, Sarawak, Malaysia. J Fish Aquat Sci 8:472
Kang YS, Huh SH, Lee SS (1990) Taxonomy and distribution of Corycaeidae (Copepoda: Cyclopoida) in the Korean waters in summer. J Oceanol Soc 25(2):49–61
Karanovic T (2003) Marine interstitial: Poicilostomatoida and Cyclopoida (Copepoda) of Australia. IDC & Martinus Nijhoff Publishers, Netherlands, pp 46–54
Kasturirangan LR (1963) Key to the identification of the more common pelagic copepods of the Indian coastal waters. INCOR, vol 2. CSIR Publications, New Delhi
Mori T (1937/1964) The Pelagic Copepoda from the neighbouring waters of Japan, 2nd edn. The Soyo Company Inc., Tokyo, 150 pp.
Motoda S (1963) Corycaeus and Farranula (Copepoda, Cyclopoida) in Hawaiian waters. Pub Seto Mar Biol Lab Kyoto Univ 11:39–92
Omori M, Ikeda T (1984) Methods in marine zooplankton ecology. Willy, New York 332 pp
Radhika R, Bijoy Nandan S, Rithin Raj M, Sanu VF (2014a) Species assemblage and community patterns of cyclopoid copepods in Kavaratti atoll along the Indian coast. Int J Curr Res 6(09):8648–8657
Radhika R, Bijoy Nandan S, Harikrishnan M (2014b) Redescription of female specimens of *Corycaeus* (*Corycaeus*) *crassiusculus* Dana and *Corycaeus* (*Onychocorycaeus*) *catus* Dahl (Poicilostomatoida: Corycaeidae) from Kavaratti Atoll, Lakshadweep Island, India. Biosystematica, 8(1&2):31–43
Robin RS, Pradipta R, Vishnu Vardhan K, Nagarjuna A, Nallathambi T, Rajanim KM, Balasubramanian T (2013) Planktonic communities and trophic interactions in the Kavaratti waters, Lakshadweep Archipelago India. Int J Ecosyst 2(2):5–18. https://doi.org/10.5923/j.ije.20120202.02
Scott A (1909) Copepoda of the Siboga expedition. Monograph, XXIV
Sewell RBS (1947) The free-swimming planktonic Copepoda. Systematic account. Sci Rep John Murray Exped 8(1):1–303
Steedman FH (ed) (1976) Zooplankton fixation and preservation. Monographs on oceanographic methodology, 4. UNESCO, Paris

Tanaka O (1957) On Copepoda of the family Corycaeidae in Japanese waters. J Fac Agric Kyushu Univ 11:77–97
Veronica F, Nagappa R (2013) Mesozooplankton community structure in the upper 1000 m along the western Bay of Bengal during the 2002 fall intermonsoon. Zool Stud 52:31
Vidjak O, Bojaniæ N (2008) Redescription of *Ditrichocorycaeus minimus* indicus M. Dahl, 1912 (Copepoda: Cyclopoida, Corycaeidae) from the Adriatic Sea. J Plankton Res 30:233–240
Wi JH, Soh HY (2013a) Two Farranula (Copepoda, Cyclopoida, Corycaeidae) species from Korean waters. J Nat Hist 47:5–12
Wi JH, Soh HY (2013b) A new species of Farranula (Copepoda, Cyclopoida, Corycaeidae) and a redescription of *Farranula carinata* from off Jeju Island, Korea. J Mar Biol Assoc UK 93(7):1813–1824
Wi JH, Kim DH, Soh HY (2013) Three species of Ditrichocorycaeus (Copepoda, Cyclopoida, Corycaeidae) from Korean waters, with new identification parameters. Ocean Sci J 48(4):419–437. http://dx.doi.org/10.1007/s12601-013-0036-8
Wilson MS (1949) A new species of copepod of the genus Corycaeus from the North American coast. Proc US Natl Mus 99(3239):321–326
Zheng Z, Li S, Li SJ, Chen B (1982) Marine planktonic copepods in Chinese waters. Shangai Science and Technology Press, Shanghai, 151 pp (in Chinese)

Sedimentological Attributes and Quartz Microtexture in the Levee Sediments of a Submarine Channel in Context of the East Antarctic Ice Sheet Fluctuations: A Study from Site U-1359 of IODP-318 Expedition

Paromita Biswas, Naresh C. Pant, Mayuri Pandey and Prakash K. Shrivastava

Abstract Sedimentological characterization of the marine sediments has been used for understanding the sediment distribution and processes occurring off the coast of East Antarctica. These sediments represent several glacial-interglacial cycles. The location of site U1359 on the eastern levee of Jussieu submarine channel in the continental rise makes it an important site for grain size analyses and quartz grains texture. The site is also important from the processes point of view as presently the area is influenced by various agents starting from the turbidity current to various bottom water currents. The grain size analyses, as well as the quartz grain textural analyses, supports the previously inferred stages of ice retreat and ice advancements in the Wilkes Land sector and displays utility of quartz microtexture and grain size study in inferring the past sedimentological processes.

Keywords Quartz microtexture · Marine sediments · East Antarctica · Ice sheet fluctuations

1 Introduction

The glacial regime determines the processes responsible for glacial debris-entrainment, transfer and sedimentation. The resultant marine sedimentary record

P. Biswas
CSIR-National Metallurgical Laboratory, Jamshedpur, India

N. C. Pant (✉)
Department of Geology, University of Delhi, New Delhi, India
e-mail: pantnc@gmail.com

M. Pandey
Department of Geology, Institute of Science, Banaras Hindu University, Varanasi, India

P. K. Shrivastava
Geological Survey of India, Gangtok, India

D. K. Pandey et al. (eds.), *Dynamics of the Earth System: Evolution, Processes and Interactions*, Society of Earth Scientists Series,
https://doi.org/10.1007/978-3-030-40659-2_8

of the Antarctic continental margin is basically a function of Cenozoic to present glacial activity (Escutia et al. 2003; Cooper et al. 2004). These sediments are particularly useful in documenting the mode of sediment transport processes related to environmental interpretations and geological characteristics of large source-rock areas in Antarctica. Since, the ice sheet is drained by a vast network of flow units, especially ice streams (Bamber et al. 2000), the composition of sediments is an average of the different lithologies eroded along the length of the flow units (Roy et al. 2007). Recent particle size analyses from the Wilkes Land offshore interpreted that the low-latitude passive margins' sequence stratigraphic models are inefficacious to polar region stratigraphy (Passchier et al. 2018). The sedimentological parameters especially clay mineralogy, grain size distribution pattern and morphological features of quartz grains are useful in deciphering the sedimentological attributes.

Grain size trends are result of sediment transport processes (Krumbein 1938; Russell 1939; Swift et al. 1972; Stapor and Tanner 1975; McCave 1978; Harris et al. 1990; Le Roux and Rojas 2007 etc.), which is primarily related to the effects of abrasion and selective sorting. The polar regime glacial and interglacial cycles are considered dominant factors controlling the supply, volume and distribution of the sediments (Escutia et al. 2003).

The microtexture study of quartz by scanning electron microscope can provide evidence of erosion, transportation and deposition (Krinsley and Doornkamp 1973). The environmental conditions in which a sand grain forms and all the procedures through which it goes, is recorded by its texture (Krinsley and Donahue 1968; Riester et al. 1982). The different morphological features like fractures, surface alteration and breakage blocks provide useful information about climate and other environmental aspects of the quartz grain's history.

Integrated Oceanic Drilling Program (IODP) Wilkes Land Expedition—318 was launched in 2010 with the major objective of examining the record of Cenozoic Antarctic glacial history and its relation with global climatic and oceanographic changes along the inshore-offshore transect (Fig. 1 and Escutia et al. 2011). Site U1359 in this expedition was selected to provide a sedimentary record which reveals a history of climate and paleo-oceanographic variability of the late Neogene to Quaternary period and to test the stability of the East Antarctic Ice Sheet (EAIS) during middle Miocene to Pleistocene Epochs. Sedimentary record of this site also aimed to provide the timing and nature of deposition of the upper seismic units (i.e., above Unconformity WL-U6) defined on the Wilkes Land margin (De Santis et al. 2003; Donda et al. 2003) (Fig. 2).

In this study, we present grain size analyses and quartz grain surface texture data from Miocene to Plio-Pleistocene sedimentary sequences from the Wilkes Land continental rise. The present work provides inferences on the depositional control and dominant sedimentary processes during Miocene to Plio-Pleistocene and its relation with the evolution of EAIS and tests the applicability of quartz grain microtexture and size as a proxy for EAIS fluctuations.

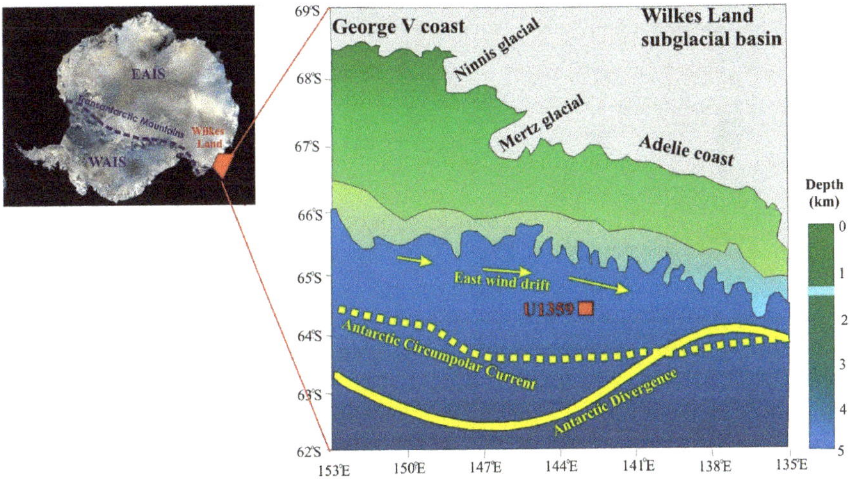

Fig. 1 The image shows the map of Antarctica in inlet (EAIS = East Antarctica Ice Sheet/WAIS—West Antarctic Ice Sheet) and the Location of the IODP Wilkes Land Expedition—318 (red box). Location of the site of U1359 (represented by red dot) is present on the continental rise area

2 Site U1359

Site U1359 of IODP Wilkes Land Expedition—318 was drilled on the eastern levee of the Jussieu submarine channel in the Wilkes Land margin, East Antarctica (Fig. 2). Site U1359 is located in an upper fan environment where the levee relief (measured from the channel thawed to the top of the levee) is ~400 m. The continental rise of East Antarctica comprises complex levee deposits and channels formed dominantly by the turbidity currents (Eittreim and Smith 1987; Hampton et al. 1987a, b; Escutia et al. 1997a, 2000). The interglacial period can be marked by the extensive rain of hemipelagic sediment and ice-rafted debris (IRD) across the continental margins which are further reworked by slumping and/or oceanic currents (Eittreim and Hampton 1987; Hampton et al. 1987a, b; Escutia et al. 1997b, 2000).

The site was selected to provide a sedimentary record which reveals a history of processes and proximity to the source area. It allows to test the inferred variability of the EAIS during the middle Miocene to Pleistocene Epoch in the Wilkes Land Sector (Escutia et al. 2011; Escutia and Brinkhuis 2014; Pant et al. 2013; Verma et al. 2014; Pandey et al. 2017). The sedimentary record was also aimed to provide the timing and nature of deposition of the upper seismic units (i.e., above Unconformity WL-U6) defined on the Wilkes Land margin (De Santis et al. 2003; Donda et al. 2003; Escutia et al. 2011). The proximity of site U1359 to the junction between East and West Antarctica makes it an important site to understand the sedimentological attributes.

Four holes were drilled (U1359A-D) within the site to depths of 193.50, 252.00, 168.70, and 602.2 m composite depth (mcd), respectively. A cumulative core of

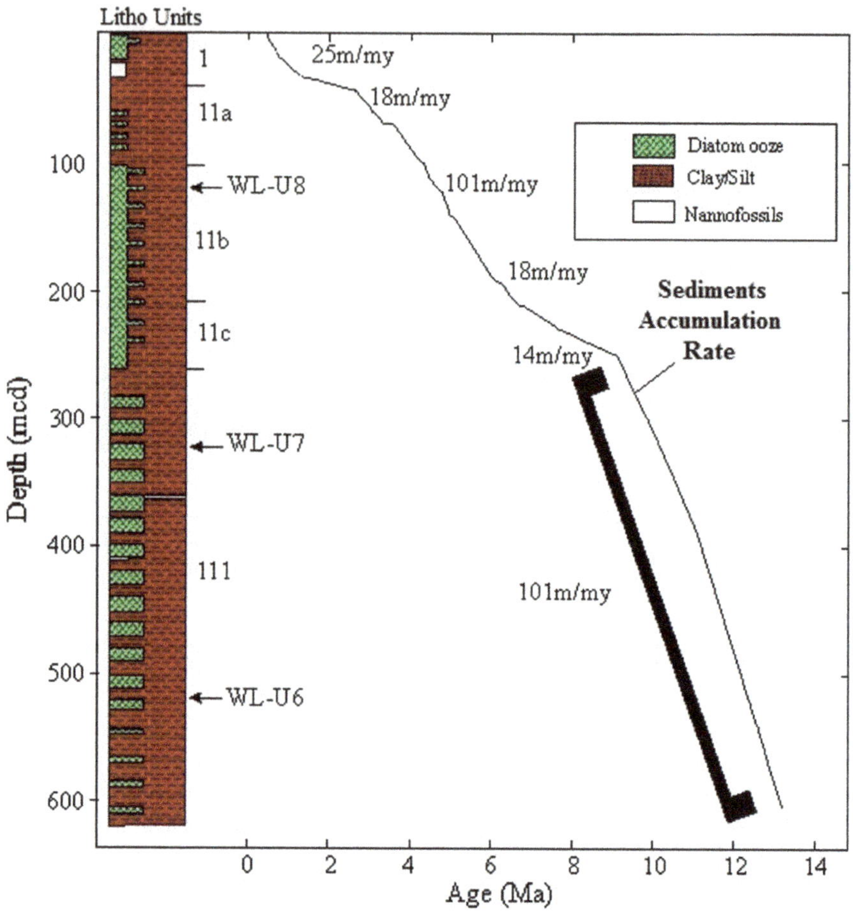

Fig. 2 Composite log of site U1359 as age-depth log which shows the seismic unconformities, different lithostratigraphic units and accumulation rates at different depths (Tauxe et al. 2012)

~600 m depth was raised from the four drilled holes within the site representing glaciomarine sediments for last ~13 million years. The raised core of 600 m sediment has been sampled at 1 m interval for the present study (Fig. 2). The age chronology based on paleomagnetic and microfossil data is adopted from Tauxe et al. (2012). The Pliocene-Pleistocene boundary is placed between 43.13 and 45.21 mcd. Based on nanofossil occurrence a hiatus or no deposition period has been reported at 1.6–2.0 Ma while the core bottom represents the age of 13.28 Ma (Tauxe et al. 2012).

Based on the nature of laminae, fossil assemblages and clast content, three lithostratigraphic units are classified within the mid-Miocene to Pleistocene sediments (Fig. 2) (Escutia et al. 2011). Unit comprising Late Miocene sediments (Unit-III) (Tauxe et al. 2012) extends from 247.1 mcd to the bottom of the cored section at 596.32 mcd and comprises bioturbated diatom-bearing silty clays and intercalations of laminated silty clays (Escutia et al. 2011). Pliocene (43.5–247.1 mcd) set of sediments represented by Unit II contains bioturbated diatom-rich silty clays with intercalations of olive-gray diatom-bearing silty clays. Pleistocene (0–43.5 mcd) lithostratigraphic (Unit-I) unit comprises yellowish brown and olive-gray diatom-rich silty clays with dispersed clasts. Sparse presence of foraminifer-bearing clayey silt and sandy silt was also reported (Escutia et al. 2011).

3 Materials and Methods

The sediment samples were treated with hydrogen peroxide, dilute HCl (10%) and SnCl (5%) separately to remove organic matter, carbonate and iron coating respectively as per the method described by Jackson (1979). Out of these chemically treated sediment samples, 2–3 g of sediments were used in Laser Particle Size Analyser, Malvern-2000 Master Size and few quartz grains ranging in size from 1 to 4Φ, were handpicked under the microscope at random to be subsequently imaged at high resolution by SEM. For the SEM studies, the quartz grains were first mounted on specially designed aluminium stubs and then coated with a 15 Å thick gold-palladium film, followed by detailed study at the Department of Geology, University of Delhi, on a Zeiss (SEM LEO 440) fitted with a Cambridge Energy Dispersive Spectrometer (EDS). The surface feature recognition and interpretation follow the concepts of previous studies (Krinsley and Doornkamp 1973; Whalley and Krinsley 1974; Marshall and Gilligan 1987; Helland and Holmes 1997; Mahaney and Kalm 2000; Damiani et al. 2006).

4 Results

4.1 Grain Size Analysis

The triangular plot based on sand, silt and clay population shows that most of the samples lie in the domain of silty sand, sandy silt and silt (Fig. 3). Few samples were gravelly sandy mud and slightly gravelly mud. Most of the silt population shows 10–20% of clay size content while sandy silt comprises less than 5% of clay size grain where (<2 μm). However, few samples classified as sandy silt also contain ~10–15% clay particles.

The downcore variations of grain size show that silt and sand has inverse relationship and silt dominates the population throughout the depth (Fig. 4a). The other plots (Fig. 4b–d) show variability of subclasses of sand and silt and define the contribution of individual groups. Medium to fine sand and coarse silt are comparable whereas

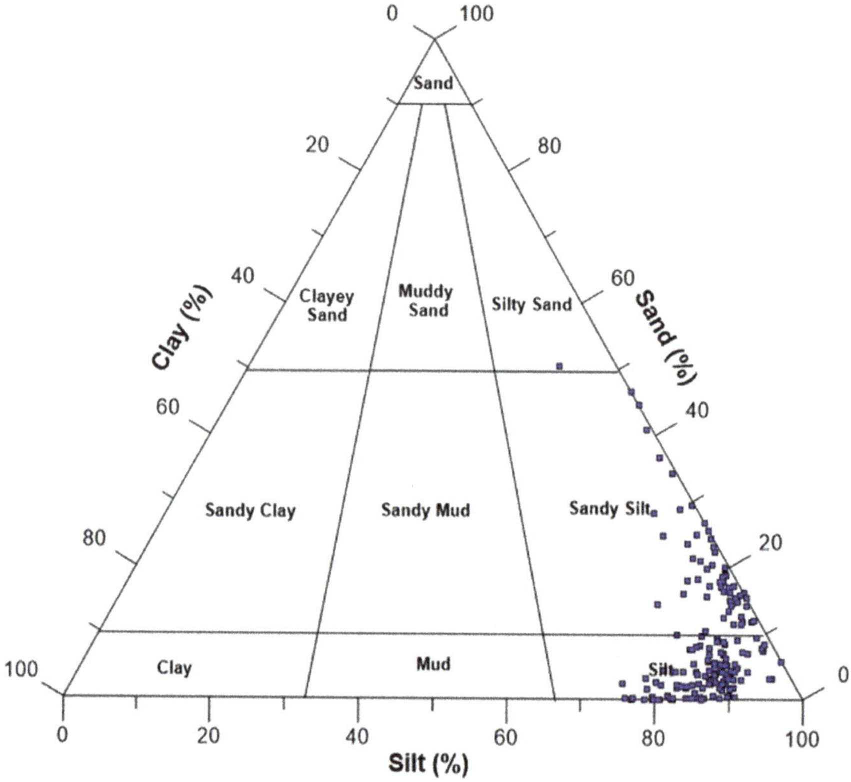

Fig. 3 Triangular plot classification of sediments of site U1359 defined in sand, silt and clay percentages (clay represented by grain size <2 μm)

medium to fine silt has an inverse relation. Coarse silt contributes the maximum to the silt population.

Silt dominates throughout the depth with an average of 82% and maximum as 94% and 41.8% as minimum values at 182.74 mcd and 590 mcd respectively (Fig. 4a). The subclasses of silt also vary significantly. The coarse silt varies from 5 (146.26 mcd) to 40% (56.13 mcd) (Fig. 4c) while fine silt varies from 7.4 (107.25 mcd) to 48% (146.44 mcd) (Fig. 4d). Sand proportion at places is present as trace and it varies up to ~50% (590.54 mcd) with an average of 9.2%.

Very fine sand contributes the maximum to sand portion with a maximum as 33.8% followed by medium sand with maximum as 20.8% and coarse sand with 16% as maximum. Clay particle content goes up to 23.9% (Fig. 4a) with an average of 8.5%.

Sample from high Ice Rafted Debris (IRD) corresponds to high sand content comparative to silt portion. Coarse silt behaves similarly to that of fine and medium-grained sand. Medium to fine silt has a negative correlation with coarse silt along with fine sand. At places, medium silt also does not follow the pattern of fine silt.

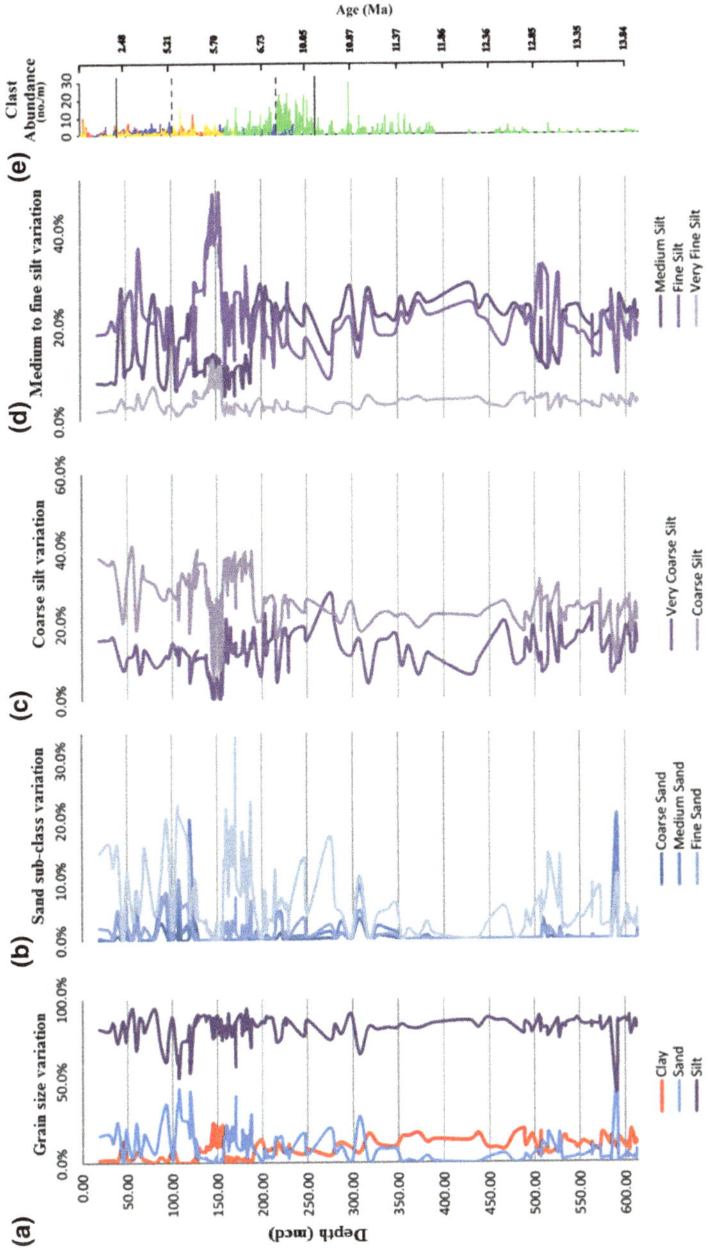

Fig. 4 Graph represents the grain size variation as a function of age of the sediments of site U1359 throughout the depth from top to bottom. **a** Sand-silt-clay proportion. **b** Coarse-medium and fine sand proportion. **c** and **d** are further division of coarse silt sediments and medium to fine silt sediments, respectively. **e** Clast abundance representing Ice Rafted Debris (from Escutia et al. 2011)

Also, higher sand content does not necessarily correspond to higher IRD. Since, discrepancies between the groups and subgroups were observed and hence subgroups of particle sizes should be considered separately.

4.1.1 Grain Size Parameters

Mean grain size is a function of the size range of available materials and amount of energy imparted to the sediment which depends on current velocity or turbulence of the transporting medium. Mean grain size of sediments of site U1359 ranges from 4ϕ to 8ϕ (Fig. 5). The maximum population of mean values lies within 6ϕ to 7ϕ range. Sediments usually become finer with a decrease in energy of the transporting medium. Sediments become finer in deeper water as in deep water the action of waves on the sea bottom is slight. Sediments of Miocene and Plio-Pleistocene are represented as two separate groups in Fig. 5. Sorting is dependent on the size range of material supplied, type of deposition, rate of supply of detritus and current characteristics. Sediments here are poorly to very poorly sorted (Fig. 5a). Currents of relatively constant strength whether low or high, will give better sorting than currents which fluctuate rapidly from almost slack to violent. For best sorting, then, currents must be of intermediate strength and also be of constant strength.

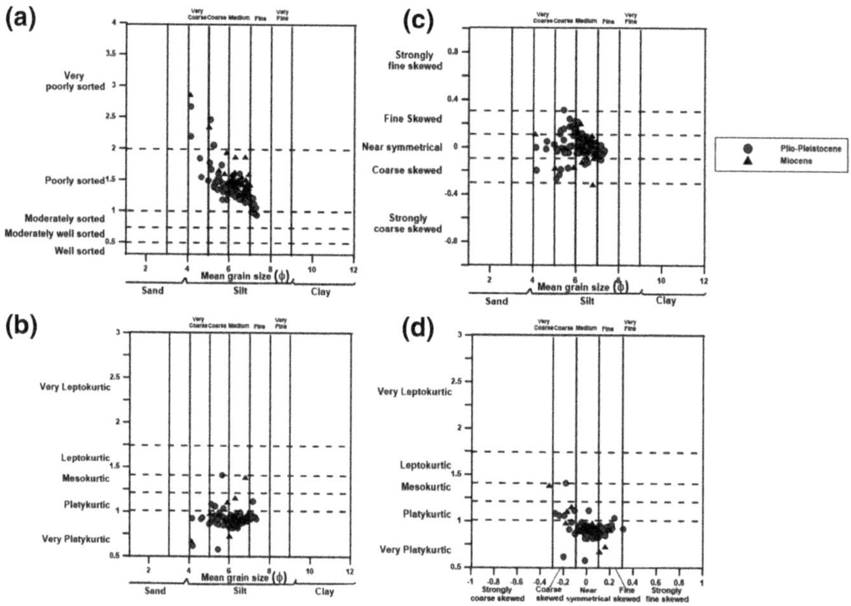

Fig. 5 Scatter diagrams explaining grain size parameters of sediments of site U1359. Black triangle represents Miocene sediment samples while grey circle are represented as Plio Pleistocene sediments. **a** Sorting versus mean grain size. **b** Skewness versus mean grain size. **c** Kurtosis versus mean grain size. **d** Kurtosis versus skewness

A negative correlation is present between the mean grain size and sorting. Very coarse to coarse silt is very poorly sorted while gradually from coarse to medium-fine silt becomes less poorly sorted. Sediments of Plio-Pleistocene age (represented by grey circle in Fig. 5) show relatively more diverse range of variability indicating higher fluctuations in terrigenous sediment supply. Skewness with respect to the mean grain size shows strong variability (Fig. 5b). The sediment population varies from coarse skewed to fine skewed. Most of the sediment samples show near symmetrical patterns indicating equal dominance of coarse as well as fine grain size sediments. Plio-Pleistocene sediments show similar pattern as of Miocene sediments suggesting pulses of mixed grain size supply throughout the core U1359. Kurtosis versus mean grain size plot (Fig. 5c) shows transition of sediment population from mesokurtic to very platykurtic. Dominantly the sediments are very platykurtic. Only a few samples fall in the domain of mesokurtic and platykurtic.

Skewness and kurtosis explain the grain-size distribution variability from the normal Gaussian probability curve, and the more extreme the values the more non-normal the size curve. It has been found that single-source sediments tend to have fairly normal curves, while sediments from multiple sources show pronounced skewness and kurtosis. Skewness as a function of kurtosis demonstrates a strong pattern in variability (Fig. 5d). Coarse skewed and fine skewed population of sediment shows maximum variability in kurtosis from very platykurtic to mesokurtic. Bimodal sediments exhibit extreme skewness and kurtosis values; extremely platykurtic (Folk 1980). All the statistical parameters point towards a although the pure end-members of such mixtures have nearly normal curves, sediments consisting dominantly of one end member with only a small amount of the other end member are extremely leptokurtic and skewed, the sign of the skewness depends on which end member dominates; sediments consisting of sub-equal amounts of the two end-members are polymodal variation which is suggestive of contribution from various processes occurring in the area. Variation of these parameters as a function of meter composite depth and respective ages are explained through a graphical presentation in Fig. 6. Sediments are very poorly sorted at 590.54, 307.4, 119.75, 107.25, 86.65 and 59.94 mcd and moderately sorted at 144.9 and ~146 mcd. It indicates that the sediment shows strong variability from moderately sorted to very poorly sort during Plio-Pleistocene while they are comparatively less poorly sorted during upper late Miocene. Cyclicity in the variability of skewness is observed throughout the depth from very finely skewed to coarsely skewed indicating either fluctuation of sediment supply or changes in the dominance of various sedimentary processes affecting the population of the fine sediment and coarse sediments. At places, symmetrical skewness is also observed indicating presence of finer as well as coarser sediments. Most of the symmetrically skewed samples are platykurtic. The IRD supply between 200 and 300 mcd corresponds to transition from Pliocene to late Miocene and these are also marked by very poorly sorted sediments. A sudden change is observable after Pliocene to Miocene transition which persisted up to WL-U6.

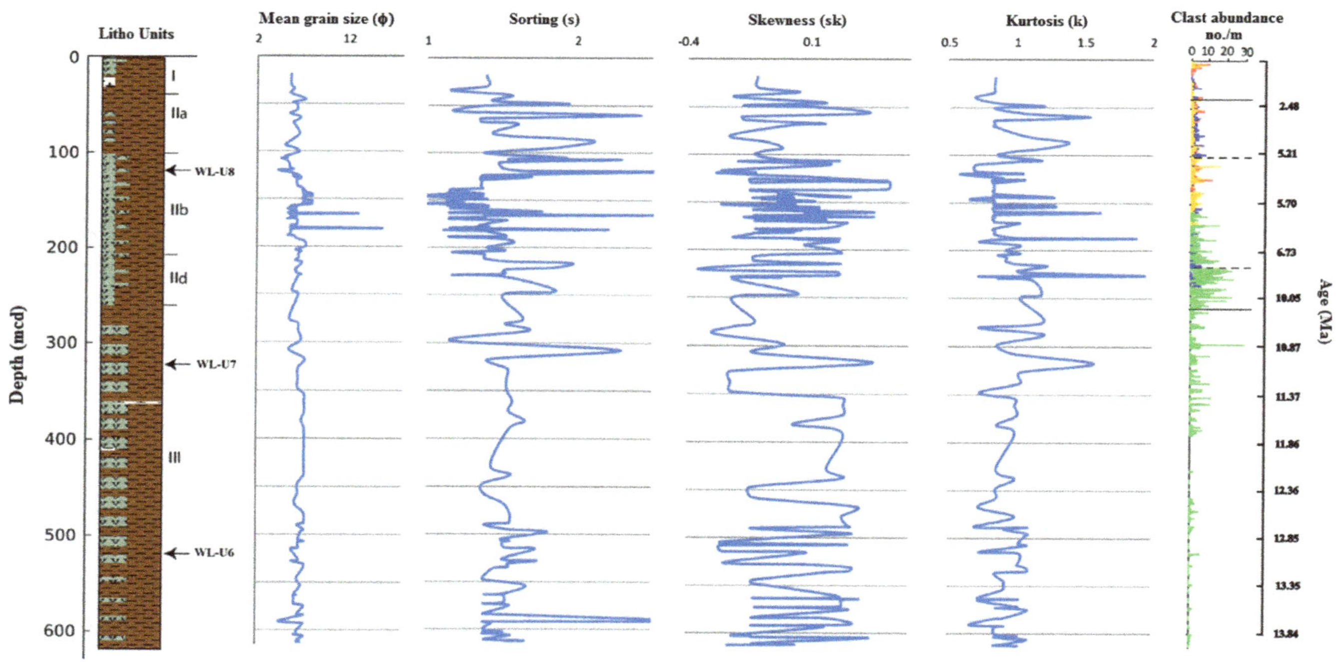

Fig. 6 Compilation of variability of various statistical parameters with respect to the meter composite depth (mcd) of the sediments of site U1359. Corresponding ages of the sediments is shown on the right hand side of Y-axis

4.2 *Quartz Grain Microtextures*

In the present study, 25 types of microstructures are identified. The modes of origin of these different types of microstructures are either mechanical or chemical or from both processes. Mechanical features identified in this study are conchoidal fractures of different sizes (5–100 μm), straight and arcuate steps, breakage blocks, fractured plate, abrasive edge, V-shaped pattern, pits of different sizes (10–150 μm). V-shaped pits also vary in size and shape. Identified chemical features are solutions pits, solutions crevasses, weathered surfaces, adhering particles and new growth. Solution pits and crevasses are different in shape and size. Selection of samples in the following sections is based on previously inferred glacial-interglacial periods during Miocene (Verma et al. 2014; Pandey et al. 2017). For clarity, ages of the sediments (Tauxe et al. 2012) are also provided along with the sediment depths.

4.2.1 Miocene Quartz Grains

Quartz grains from 508 to 516 mcd (12.25–12.33 Ma) are strongly angular. Almost 80% of the population shows carved surfaces, sharp edges, U shaped gullies, pitted surfaces indicating strong glacial activity (Fig. 7a and b).

Quartz grain samples from 379 to 384 mcd (~10.9–11 Ma) represent two distinct characteristics of glacial as well as fluvial input (Fig. 7c–f). Almost 40% of grain shows angular to sub-angular outline (Fig. 7e). These grains bear mechanical features like conchoidal fractures, parallel to sub-parallel steps and breakage blocks. The other 50% represents the extensively eroded quartz grains which are quite well rounded and are spherical in shape. These grains also have experienced mechanical breakage as well (Fig. 7c). Few grains show sub-rounded texture and mechanical breakages like smooth conchoidal surface, pits, striations and arcuate steps indicating glaciofluvial influence along with chemical weathering (Fig. 7d). Presence of discrete types of quartz grain population indicates reworking as well as mixing of the sediments especially at this depth. Interestingly the fine-grained quartz population comprises grains which are very well rounded inferring the aeolian input to the sediments.

Quartz grains between depths of 343 and 347 mcd (~10.46–10.52 Ma) show mostly sub-rounded to rounded nature with medium to low relief. Almost 80–90% grains bear V-shaped breakage and smoother edges indicating the fluvial activity however, irregular impact pits are commonly present representing mechanical breakage (Fig. 7g and h). Sphericity of grains varies from elongated to highly spherical.

Quartz grains from 313 to 317 mcd (~10–10.10 Ma) show mostly two different type of population; one is sub-rounded with pitted surface while other is sub angular indicating fluvio-glacial influence. Almost 80–90% of grains bear mechanical V-shaped, irregular impact pits are commonly present (Fig. 7i and j).

The 70–80% quartz grains s from depth between 224.24 and 226.84 mcd (~7.43–7.55 Ma) show sharp angular to sub-angular outline with high relief, arcuate and/or

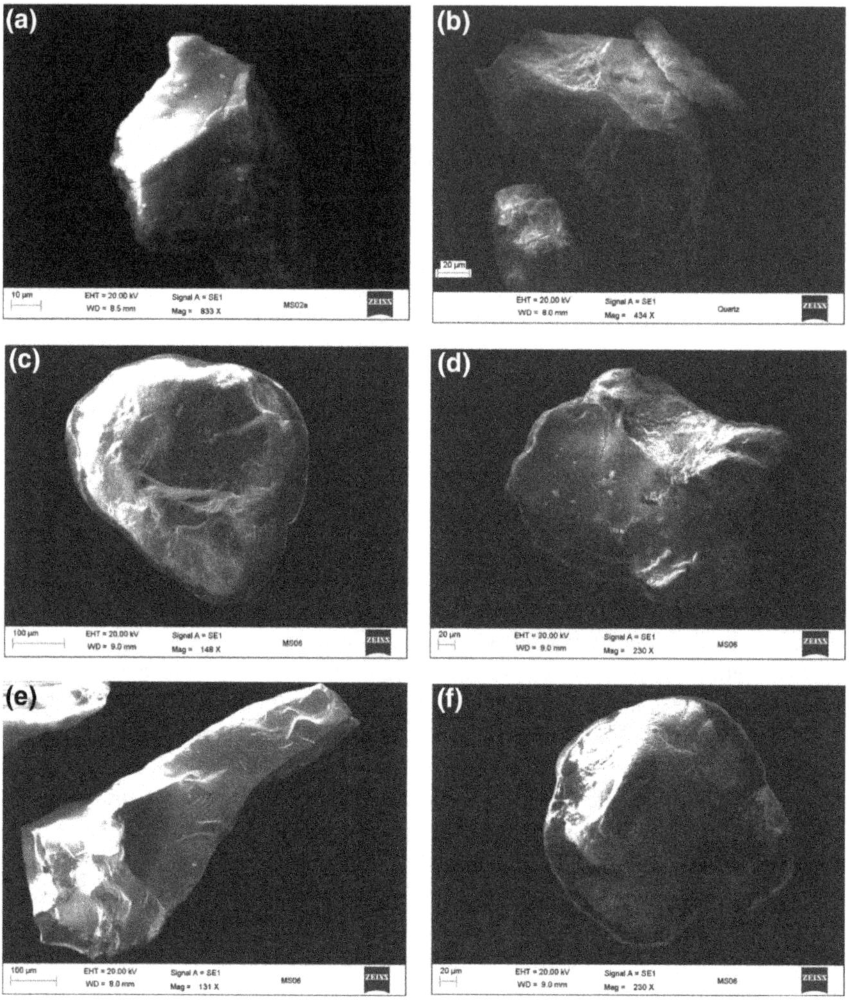

Fig. 7 Selected SEM micrographs showing typical morphological and mechanical textures of quartz grains representing late Miocene to mid Miocene sediments from U1359. **a** and **b** are highly angular quartz grains with sharp edges and U shaped carved surfaces, **c** well rounded appearance with block breakage, carved surfaces, **d** irregular quartz grain with rounded edges and carved surface, **e** strongly angular grain with irregular depressions, conchoidal fractures and arc steps/straight steps, **f** a ~10 μm grain with highly rounded texture, **g** smoother and rounded edges in spherical as well as elongated grains, **h** subrounded quartz grains with conchoidal fractures, **i** conchoidal fractures, subrounded elongated grain, striations present at the surface arcuate steps present at one edge, **j** smooth surface, and rounded edges, **k** presence of U shaped carving with conchoidal striations on the surface, sharp edges and arcuate steps, **l** U shaped gully present with sharp edges, fine layer of flakes above the surface, block breakage and pits are also present

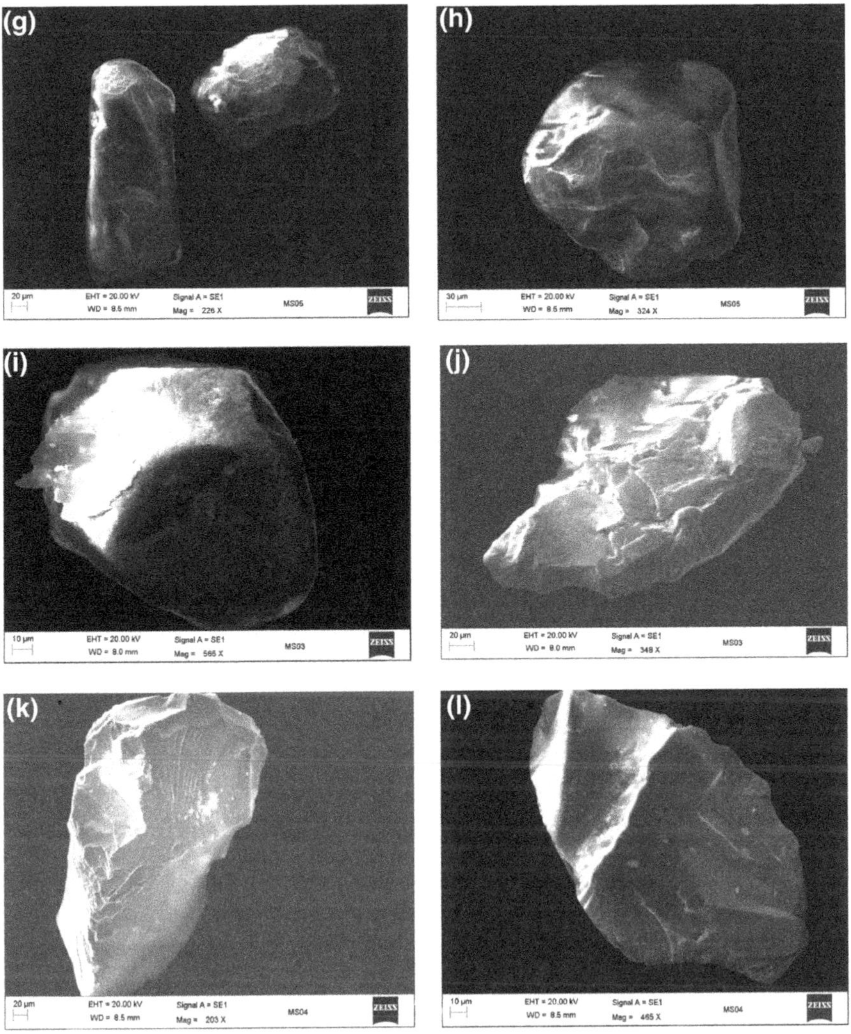

Fig. 7 (continued)

straight steps, conchoidal fractures, and abrasive edge (Fig. 7k and l). Only 10–20% grains show sub-rounded outline with a partially smoothed edge with features like arcuate and/or straight steps, conchoidal fractures. The grains are characterised by typical mechanical U-shaped feature and abrasive surfaces (Fig. 7l).

However, the dominance of the fluvioglacial and glaciofluvial input was observed in Miocene quartz grains when compared to the Pleistocene grains. Since the accumulation rate was quite high (~100 m/year; Fig. 2) during upper to mid-Miocene, representative samples were considered for the quartz textural features.

4.2.2 Plio-Pleistocene Quartz Grains

Quartz grains from depth intervals of 107–111.99 mcd (~4.4–4.5 Ma), 143.68–150.48 mcd (4.9–5.28 Ma) contain some highly rounded, often spherical quartz grains (Fig. 8a). The surfaces show abundant upturned plates and some dish-shaped concavities, but mechanical breakage features are missing. At the depth of 84.39 (~3.8 Ma) and 95.64 mcd (~4.1 Ma) and 107.49–153.2 mcd (4.4–5.3 Ma) mix population along with mechanical features as well as chemical alteration features is observed. These depths mainly describe the fluvioglacial input.

The quartz grain samples from the depth between 78.2 to 82.72 mcd (3.6–3.7 Ma) and 96.14 to 106.99 mcd (4.2–4.4 Ma) represents typical glacial characteristics (Fig. 8b and c). In these, 60–80% grains shows angular to sub-angular outline. Almost all grains possess mechanical features like conchoidal fractures, parallel to sub-parallel steps and breakage blocks. Chemical alteration features are present in very few grains (5–10%).

Quartz grains from 72.56 mcd (~3.5 Ma) shows mostly sub-rounded to rounded features with medium to low relief. Almost 80–90% of grains have chemical alteration signatures. Mechanical V-shaped, irregular impact pits are commonly present (Fig. 8e–g, i and k). These depths appear to be influenced by an increase in fluvial activity.

Quartz grains from ~48 to 57 mcd (2.7–3.0 Ma) are mostly angular to sub-angular with a sharp edge. Micro features like conchoidal fractures, arc and straight steps, parallel to semi-parallel striations are mostly present. The grains mainly bear imprints of mechanical processes (Fig. 8b and c). Very few grains show chemical alteration, rounded outline and V-shaped mechanical cracks. In quartz grains from ~43 to 47 mcd (~2.6–2.7 Ma), 50–60% grains show sub-angular to sub-rounded outline with medium relief. The surfaces show precipitation/solution features with a thick silica coating and mechanical V-shaped imprints (Fig. 8e and f). In few grains, mechanical features are partially obliterated by chemical precipitation/solution activity (Fig. 8g and i). Quartz grains are influenced mainly by the glaciofluvial activity with minor fluvioglacial influence.

The 50–60% quartz grain samples from the depth between 15 and 41.32 mcd (~0.6–1.6 Ma) show angular to sub-angular outline with high to medium relief, arcuate and/or straight steps, conchoidal fractures, and abrasive edge (Fig. 8b and c). 30–40% grains show sub-rounded outline with smoothed edge and medium to low relief with features like arcuate and/or straight steps, conchoidal fractures, and abrasive edge (Fig. 8b). 10–15% grains are characterised by typical mechanical V-shaped feature and abrasive surfaces (Fig. 8e and f). Glacial influence is indicated for these youngest sediments.

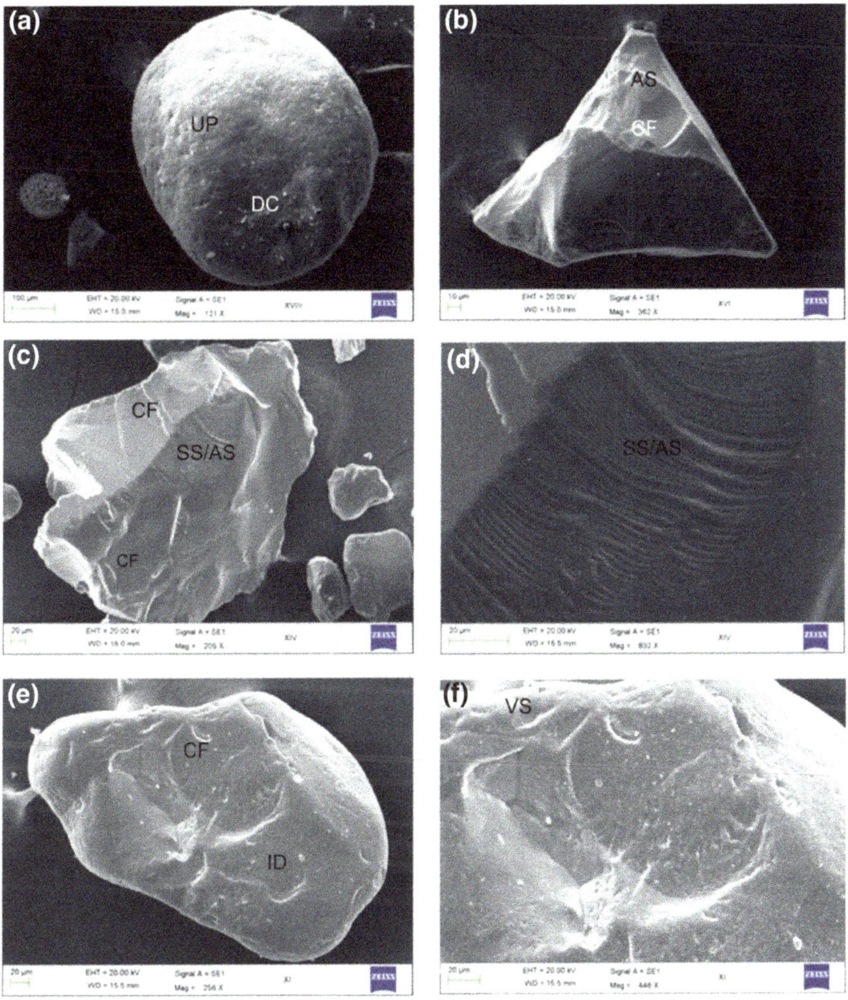

Fig. 8 Selected SEM micrographs of quartz grains showing typical morphological and mechanical textures of quartz grains representing Plio-Pleistocene sediments from U1359. **a** Rounded grains with dish-shaped concavities and upturned plates, **b** sub-angular grain with CF, **c** angular grains with conchoidal fracture (CF), arc steps/straight steps (AS/SS), **d** close up view of AS/SS, **e** and **f** rounded grains with V-shaped mechanical features and irregular depression (ID), **g** angular grains with irregular depression and new silica growth, **h** fracture planes and abrasive edge, **i** silica precipitation or silica re crystallization, **j** close up view of the same grain, **k** sub-angular grains showing straight and arc step features, silica globules and various stages of crystal overgrowth, **l** close up view of silica globules

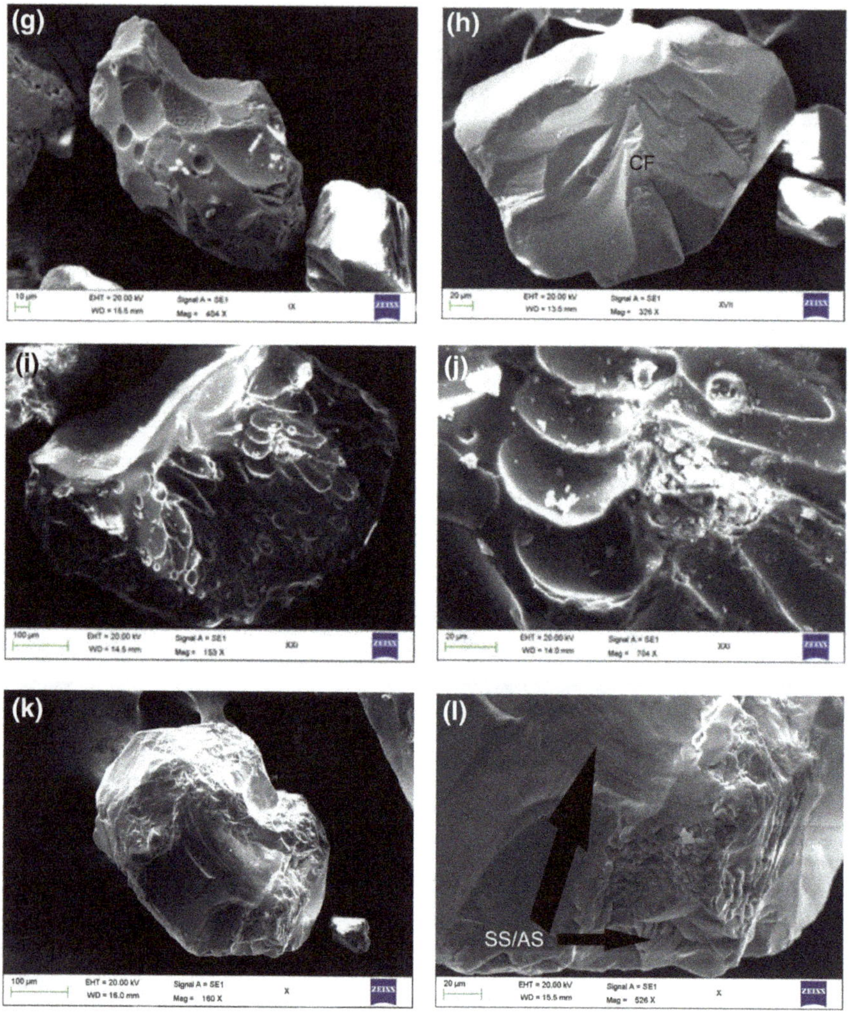

Fig. 8 (continued)

5 Discussion

5.1 Sedimentary Processes

Long term variations in the polar regime of the continent and the glacial-interglacial cycle are the two significant factors controlling the type, volume, and distribution of sediment in the Wilkes Land margin during Miocene to Pleistocene period. Using independent proxies of clay minerals as well heavy minerals several retreat and advancement of East Antarctic Ice Sheet were earlier inferred from U1359 sediments

(Fig. 9; Verma et al. 2014; Pandey et al. 2017). Continental rise sediment consists of mud and silty mud with alternating massive intervals with more abundant IRD, and laminated silty to sandy mud. Outlet glaciers such as the Ninnis and the Mertz have an important role in ice drainage and sediment delivery to the ocean (Drewry and Cooper 1981; Lindstrom and Tyler 1984; MacDonald et al. 1989; Anderson 1999), and the glaciers may have changed position during different glacial-interglacial periods (Eittreim et al. 1995; Escutia et al. 2000). Bottom contour-current deposition may also be an active process on the continental slope and rise area (Hayes and Frakes 1975; Hampton et al. 1987b; Escutia et al. 2000, 2002). Presence of laminated silts is indicative of Contourite deposition (Fig. 10). The bottom contour-current deposition has been a long-lasting process in this margin, as evidenced by the sediment mounds that developed in the continental rise interchannel areas (Escutia et al. 1997b, 2000, 2002). Deposition of clasts can take place due to higher melting of ice cover, occurrences of high energy condition for sediment transport and/or proximity of source. Therefore, a higher concentration of clasts seems linked to interglacials rather than to the periods of glacial maxima.

5.1.1 Sedimentary Processes During the Miocene

Principally, the turbidity currents are considered as the main sedimentary process which was active during the Miocene (Escutia et al. 2003). Laminated intervals in drillcore probably represent sediments deposited from turbidity currents which are during Miocene (Fig. 10) (Payne and Conolly 1972; Hayes and Frakes 1975; Hampton et al. 1987b; Escutia et al. 2002, 2003). Besides, the bottom contour-current deposition is also considered an important depositional process at the continental rise (Fig. 10) (Hayes and Frakes 1975; Hampton et al. 1987b; Escutia et al. 2000). Presence of laminated silts is indicative of Contourite deposition (Fig. 10).

5.1.2 Sedimentary Processes During the Plio-Pleistocene

Plio-Pleistocene sediments of site U1359 are also characterized by low-density turbidity currents. The Pleistocene interglacial sedimentation was characterized by massive and more IRD abundant hemipelagic sediments between the glacial stratified to laminated intervals (Verma et al. 2014; Escutia et al. 2003, 2011). During interglacials, the supply of IRDs along with sand-size grains to the site appears to have been greater relative to the fine-grained terrigenous contribution (Figs. 4 and 6). Therefore, the deposition of IRDs seems to be a result of interglacials rather than the periods of glacial maxima. Erosion and redeposition of fine-grained sediment by bottom contour currents is another important and likely process during interglacials of Pleistocene and upper late Miocene.

Influxes of coarser to medium sand are possibly the result of turbidity currents triggered by movements of the ice sheet and hence can be considered the main process for sediment supply from the shelf and slope to the continental rise. But the presence

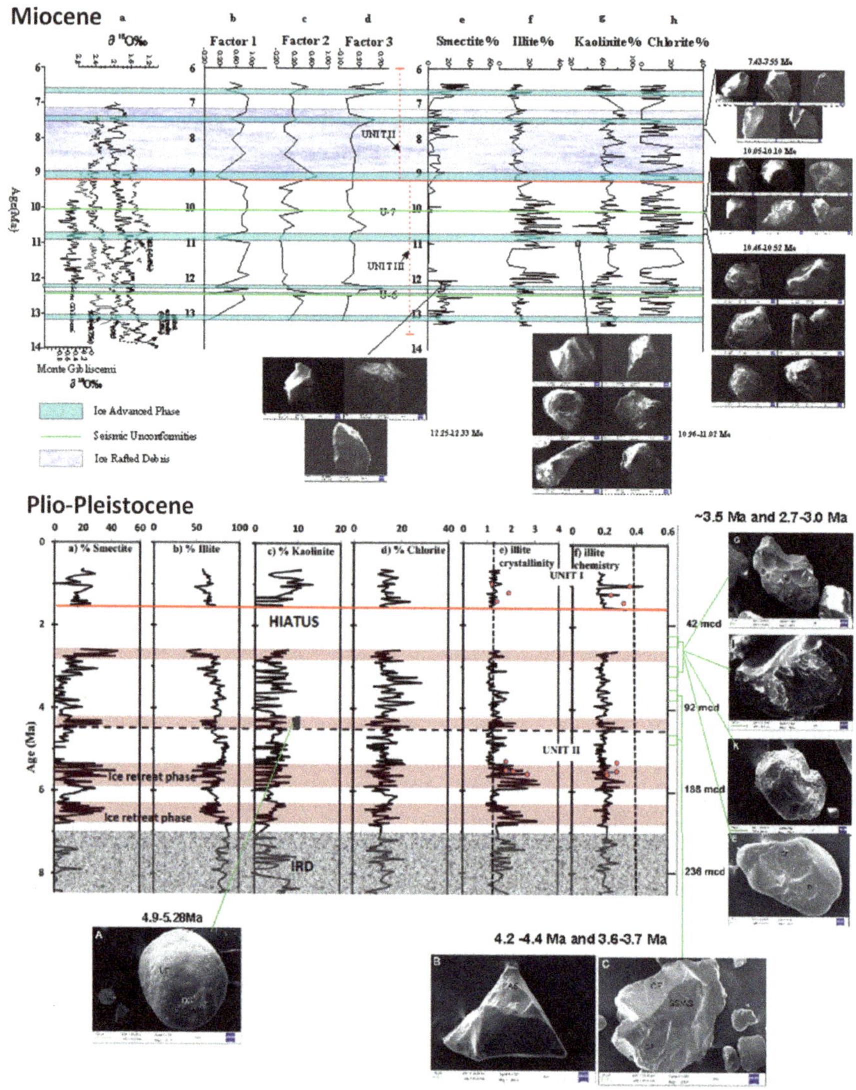

Fig. 9 Compilation of clay minerals, heavy minerals and quartz textural behaviour of sediments with respect to the age/depth wise core log of site U1359. The first graph shows downcore variability of the heavy mineral population and clay minerals during Pliocene to Miocene (modified from Pandey et al. 2017). The blue rectangles represent the previously inferred ice advance stages during Pliocene to Miocene and its correspondence with the global high-resolution oxygen isotope record of benthic foraminifers. Miocene quartz texture data in conjunction with earlier inferences shows coherent behaviour with ice advance and ice retreat stages previously reported through other proxies. The second graph shows downcore variation of clay minerals and corresponding quartz textures of Plio-Pleistocene sediments of site U1359 (modified from Verma et al. 2014). The pink and white shaded rectangles indicate the ice retreat and waxing ice sheet condition, respectively (Verma et al. 2014). Textured region shows the abundance of clasts (IRDs) in both the graphs representing the common depth and comparatively warmer conditions. The incorporated quartz textures of Plio-Pleistocene are also in accordance with the previously reported ice retreat stages during Plio-Pleistocene

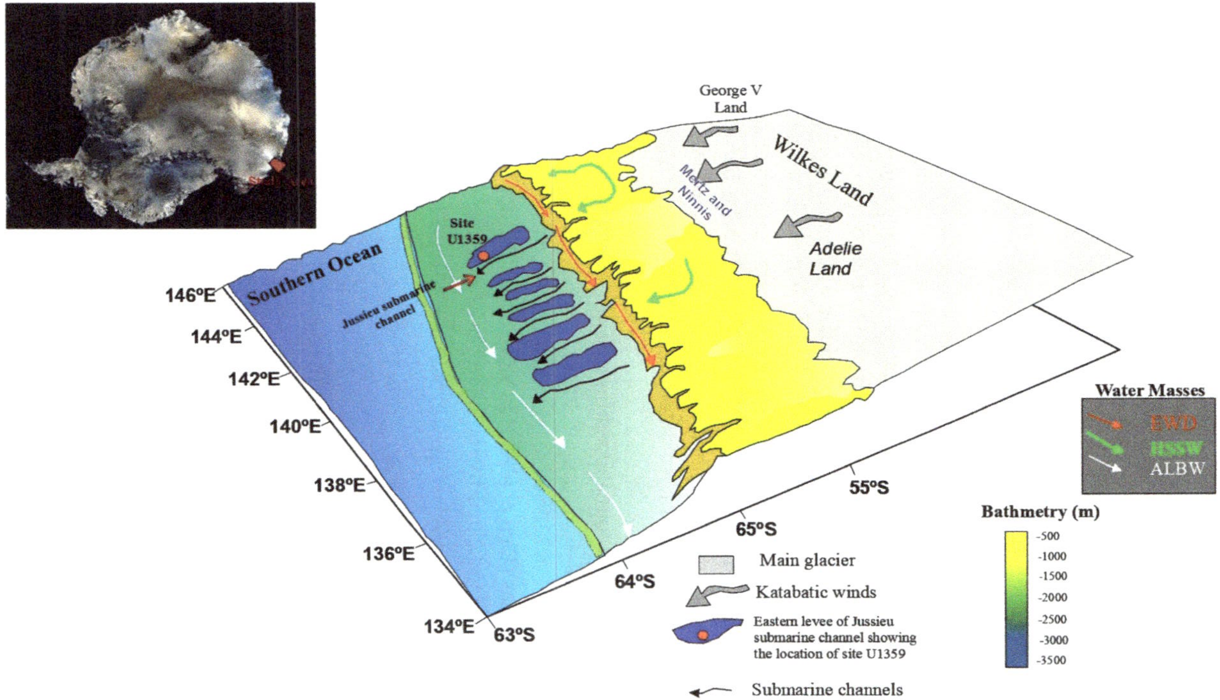

Fig. 10 Map of the study area, showing the location of sediment site U1359, bathymetry, main glaciers, detail of oceanographic currents and different water masses and katabatic wind directions. EWD, East Wind Drift; HSSW, high-salinity shelf water; ALBW, Adélie Land bottom water; ACC: Antarctic Circumpolar Current. Bathymetry (after Presti et al. 2011)

of fine-grained clay throughout the depth indicates the erosion and the redeposition through bottom currents in the Wilkes Land continental rise (Donda et al. 2003; Escutia et al. 1997a, b, 2003, 2005). This results in the polymodal behaviour of the sediments. The Antarctic Bottom Water (ABW, Fig. 10) produced in the Ross Sea flows westward along the Wilkes Land continental slope and rise (Gordon and Tchernia 1972), where it mixes with the Antarctic Bottom Water produced in the Adelie Depression (Gordon and Tchernia 1972; Bindoff et al. 2000).

5.2 *Depositional Environment*

The increased terrigenous supply may have resulted from ice advance across the shelf or increase in sedimentation from bottom currents or by a change in energy conditions of bottom currents or by the combination of all these three processes. The above interpreted processes are diagrammatically explained by Presti et al. (2011) (Fig. 10). This figure explains the present glaciers, channels, water masses and currents as major sedimentary and oceanographic processes active on the continental shelf, slope and rise. Present course of the current may or may not have been active during Pleistocene, Pliocene and Miocene but the processes explained would have been mostly similar. However, paths of sediment supply might have changed from mid-Miocene to Pleistocene.

Icesheet fluctuations with retreat conditions at 4.5, 5.5–5.8 and 6.2–6.8 Ma and ice sheet advances at 6.62–6.72 (minor stage), 7.43–7.60, 8.04–8.42, 9.06–9.13, 10.83–10.96 and 12.06–12.43 Ma were previously inferred using heavy mineral and clay mineral assemblages (Pant et al. 2013; Verma et al. 2014; Pandey et al. 2017). Depositional conditions indicated by the quartz microtextures in the present study are discussed in relation to these fluctuations of the ice sheet during Miocene and Plio-Pleistocene.

5.2.1 Depositional Environment During the Miocene

For Miocene, quartz grains were selected from three ice advance stages (7.43–7.60, 10.96–11.02 and 12.25–12.33 Ma) and two ice retreat stages (10.05–10.1 and 10.46–10.52 Ma). Quartz grains from the 224 to 227 mcd representing the first stage of glacial advancement preserve angular grains with mechanical breakage indicating the supply through glacial or glaciofluvial (occasionally) activity. Quartz grains from two depth ranges, from 313.04 to 317.34 and 343.34 to 347.55 mcd, shows the textural features indicating sub-aqueous sediment supply and fluvial to glaciofluvial input which represents the retreat stage during late Miocene.

Quartz grains from 379 to 384 mcd (~10.96–11.02 Ma) depicted mix population of grains transported by the glacial activity and grains flushed by fluvial activity. As the depth marks the end of the glacial advance stage (10.83–10.96 Ma); the presence of two end members of grain shape can be explained by the reworking and mixing of the sediments. Quartz grains from 508.44 to 516.54 mcd are the result of glacial and glaciofluvial input and also correlate directly with the inferred advance stage during 12.25–12.33 Ma.

5.2.2 Depositional Environment During Plio-Pleistocene

Quartz grains from the column representing the Pleistocene (0.6–3 Ma) time starting from 15 to 57 mcd exhibits numerous mechanical breakage features of various sizes that include conchoidal fractures, arc and straight steps, parallel to semi-parallel striations, angularity; fresh surface and sharp edge indicating typical glacial origin. Quartz grain sample of U1359 show dominance of glacial origin surface textures due to basal transport of the continental glacier similar to those described by Mahaney and Kalm (1995) and Mahaney et al. (1996) indicating an extension of icesheets to the shelf areas in the Wilkes Land sector.

Presence of rounded-subrounded quartz grains, thick coat of silica and V-shaped features marks the glacial subaqueous origin, exposure to high energy marine transport during the time close to 4.9–5.3 Ma (Fig. 8g and k) and the initiation of warming event leading to glacial retreat. This is supported by the grain size data also where the supply of fine grained sediments overcame the coarser population and was accompanied by increased accumulation rates and IRDs. Hence, studied quartz grains indicate dominance of fluvial and fluvioglacial input and support the previously inferred retreats during 5.5–5.8 Ma (Verma et al. 2014) during the Pliocene warming stage. The usefulness of the quartz microtexture data in paleoclimatic interpretations is well illustrated in Fig. 9 where multiproxy data in two plots (representing Miocene and Plio-Pleistocene sedimentation) is compared with that of the representative quartz microtextures. The inferred paleoclimate signals in terms of waning and waxing of EAIS show significant correlation between the data sets.

A few grains of aeolian quartz grain were characterized at depths of 107–111.99, 143.68–150.48 and 343–347 mcd (Fig. 8e and f). Aeolian quartz grains have been reported to occur in polar marine sediment (Rex et al. 1970; Hill and Nadeau 1984; Hodel et al. 1988; Damiani et al. 2006) as a result of katabatic winds which dominantly flow towards the ocean from the land (Fig. 10). Those grains may have formed under arid conditions and probably were derived from dunes deposited in deglaciated areas. These grains did not suffer glacial abrasion and thus rule out a reworking of old sediment strata by glacial grinding. Aerial transport on to the ice instead is, thus, inferred instead of overridden by ice. Successively aeolian grains can be transported by icebergs or by wind, sedimented as the ice melts and dropped through the water column with no further abrasion by marine currents.

6 Conclusions

The pulses of sand size corroborates with rounded to subrounded grains indicating high influx of terrigenous input possibly due to extensive melting during ice retreat stage point towards warmer conditions. The present inference however, contradicts the suggested behaviour of polar regime from global scenario by Passchier et al. (2018). The rounded to sub-rounded quartz grains would have also been derived from a distal sediment provenance but angular and subangular grains are direct result of proximal glacial activity supporting the multiple sourcing of sediments from at least two distinct terrain (Pant et al. 2013).

Compilation of processes explains that grain size data cannot be directly used to interpret the paleo-environment especially at continental rise as the grain size population is affected by channel flows from the hinterland, bottom current flow contributing sediments from Ross sea and Adelie depression, katabatic wind supply and bioturbations (Fig. 10). The interplay of processes and change in dominance of particular sedimentary process are the two main factors which explain the variability of depositional parameters during Miocene. Less sorted sediments during upper late Miocene indicate that either the fluvial environment was playing a key role, possibly due to increased melting or there was intensification of Antarctic bottom water supply. The relative contribution of the two cannot be assessed in the present study.

A number of grains show surfaces extensively modified by high chemical activity indicating a high-chemical diagenetic environment possibly as a result of interglacial climate influences and reworking of grains during Plio-Miocene and Miocene.

Acknowledgements This work is a result of analyses on sediments of the drill core U1359 provided by the IODP Expedition 318. The National Centre for Polar and Ocean Research funded this project. Dr. N. C. Mehra, Department of Geology, University of Delhi is acknowledged for his support in SEM-EDS analysis. SEM-EDS support from MoES funded project (**MoES/PAMC/H&C/51/2013-PC-II**) is also thankfully acknowledged. We are grateful to Professor Partha Pratim Chakraborty, Department of Geology, University of Delhi and Rashmi Gupta (CSIR-JRF), Department of Geology, Banaras Hindu University, Varanasi for constructive discussions and support.

References

Anderson JB (1999) Antarctic marine geology. Cambridge University Press, Cambridge, UK, 289 pp.

Bamber JL, Vaughan DG, Joughin I (2000) Widespread complex flow in the interior of the Antarctic Ice Sheet. Science 287:1248. https://doi.org/10.1126/science.287.5456.1248

Bindoff NL, Rosenberg MA, Warner MJ (2000) On the circulation and water masses over the Antarctic continental slope and rise between 80 and 150 E. Deep Sea Res Part II 47(12–13):2299–2326

Cooper AK, O'Brien PE, Richter C (eds) (2004) Proceedings of the Ocean Drilling Program, scientific results, College Station, TX, Ocean Drilling Program. https://doi.org/10.2973/0dp.proc.sr.188.2004

Damiani D, Giorgetti G, Turbanti IM (2006) Clay mineral fluctuations and surface textural analysis of quartz grains in Pliocene-Quaternary marine sediments from Wilkes Land continental rise (East-Antarctica): paleoenvironmental significance. Mar Geol 226(3–4):281–295
De Santis L, Brancolini G, Donda F (2003) Seismic-stratigraphic analysis of the Wilkes Land continental margin (East Antarctica). Influence of glacially-driven processes on the Cenozoic deposition. Deep-Sea Res II 50:1563–1594
Donda F, Brancolini G, De Santis L, Trincardi F (2003) Seismic facies and sedimentary processes on the continental rise off Wilkes Land (East Antarctica): evidence of bottom current activity. Deep-Sea Res II 50:1509–1528
Drewry DJ, Cooper APR (1981) Processes and models of Antarctic glaciomarine sedimentation. Ann Glaciol 2:117–122
Eittreim SL, Hampton MA (eds) (1987) The Antarctic continental margin: geology and geophysics of offshore Wilkes Land. Earth science series (N. Y.), 5A
Eittreim SL, Smith GL (1987) Seismic sequences and their distribution on the Wilkes Land margin. In: Eittreim SL, Hampton MA (eds) The Antarctic continental margin: geology and geophysics of offshore Wilkes Land. Earth science series (N. Y.), 5A, pp 15–43
Eittreim SL, Cooper AK, Wannesson J (1995) Seismic stratigraphic evidence of ice-sheet advances on the Wilkes Land margin of Antarctica. Sed Geol 96(1–2):131–156
Escutia C, Brinkhuis H (2014) Greenhouse to Icehouse at the Wilkes land Antarctic margin: IODP expedition 318 synthesis of results. In: Blackman DK, Inagaki F, Larsen H-C, Stein R (eds) Developments in marine geology. Elsevier, Amsterdam, The Netherlands, pp 295–328
Escutia C, Eittreim SL, Cooper AK (1997a) Cenozoic sedimentation on the Wilkes Land continental rise, Antarctica. In: Ricci CA (ed) The Antarctic region: geological evolution and processes. Proceedings of international symposium on Antarctic earth science, vol 7, pp 791–795
Escutia C, Eittreim SL, Cooper AK (1997b) Cenozoic glaciomarine sequences on the Wilkes Land continental rise, Antarctica. In: Proceedings, vol VII. International symposium on Antarctic earth sciences, pp 791–795
Escutia C, Eittreim SL, Cooper AK, Nelson CH (2000) Morphology and acoustic character of the Antarctic Wilkes Land turbidite systems: ice-sheet-sourced versus river sourced fans. J Sediment Res 70(1):84–93. https://doi.org/10.1306/2DC40900-0E4711D7-8643000102C1865D
Escutia C, Nelson CH, Acton GD, Cooper AK, Eittreim SL, Warnke DA, Jaramillo J (2002) Current controlled deposition on the Wilkes Land continental rise. In: Stow D et al (eds) Deep-water Contourite systems: modern drifts and ancient series, seismic and sedimentary characteristics. The Geological Society of London, Memoirs 22, pp 373–378. https://doi.org/10.1144/gsl.mem.2002.022.01.26
Escutia C, Warnke D, Acton GD, Barcena A, Burckle L, Canals M, Frazee CS (2003) Sediment distribution and sedimentary processes across the Antarctic Wilkes Land margin during the Quaternary. Deep-Sea Res Part II 50(8–9):1481–1508. https://doi.org/10.1016/s0967-0645(03)00073-0
Escutia C, De Santis L, Donda F, Dunbar RB, Cooper AK, Brancolini G, Eittreim SL (2005) Cenozoic ice sheet history from East Antarctic Wilkes Land continental margin sediments. Global Planet Change 45(1–3):51–81. https://doi.org/10.1016/j.gloplacha.2004.09.010
Escutia C, Brinkhuis H, Klaus A, Expedition 318 Scientists (2011) IODP Expedition 318: Cenozoic East Antarctic ice sheet evolution from Wilkes Land margin sediment. In: Proceedings of the Integrated Ocean Drilling Program, vol 318, College Station, TX (Integrated Ocean Drilling Program Management International, Inc.)
Folk RL (1980) Petrology of Sedimentary Rocks. Hemphill Publishing Company, Austin
Gordon AL, Tchernia P (1972) Waters of the continental margin off the Adelie Coast, Antarctica. In: Hayes DE (ed) Antarctic oceanology II. The Australian-New Zealand sector. Antarctic research series, vol 19, pp 59–70
Hampton MA, Eittreim SL, Richmond BM (1987a) Postbreakup sedimentation on the Wilkes Land Margin, Antarctica. In: Eittreim SL, Hampton MA (eds) The Antarctic continental margin: geology and geophysics of offshore Wilkes Land. Circum-Pacific Council for Energy and Mineral Resources Earth Sciences Series, 5A, Houston, pp 75–88

Hampton MA, Eittreim SL, Richmond BM (1987b) Geology of sediment cores from the George V continental margin. In: Eittreim SL, Hampton MA (eds) The Antarctic continental margin: geology and geophysics of offshore Wilkes Land. Circum-Pacific Council for Energy and Mineral Resources Earth Sciences Series, 5A, Houston, pp 75–88

Harris PT, Keene JB, Cole A, Pattiaratchi CB (1990) Modelling the evolution of a linear sandbank field, Moreton bay, Queensland: report of results obtained during the cruise of AM brolga in July, 1989. Ocean Sciences Institute, University of Sydney

Hayes DE, Frakes LA (1975). Initial reports of the Deep Sea Drilling Project, 28. U.S. Govt. Printing Office, Washington, DC. https://doi.org/10.2973/dsdp.proc.28.1975

Helland PE, Holmes MA (1997) Surface textural analysis of quartz sand grains from ODP Site 918 off the southeast coast of Greenland suggests glaciation of southern Greenland at 11 Ma. Palaeogeogr Palaeoclimatol Palaeoecol 135(1):109–121

Hill PR, Nadeau OC (1984) Grain-surface textures of late Wisconsinan sands from the Canadian Beaufort Shelf. J Sediment Res 54(4):1349–1357

Hodel KL, Reimnitz E, Barnes PW (1988) Microtextures of quartz grains from modern terrestrial and subaqueous environments, north slope of Alaska. J Sediment Res 58(1):24–32

Jackson ML (1979) Soil chemical analysis—advanced course, 2nd edn. Published by the Author, Madison, Wisconsin

Krinsley DH, Donahue J (1968) Environmental interpretation of sand grain surface textures by electron microscopy. Geol Soc Am Bull 79(6):743–748

Krinsley DH, Doornkamp JC (1973) Atlas of sand grain surface textures. Cambridge University Press, Cambridge

Krumbein WC (1938) Pettijohn. FJ-1938—manual of sedimentary petrography, New York, London

Le Roux JP, Rojas EM (2007) Sediment transport patterns determined from grain size parameters: overview and state of the art. Sediment Geol 202(3):473–488

Lindstrom D, Tyler D (1984) Preliminary results of Pine Island and Thwaites glaciers study. Antarct JUS 19(5):53–55

MacDonald TR, Ferrigno JG, Williams RS Jr, Lucchitta BK (1989) Velocities of Antarctic outlet glaciers determined from sequential Landsat images. Antarct JUS 24:105–106

Mahaney WC, Kalm V (1995) Scanning electron microscopy of Pleistocene tills in Estonia. Boreas 24:13–29

Mahaney WC, Kalm V (2000) Comparative scanning electron microscopy study of oriented till blocks, glacial grains and Devonian sands in Estonia and Latvia. Boreas 29(1):35–51

Mahaney WC, Claridge GC, Campbell I (1996) Micro texture on quartz grains in tills from Antarctica. Paleogeogr Palaeoclimatol Palaeoecol 121:89–103

Marshall B, Gilligan LB (1987) An introduction to remobilization: information from ore-body geometry and experimental considerations. Ore Geol Rev 2(1–3):87–131

McCave IN (1978) Grain-size trends and transport along beaches: example from eastern England. Mar Geol 28(1–2):M43–M51

Pandey M, Pant NC, Biswas P, Shrivastava PK, Joshi S, Nagi N (2017) Heavy mineral assemblage of marine sediments as an indicator of provenance and East Antarctic ice sheet fluctuations. Geological Society, London, Special Publications, 461, p SP461-2

Pant NC, Biswas P, Shrivastava PK, Bhattacharya S, Verma K, Pandey M, IODP Expedition 318 Scientific Party (2013) Provenance of Pleistocene sediments from site U1359 of the Wilkes Land IODP Leg 318—evidence for multiple sourcing from the East Antarctic Craton and Ross Orogen. Geological Society of London, Special Publications, 381. https://doi.org/10.1144/sp381.11

Passchier S, Ciarletta DJ, Henao V, Sekkas V (2018) Sedimentary processes and facies on a high-latitude passive continental margin, Wilkes Land, East Antarctica. Geological Society, London, Special Publications, 475, p SP475-3

Payne RR, Conolly JR (1972) Turbidite sedimentation off the Antarctic continent. Antarct Oceanol II: Aust-N Z Sect 19:349–364

Presti M, Barbara L, Denis D, Schmidt S, De Santis L, Crosta X (2011) Sediment delivery and depositional patterns off Adélie Land (East Antarctica) in relation to late Quaternary climatic cycles. Mar Geol 284(1–4):96–113

Rex RW, Margolis SV, Murray B (1970) Possible interglacial dune sands from 300 meters water depth in the Weddell Sea, Antarctica. Geol Soc Am Bull 81:3465–3472

Riester DD, Shipp RC, Ehrlich R (1982) Patterns of quartz sand shape variation, Long Island littoral and shelf. J Sediment Res 52(4):1307–1314

Roy M, Van de Flierdt T, Hemming SR, Goldstein SL (2007) 40Ar/39Ar ages of hornblende grains and bulk Sm/Nd isotopes of circum-Antarctic glacio-marine sediments: implications for sediment provenance in the Southern Ocean. Chem Geol 244:507–519

Russell RD (1939) Effects of transportation of sedimentary particles. In: Trask PD (ed) Recent marine sediments. Society of Economic Paleontologists and Mineralogists, T&A, Oklahoma, p 3247

Stapor FW, Tanner WF (1975) Hydrodynamic implications of beach, beach ridge and dune grain size studies. J Sediment Res 45(4):926–931

Swift DJP, Ludwick JC, Boehmer WR (1972) Shelf sediment transport: a probability model. In: Swift DJP, Duane DB, Pilkey OH (eds) Shelf sediment transport: process and pattern. Dowden, Hutchingson and Ross, Stroudsburg, pp 195–223

Tauxe L, Stickley C, Sugisakil S, Bij PK, Bohaty S, Brinkhuis H, Escutia C, Flores JA, Iwai M, McKay R, Passchier S, Pross J, Riesselman C, Röhl U, Sangiorgi F, Welsh K, Williams T (2012) Magneto and biostratigraphic constraints for the paleoceanographic record of the Wilkes Land Margin cores: IODP Expedition 318. American Geophysical Union, Paleocenography 27:PA2214. https://doi.org/10.1029/2012pa002308

Verma K, Bhattacharya S, Biswas P, Shrivastava PK, Pandey M, Pant NC, I.O.D.P Expedition Scientific Party (2014) Clay mineralogy of the ocean sediments from the Wilkes Land margin, East Antarctica: implications on the paleoclimate, provenance and sediment dispersal pattern. Int J Earth Sci 103:2315–2326

Whalley WB, Krinsley DH (1974) A scanning electron microscope study of surface textures of quartz grains from glacial environments. Sedimentology 21(1):87–105

Late Quaternary Sedimentation and Slope Failure Events on the Costa Rican Margin

B. Nagender Nath, V. Yatheesh, A. Kazip Yazing, Surabhi R. Prabhu, Avinash Borker, D. K. Pandey and K. A. Kamesh Raju

Abstract The Costa Rican margin is an erosional convergent margin formed by subduction of the Cocos Plate under the Caribbean Plate along the Middle America Trench. The Integrated Ocean Drilling Program (IODP) Expedition 334 was conducted for drilling long sediment cores from the upper Caribbean Plate and the subducting Cocos Plate. Using the first 30 mbsf of the two drill cores retrieved from upper slope (U1379) and middle slope (U1378) of the Costa Rican margin, we attempted to decipher slope failure induced events based on combined geochemical, sedimentological, mineralogical and isotope chemical proxy analyses. The downhole variation data of clay minerals indicates that the source rocks are mainly the young volcanic rocks in central Costa Rica region. Poor crystallinity of clay minerals with fining upward sequence in the middle slope indicates at least two major episodes of slumping/mass deposition from the shallower areas during the glacial period. Major rock forming elements such as SiO_2, TiO_2, CaO, MgO and Na_2O were distinctly different in the cores from upper and middle slopes. C and N isotopes were also found to be different at depths with lithological changes. Our results suggest that slope failure induced events such as turbiditic sedimentation and mass flows mainly during the last glacial period are evident in the Costa Rican margin.

Keywords Costa Rican margin · Erosional convergent margin · Slope failure · IODP Expedition · Clay minerals · Turbiditic sedimentation

1 Introduction

The Integrated Ocean Drilling Programme (IODP) Expedition 334 was planned to retrieve long sediment cores from the Costa Rican margin (Fig. 1), where the Cocos

B. Nagender Nath · V. Yatheesh (✉) · A. Kazip Yazing · Surabhi R. Prabhu · Avinash Borker · K. A. Kamesh Raju
CSIR - National Institute of Oceanography, Dona Paula, Goa, India
e-mail: yatheesh@nio.org

D. K. Pandey
ESSO - National Centre for Polar and Ocean Research, Vasco da Gama, Goa, India

D. K. Pandey et al. (eds.), *Dynamics of the Earth System: Evolution, Processes and Interactions*, Society of Earth Scientists Series,
https://doi.org/10.1007/978-3-030-40659-2_9

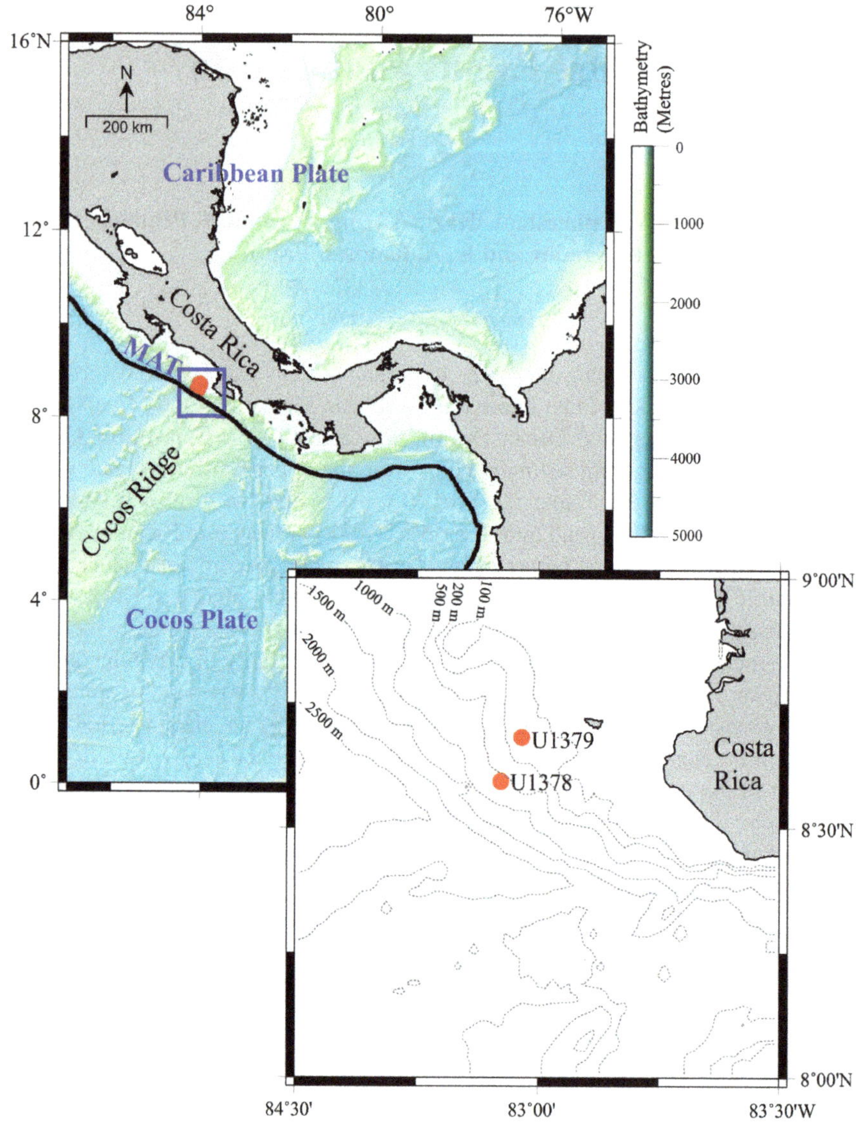

Fig. 1 Bathymetric maps showing the locations of sites U1379 and U1378 drilled during the Integrated Ocean Drilling Programme Expedition 334. Thick black line represents the location of Middle America Trench (MAT). Bathymetry data is from Smith and Sandwell (1997)

Plate subducts beneath the Caribbean Plate along the Middle America Trench (von Huene et al. 1995). This expedition, also known as the Costa Rica Seismogenesis Project (CRISP), was designed to understand the processes that control nucleation and seismic rupture of large earthquakes at erosional subduction zones. During this expedition, drilling was carried out over two sites (U1378 and U1379) in the upper plate (Caribbean Plate) and one site (U1381) over the subducting plate (Cocos Plate). Both the sites U1379 and U1378 are located (Fig. 1) in the upper plate and the drilling operations could retrieve ~890 m and ~524 m long sediment cores, respectively (see Expedition 334 Scientists 2011, 2012a, b, c; Vannucchi et al. 2013a, b).

We have studied two core sediments from sites U1379 and U1378 drilled on the Costa Rican continental margin with the objectives to decipher the sediment provenance, timing of late Quaternary slope failure events and evaluation of role of tectonics in altering the sediment characteristics. A diverse oceanic crust comprising of high-standing Cocos Ridge, flanked by chains of seamounts subducts along the Middle America Trench off Panama and Costa Rica. The subducting seamounts destabilizes the overlying seafloor, sometimes leading to occurrence of tsunamigenic earthquakes (von Huene et al. 2004). The effect of oceanic crust subduction on continental margin structure is most evident in the middle and upper slope, erosion being more pronounced in these areas relative to the stable slope where smooth lower plate subducts (von Huene et al. 2000). Past slope failure events (e.g., 1.1 Myr; 0.1 Myr BP and a recent one) were reported in the region (von Huene et al. 2004 and references therein), some of them would have left their traces in the slope sediments. With a view of deciphering the late Quaternary slope failure events recorded if any, the top 30 m of drilled cores of sites U1379 and U1378 representing upper and middle slope regions were studied (Fig. 1) employing the geochemical (major elements), clay mineralogical and carbon and nitrogen isotope chemical proxies, diagnostic of provenance and sedimentary processes.

2 Materials and Methods

2.1 *Samples and Data*

The core samples used for the present study were retrieved by drilling carried out under the Integrated Ocean Drilling Program Expedition 334. As a part of this expedition, drilling was conducted at two sites, U1378 and U1379, which are located in the Costa Rican margin, offshore Osa Peninsula, on the Caribbean Plate (Fig. 1). The Site U1379 is on the upper slope at 125.9 m water depth, while the Site U1378 is on the middle slope at 521.8 m water depth. The drilling operations in these sites U1379 and U1378 recovered ~890 m and ~524 m long sediment cores, respectively. For this work, one sample every 1 m from the top 30 m totalling to 60 sediment samples for both the sites were studied. The sediments recovered from 0 to 30 mbsf at U1378 is composed mainly of soft, dark greenish gray, terrigenous silty clay. Two thick fining-upward sequences consisting of lithic sands and tephra layers are observed as intercalated in this silty clay (Expedition 334 Scientists 2011, 2012a, b, c). The sediments recovered from 0 to 30 mbsf at U1379 is composed of mainly olive-green clayey silt and silty clay with minor layers of sandy silty clay, with its topmost part (0.0–0.93 mbsf) consisting of medium to coarse grained sand with abundant shell fragments. Superimposed on this main lithostratigraphic unit, fining and coarsening upward decimeter-scale sequences of silty sands and sand stone are observed. This unit is massive with minor changes in the proportions of clay, silt and sand grain sizes occurring gradually over <1 m intervals (Expedition 334 Scientists 2011, 2012a, b, c). In addition to the sediment samples, we also used digital physical property data, consisting of bulk density and magnetic susceptibility, which were measured using the Whole-Round Multisensor Logger.

2.2 Sample Processing for XRD Analysis

For the clay mineral studies, all the samples collected from 0 to 30 mbsf of cores retrieved from sites U1379 and U1378 were made free of salt by washing repeatedly with distilled water followed by centrifuging. These salt-free samples were sieved carefully using a 230 mesh (63 μm size). The sediments retained on the sieve represent the sand-sized fraction, which were collected and dried. The silt and clay fractions that passed through the sieve were transferred into 1000 ml measuring cylinders and made up to the 1000 ml mark by adding Milli-Q water. From this sediment suspension, different size fractions of <2, 2–4, 4–8, 8–16 and 16–32 μm were collected at a depth of 10 cm, at timings which were calculated for a room temperature of 29 °C. Further, the clay fraction (<2 μm) has been treated with acetic acid and hydrogen peroxide to remove carbonates and organic matter respectively and washed several times to make them free of these chemicals. The clay suspension was deposited on a glass slide with a pipette to prevent non-uniform sample surface roughness or flatness. After the deposition, all the slides were air-dried at room temperature. The air-dried clay slides were then placed in the desiccator containing ethylene glycol in the oven at 100 °C for about an hour. These samples were then used for the analysis of clay minerals using X-Ray Diffractometer.

2.3 Quantification of Clay Minerals

While carrying out XRD analysis, we used two types of scanning speeds, normal scanning speed and slow scanning speed. In normal scanning speed, the samples were scanned between 3° and 22° 2θ with a scanning speed of 0.02° 2θ/s and chart speed of 10 mm/°2θ and the range of intensity of diffraction beam was 2000 cps (counts per seconds). Between 24° and 26° 2θ, the scanning speed was 0.01° 2θ/s with a chart speed of 20 mm/°2θ. Among the clay minerals, smectite is expanding type and therefore, clay mounts were glycolated to expand smectite. After glycolation, smectite shows peak at 5.3° 2θ. Illite and chlorite were identified from their characteristic peaks commonly used in the literature. Slow scanning has not revealed any kaolinite in the samples.

After identifying each clay mineral, the integrated intensity for the clay minerals was calculated using the RIGAKU software. The glycolation for smectite is a process to artificially expand the mineral and therefore, some compensation for other minerals is also needed. Hence, the following compensation factor was used to obtain percentage of clay minerals:

$$\text{smectite} \times 1 = \text{illite} \times 4 = \text{kaolinite} + \text{chlorite} \times 2.5$$

The crystallinity of two minerals, smectite and illite were determined using formulae from the published literature. The crystallinity of smectite was measured based on the ratio of the height of the peak above the baseline (P) and the depth of the valley (V) on the low angle side of the peak. Crystallinity of illite was calculated by measuring width (Δ 2θ) at half peak height.

2.4 *Sample Processing for XRF Analysis*

X-ray fluorescence (XRF) spectrometry was used for determining the major element composition. For the analyses, 10 g of the bulk sediment samples were transferred to a beaker and treated with acetic acid till all the $CaCO_3$ was completely dissolved, until the time the effervescence had ceased. The samples are then washed repeatedly with distilled water till the acetic acid is washed off. The samples were dried in the oven at a temperature of 60 °C and then were powdered. The acetic acid-treated sediments thus represent the lithogenic fraction. Approximately 1.5 g of this lithogenic fraction was heated in the oven at 110 °C for at least 12 h to remove the moisture. From this, 0.55 g of each sample was weighed and kept in the desiccator for XRF analysis. For this, Spectromelt A12 of MERCK (di-lithium tetra-borate, 66% and lithium meta-borate 34%) was added in a ratio of 1:10 to the sample. This was then mixed thoroughly in a platinum crucible and placed in a bead making Minifuse-2 machine. The samples were fused/ melted with the flux in the crucible using the furnace. The composite red hot melt was poured in a casting dish and cooled into a glass bead which was then used for XRF analysis.

2.5 *Sample Processing for IRMS Analysis*

Isotope Ratio Mass Spectrometry (IRMS) was used for isotope ratio analysis of carbon and nitrogen in the organic matter of the samples collected from 0 to 30 mbsf of cores retrieved from sites U1379 and U1378. For this analysis, sediment samples were treated with 10% HCl to remove carbonates (Schubert and Nielsen 2000). Excess acid was removed by repeated washing with Milli-Q water. The decarbonated sediments were then dried in a freeze dryer and were ground to fine powder using agate mortar. These sample powders were then used for the analysis of carbon and nitrogen isotope ratios using IRMS. Analysis was performed with the Thermo Finnigan Flash 1112 elemental analyzer, linked with a Thermo Finnigan Delta V plus IRMS at CSIR-NIO. IAEA grade standards Sucrose ($\delta^{13}C = -10.449‰$) and Ammonium Sulphate ($\delta^{15}N = 20.3‰$) were used for instrument calibration for stable isotope analyses of carbon and nitrogen respectively. Stable isotopic ratio of OC and TN were represented using delta notation relative to standard as,

$$\delta\ (‰) = \{(R_{Sample}/R_{Standard}) - 1\} \times 1000$$

where R represents $^{13}C/^{12}C$ for C isotopes and $^{15}N/^{14}N$ for N isotopes. The standards used for $\delta^{13}C_{org}$, and $\delta^{15}N$ were PDB and atmospheric N_2 respectively. The overall analytical precision for replicate samples was within ±0.2‰ for $\delta^{13}C_{org}$ and ±0.1‰ for $\delta^{15}N$.

2.6 *Age Model*

The biostratigraphic age model generated onboard by the Shipboard participants (Fig. 2a and b) has been used for depth to age conversion for the top 30 m studied here. For the sites U1379 and U1378, the sedimentation rates for the top sections (applicable for top 30 m) estimated were 1035 m/m.y. and 516 m/m.y., respectively

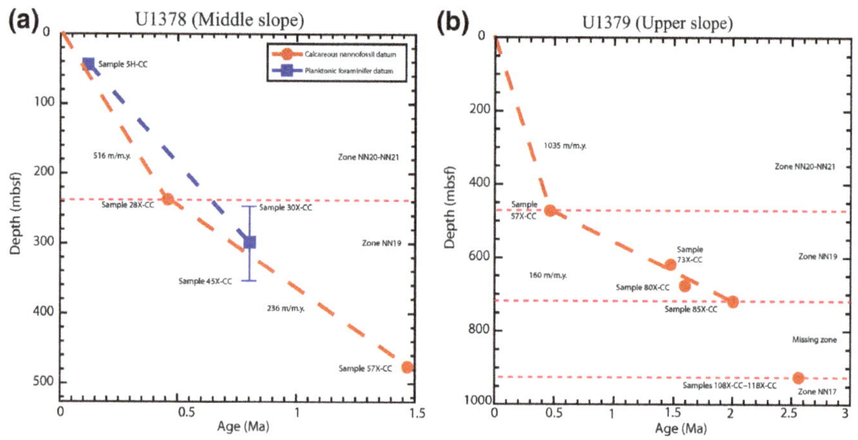

Fig. 2 Age-depth plot based on first and last occurrence datums of microfossils in **a** Hole U1378 and **b** Hole U1379 (after Expedition 334 Scientists 2012b)

(Fig. 2a and b). From these sedimentation rates, relative downcore ages were assigned and the clay mineral, elemental and isotopic data were plotted with age to compare the coeval events in both upper and middle slope if any.

3 Results and Discussion

3.1 Clay Minerals

A total number of 58 sediment samples together from sites U1379 and U1378 were analyzed as above for quantifying the clay minerals and these data are listed in Tables 1 and 2. Our analysis show that the clay mineral kaolinite is absent in the analyzed sediment samples. The downhole variations of percentages of smectite, chlorite and illite as well as smectite and illite crystallinity for the cores from sites U1379 and U1378 are shown in Figs. 3 and 4, respectively. For comparison, we also plotted the concentration of clay minerals for the sites U1379 and U1378 together as a function of depth (Fig. 5) as well as age (Fig. 6).

3.1.1 Site U1379

In this core, clay minerals smectite, chlorite and illite are present in order of abundance (Fig. 3). Smectite is the most dominant mineral. The smectite content ranges from 34 to 73% (Av. 52%). Highest percentage of smectite occurs at 1.92 m and least at 7.71 m. The smectite content is uniform within the depth of 9.21–10.71 and 18.71–27.51 m, and a peak is found at 28.37 m. As they are calculated as relative percentages, whenever smectite concentration is low, illite and chlorite contents were high. Chlorite ranges from 14 to 34% (av. 26%) with maximum occurring at 26.1 mbsf and least at 1.92 mbsf. Chlorite content increases with the depth until about 5 mbsf but nearly uniform thereafter throughout the core. Until about 15 mbsf, illite is higher with two subsurface peaks (maximum occurring at 7.71 mbsf) with uniform values throughout. The change in smectite content at 12 mbsf with a concomitant change in other clays, which also coincides with lithological changes such as the presence of mineral grains (onboard description) and a peak in magnetic susceptibility (Fig. 3) can be inferred to represent an episode of increased run-off at about 12 kyr BP (Fig. 6). Other susceptibility peaks can indicate the periods of increased terrestrial input with the peak at ~26 mbsf coinciding with the recovery of coarser mineral grains and increased illite crystallinity values (Fig. 3) may represent another major episode of runoff of high energy during the last glacial period (Fig. 6). High chemical weathering would have shown low-crystallized illite and a fining grain-size sequence (Lamy et al. 1999).

Table 1 Clay mineral content and calculated crystallinity index for smectite and illite for Site U1379 (upper slope)

Sample ID	Depth (mbsf)	Mean depth (mbsf)	Smectite (%)	Illite (%)	Chlorite (%)	(K + C) (%)	Smectite crystallinity	Illite crystallinity
1H1: 126–128	1.26–1.28	1.26	58.69	24.17	17.14	17.14	0.56	0.60
1H2: 24–26	1.74–1.76	1.75	70.38	13.00	16.61	16.62	0.51	0.50
1H2: 40–43	1.90–1.93	1.92	72.58	13.18	14.24	14.25	0.62	0.50
1H3: 40–43	3.40–3.43	3.42	56.19	25.46	18.35	18.35	0.48	0.90
1H4: 40–43	4.90–4.93	4.92	41.72	24.57	33.70	33.71	0.55	1.00
1H5: 40–43	6.40–6.43	6.42	53.49	15.76	30.74	30.75	0.61	0.70
2H1: 100–102	7.70–7.72	7.71	33.88	39.35	26.77	26.77	0.42	0.60
2H2: 100–102	9.20–9.22	9.21	56.68	22.43	20.89	20.90	0.55	0.55
2H3: 50–52	10.20–10.22	10.21	59.83	16.89	23.27	22.28	0.54	0.50
2H3: 100–102	10.70–10.72	10.71	59.28	22.15	18.56	18.57	0.65	0.60
2H4: 30–32	11.50–11.52	11.51	51.55	22.42	26.03	26.03	0.51	0.45
2H4: 100–102	12.20–12.22	12.21	56.77	19.59	23.64	23.64	0.54	0.50
2H5: 30–32	13.00–13.02	13.01	39.04	30.28	30.67	30.68	0.55	0.50
2H6: 20–22	14.40–14.42	14.41	48.68	31.13	28.18	28.19	0.40	0.60
3H1: 100–102	17.20–17.22	17.21	62.98	16.74	20.28	20.28	0.61	0.50
3H2: 100–102	18.70–18.72	18.71	51.80	21.90	26.30	26.30	0.50	0.60
3H3: 100–102	20.20–20.22	20.21	47.18	19.88	32.93	32.94	0.45	0.50
3H4: 100–102	21.70–21.72	21.71	47.50	20.72	31.77	31.78	0.55	0.60
3H5: 100–102	23.20–23.22	23.21	42.73	23.98	33.29	33.98	0.54	0.55

(continued)

Table 1 (continued)

Sample ID	Depth (mbsf)	Mean depth (mbsf)	Smectite (%)	Illite (%)	Chlorite (%)	(K + C) (%)	Smectite crystallinity	Illite crystallinity
3H6: 30–32	24.00–24.02	24.01	50.58	20.25	29.16	29.17	0.57	0.80
4H1: 100–102	25.20–25.22	25.21	45.80	24.59	29.60	29.61	0.47	1.00
4H2: 30–32	26.00–26.02	26.01	42.34	23.33	34.33	34.34	0.53	1.10
4H2: 100–102	26.70–26.72	26.71	38.63	28.44	32.92	32.93	0.48	1.10
4H3: 30–32	27.50–27.52	27.51	49.45	22.03	28.51	28.52	0.56	0.60
4H3: 116–118	28.36–28.38	28.37	66.96	12.68	20.35	20.36	0.53	0.85
4H4: 30–32	29.00–29.02	29.01	42.80	26.02	31.18	31.18	0.51	0.65

Crystallinity Index:

Crystalline index can help in qualitatively assessing the proximity or distance to the source. Lower the index value, better is the degree of crystallinity and vice versa. Smectite crystallinity ranges from 0.4 to 0.65 (av. 0.5). Higher the smectite concentration, poorer is the crystallinity. Smectite shows good crystallinity within 18.71–26.71 mbsf. Illite crystallinity ranges from 0.4 to 1.1 (av. 0.7). Illite shows good crystallinity throughout the core whereas distinctly poor crystallinity is seen at two sub-surface depths; one at ~5 mbsf and the second between 24.01 and 26.71 mbsf (Fig. 3). The depths between 7.71 and 24 mbsf show good illite crystallinity. Occurrence of a distinct mineral grain layer was logged in the onboard description at the depth (~26 mbsf) where the illite crystallinity was found to be poor, which was also coincident with fining upward sequence.

3.1.2 Site U1378

Similar to the core from the upper slope, clay minerals smectite, chlorite and illite are present in order of abundance (Fig. 4) in this site. The smectite content ranges from 42 to 73% (av. 57%). A change at about 10 m is seen with lower smectite content grading to higher content which is nearly similar thereafter, except two troughs in the deeper section. When smectite concentration is low, chlorite was found to be high and vice versa. Illite content is nearly similar with only minor subsurface peaks. Near similar illite and chlorite downcore trend may suggest the changes in physical weathering in the source areas. Illite and chlorite mainly result from physical weathering of crystalline (i.e., acidic plutonites for illite and basic igneous rocks for chlorite) and low-grade metamorphic rocks (e.g., Chamley 1989).

Table 2 Clay mineral content and calculated crystallinity index for smectite and illite for Site U1378 (middle slope)

Sample ID	Depth (mbsf)	Mean depth (mbsf)	Smectite (%)	Illite (%)	Chlorite (%)	(K + C) (%)	Smectite crystallinity	Illite crystallinity
1H1: 68–70	0.68–0.70	0.69	61.62	13.69	24.70	24.70	0.52	0.50
1H2: 68–70	2.18–2.20	2.19	59.66	13.78	26.56	26.56	0.42	0.53
1H3: 68–70	3.68–3.70	3.69	51.88	18.10	30.02	30.02	0.46	0.53
2H1: 68–70	5.98–6.00	5.99	57.98	16.21	25.81	25.81	0.50	0.70
2H2: 68–70	7.48–7.50	7.49	53.95	18.05	28.00	28.00	0.39	0.63
2H3: 68–70	8.98–9.00	8.99	55.72	20.93	23.35	23.35	0.51	0.65
2H3: 96–98	9.26–9.28	9.27	44.41	15.91	39.67	39.68	0.54	0.50
2H4: 32–34	10.12–10.14	10.13	47.37	17.22	35.40	35.41	0.54	0.55
2H4: 39–41	10.19–10.21	10.20	41.77	19.14	39.09	39.10	0.56	0.70
2H4: 50–52	10.30–10.32	10.31	49.47	16.30	34.22	34.23	0.57	0.55
2H4: 68–70	10.48–10.50	10.49	65.84	16.74	17.42	17.42	0.45	0.50
2H4: 99–101	10.79–10.81	10.80	48.53	14.34	37.12	37.12	0.52	0.60
2H4: 108–110	10.88–10.90	10.89	44.94	19.09	35.96	35.96	0.59	0.90
2H4: 118–120	10.98–11.00	10.99	43.62	23.37	33.00	33.00	0.61	0.50
2H5: 40–42	11.70–11.72	11.71	44.45	16.26	39.28	39.29	0.65	0.75
2H5: 68–70	11.98–12.00	11.99	69.15	13.41	17.44	17.44	0.45	0.50
2H5: 81–83	12.11–12.13	12.12	50.65	12.91	36.43	36.44	0.52	0.70
2H5: 109–111	12.39–12.41	12.40	47.69	13.86	38.44	38.45	0.55	0.50
2H6: 48–50	13.28–13.30	13.29	52.52	14.95	32.52	32.52	0.51	0.60

(continued)

Table 2 (continued)

Sample ID	Depth (mbsf)	Mean depth (mbsf)	Smectite (%)	Illite (%)	Chlorite (%)	(K + C) (%)	Smectite crystallinity	Illite crystallinity
2H6: 68–70	13.48–13.50	13.49	64.74	14.15	21.11	21.11	0.44	0.35
3H1: 65–67	15.45–15.47	15.46	72.87	10.13	17.00	17.00	0.38	0.25
3H2: 68–70	16.98–17.00	16.99	61.65	16.57	21.78	21.78	0.52	0.55
3H3: 68–70	18.48–18.50	18.49	65.03	16.25	18.72	18.72	0.51	0.40
3H3: 116–118	18.96–18.98	18.97	44.05	20.90	35.04	35.05	0.48	0.60
3H4: 68–70	19.98–20.00	19.99	67.67	13.28	19.05	19.05	0.44	0.50
3H5: 72–74	21.52–21.54	21.53	71.93	9.85	18.22	18.22	0.48	0.40
3H6: 67–69	22.97–22.99	22.98	70.82	13.51	15.67	15.67	0.41	0.20
4H2: 67–69	25.58–25.60	25.59	66.18	17.30	16.51	16.51	0.46	0.45
4H3: 65–67	27.06–27.08	27.07	67.80	8.32	23.88	23.88	0.43	0.30
4H3: 119–121	27.60–27.62	27.61	43.93	15.10	40.97	40.97	0.55	0.40
4H4: 66–69	28.57–28.60	28.58	73.25	9.56	17.19	17.19	0.47	0.40
4H5: 66–69	30.07–30.10	30.08	63.32	12.92	23.76	23.76	0.54	0.30

Rapid mineralogical changes noticed between 10 and 13 mbsf distinctly coincide with the sediment with coarsening upward sequence and the occurrence of tephra (Fig. 4), possibly indicating the deposition of tephra containing turbidite at about 23–20 kyr when peak glacial conditions would have prevailed. A cooling of 7–8°C during the Last Glacial Maximum (LGM) for montane Costa Rica was reported (Islebe and Hooghiemstra 1997). The lowered sea level during LGM would have led to the turbidite deposition. Clay mineralogical changes are also noticed at the depths at which presence of visual mineral grain layers (Fig. 4) were noted in the onboard lithological description. When both the cores are compared, broadly illite content is higher in upper slope with a concomitant decrease in smectite (Fig. 5). Smectites typically are the products of erosion of basic rocks such as basalts or other young volcanics. Smectites are abundant clay mineral in both the sites and the primary source could be the elevated portions with volcanic rocks of central Costa Rica. In Costa Rica, the oldest rocks are around 180 million years old and are chunks of ophiolites, overlain by younger volcanic rocks and recent deposits. The Sierra de

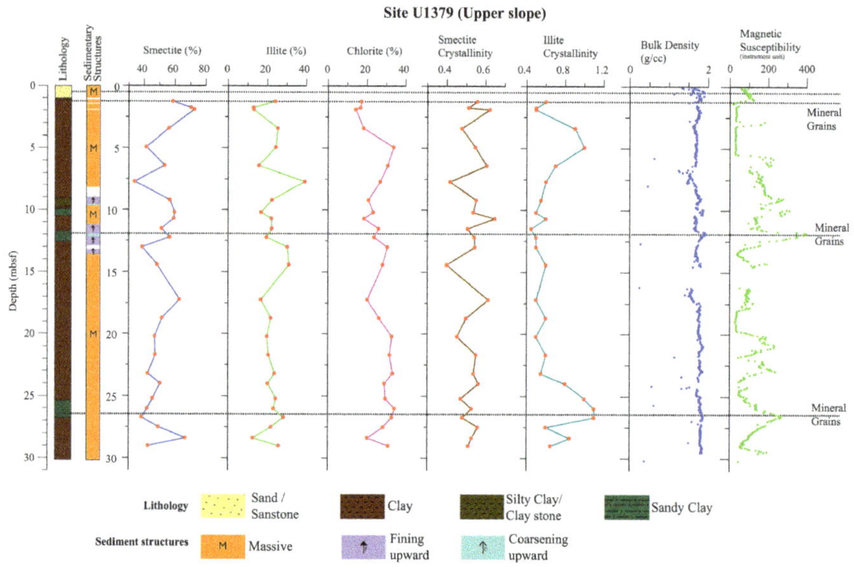

Fig. 3 Downhole variations of clay minerals smectite, chlorite and illite and smectite crystallinity and illite crystallinity for the Site U1379. For the correlation, also plotted the lithology, sedimentary structures and physical property parameters (bulk density and magnetic susceptibility)

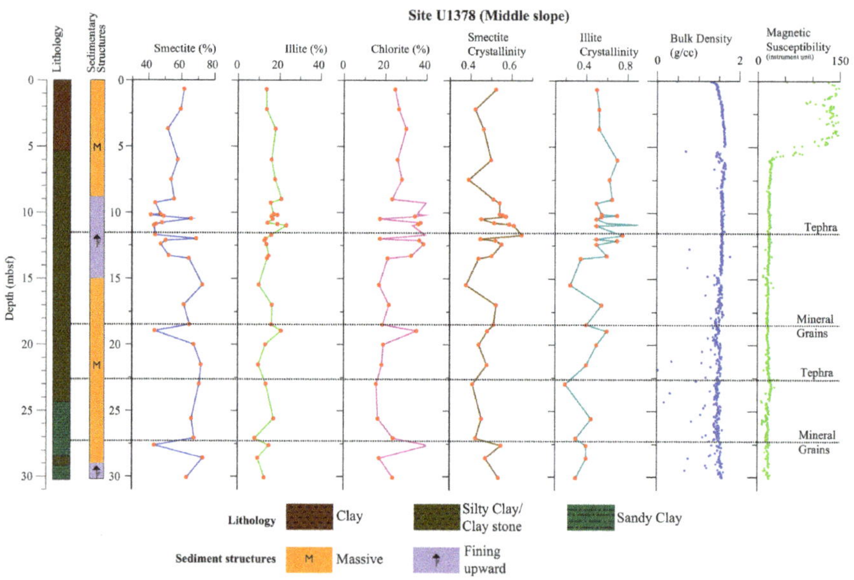

Fig. 4 Downhole variations of clay minerals smectite, chlorite and illite and smectite crystallinity and illite crystallinity for the Site U1378. For the correlation, also plotted the lithology, sedimentary structures and physical property parameters (bulk density and magnetic susceptibility)

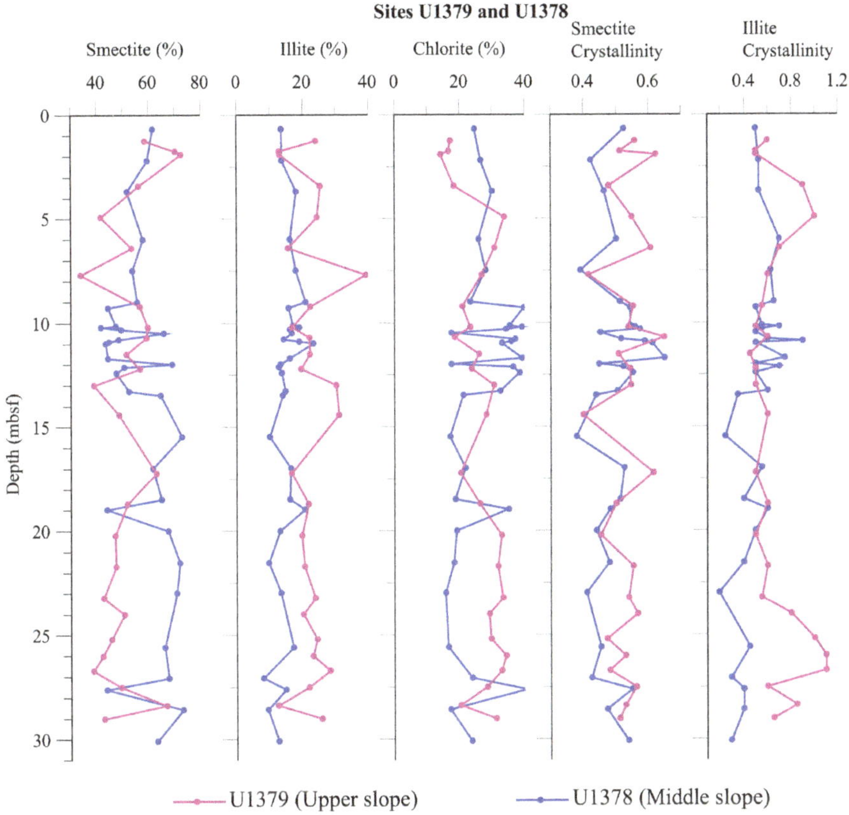

Fig. 5 Downhole variations of clay minerals smectite, chlorite and illite and smectite crystallinity and illite crystallinity together for the sites U1379 and U1378

Talamanca exposes the eroded roots of a former volcanic chain and to their east lie the oceanic sedimentary rocks (e.g., Dutch 2009).

Crystallinity Index:

Smectite crystallinity ranges from 0.38 to 0.65 (av. 0.5). Smectite shows a better crystallinity with an increase in depth. In contrast to the upper slope, smectite shows better crystallinity when the smectite concentration is high. Poor crystallinity is shown within the depth of 5.99 and 10.89 mbsf. Illite crystallinity ranges from 0.2 to 0.9 (av. 0.5). Higher the illite %, poorer is the crystallinity. Illite in general shows better crystallinity in the deeper portion of the core compared to the top section (Fig. 4). Similar to mineralogical changes, distinctly poor clay mineral crystallinity was found in the sediments with fining upward sequence (Fig. 4) suggesting a distal source. When both the cores are compared, smectite crystallinity was nearly similar whereas the lower slope core has shown better illite crystallinity at two depths (Fig. 5).

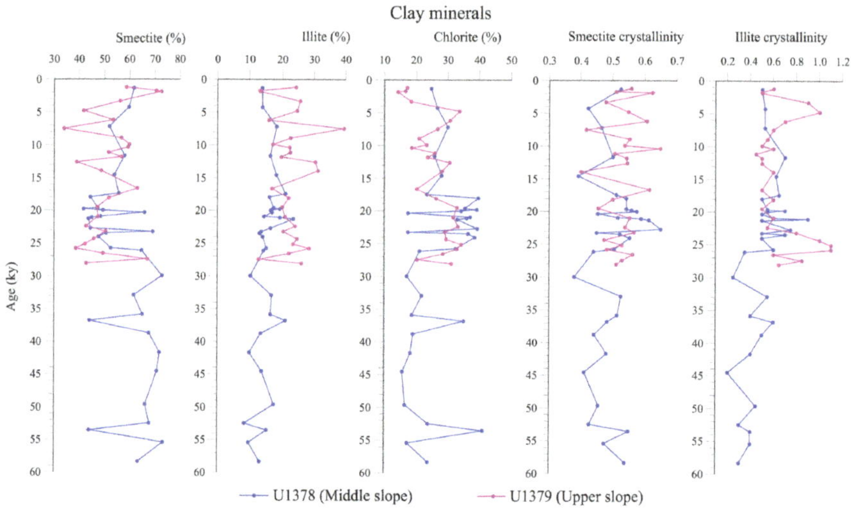

Fig. 6 Concentration of clay minerals for the sites U1379 and U1378 plotted together as a function of age

While the ranges of crystallinity index values of both the sites are nearly similar, a distinct contrast in relation between smectite abundance and the crystallinity is seen in these two sites. Poor crystallinity when smectite is high in upper slope could indicate the normal sedimentary process of alteration, transport and deposition. On the other hand, better crystallinity when smectite is abundant in middle slope could indicate periods of rapid runoff from the terrestrial and shelf areas and their burial without being altered much, possibly through turbiditic currents. Incidentally, the two main episodes of turbiditic sedimentation at about 10 and 12 mbsf occurring in middle slope (U1378) roughly correspond to the Last Glacial Period (~25 kyr BP—Fig. 6) when the sea level would be lower inducing a slope instability. Sandwiched between two layers has sediment with poor smectite crystallinity which is coeval with the presence of tephra layer (Fig. 4). At about same time period (Fig. 6), depositional change with coarser grain deposition, increased susceptibility and deposition of sediments with high crystallinity is noticed in upper slope site too (U1379) suggesting the role of glacial climate driven slope instability/turbiditic sedimentation.

3.2 Quantification of Major Elements

A total number of 24 sediment samples together from sites U1379 and U1378 were analyzed for determining the major elements and these data are listed in Tables 3 and 4. The downhole distribution of percentages of major element oxides (SiO_2, Al_2O_3, TiO_2, Fe_2O_3, MnO, MgO, CaO, Na_2O, K_2O and P_2O_5) for the cores from sites U1379 and U1378 are shown in Figs. 7 and 8 respectively.

Table 3 Major element content for the core retrieved from Site U1379 (upper slope)

Sample ID	Depth (mbsf)	Mean depth (mbsf)	Sum (%)	SiO_2 (%)	Al_2O_3 (%)	TiO_2 (%)	Fe_2O_3 (%)	MnO (%)	MgO (%)	CaO (%)	Na_2O (%)	K_2O (%)	P_2O_5 (%)
2H2: 100–102	9.2–9.22	9.21	90.012	55.713	16.512	1.082	8.159	0.028	2.768	2.409	1.626	1.566	0.151
2H3: 50–52	10.2–10.22	10.21	88.927	55.060	16.034	1.083	8.413	0.028	2.721	2.343	1.575	1.507	0.163
2H3: 100–102	10.7–10.72	10.71	88.546	55.076	16.692	1.092	8.166	0.022	2.503	1.906	1.440	1.516	0.135
2H4: 100–102	12.2–12.22	12.21	89.845	56.153	16.146	1.094	8.055	0.029	2.638	2.374	1.644	1.559	0.153
2H6: 20–22	14.4–14.42	14.41	89.034	55.399	16.307	1.083	8.045	0.023	2.538	2.320	1.604	1.572	0.143
3H2: 100–102	18.7–18.72	18.71	89.035	55.417	16.601	1.098	7.784	0.024	2.654	2.175	1.590	1.557	0.136
3H3: 100–102	20.2–20.22	20.21	89.316	55.527	16.234	1.102	7.877	0.025	2.715	2.472	1.658	1.554	0.150

(continued)

Table 3 (continued)

Sample ID	Depth (mbsf)	Mean depth (mbsf)	Sum (%)	SiO_2 (%)	Al_2O_3 (%)	TiO_2 (%)	Fe_2O_3 (%)	MnO (%)	MgO (%)	CaO (%)	Na_2O (%)	K_2O (%)	P_2O_5 (%)
3H4: 100–102	21.7–21.72	21.71	88.812	54.408	16.826	1.143	8.285	0.026	2.843	2.183	1.460	1.494	0.144
3H5: 100–102	23.2–23.22	23.21	88.557	54.553	16.063	1.113	8.443	0.027	2.746	2.426	1.531	1.484	0.171
3H6: 30–32	24–24.02	24.01	87.909	54.272	16.802	1.118	8.223	0.019	2.597	1.928	1.333	1.476	0.141
4H1: 100–102	25.2–25.22	25.21	88.547	54.951	16.297	1.113	8.095	0.023	2.680	2.250	1.519	1.471	0.147
4H2: 30–32	26–26.02	26.01	88.562	54.340	16.337	1.126	8.454	0.024	2.884	2.330	1.463	1.462	0.144
4H2: 100–102	26.7–26.72	26.71	88.941	54.723	15.118	1.103	8.423	0.032	3.159	2.961	1.829	1.429	0.162
4H3: 30–32	27.5–27.52	27.51	88.537	53.803	16.486	1.159	8.508	0.028	3.086	2.250	1.557	1.515	0.146

Table 4 Major element content for the core retrieved from Site U1378 (middle slope)

Sample ID	Depth (mbsf)	Mean depth (mbsf)	Sum (%)	SiO_2 (%)	Al_2O_3 (%)	TiO_2 (%)	Fe_2O_3 (%)	MnO (%)	MgO (%)	CaO (%)	Na_2O (%)	K_2O (%)	P_2O_5 (%)
2H3: 68–70	8.98–9.00	8.99	86.384	51.405	16.014	1.159	9.156	0.029	3.600	2.248	1.265	1.367	0.140
3H1: 65–67	15.45–15.47	15.46	86.273	51.844	16.307	1.148	8.822	0.021	3.256	2.257	1.141	1.318	0.159
3H3: 68–70	18.48–18.50	18.49	86.608	52.825	16.418	1.136	8.382	0.023	3.273	1.908	1.170	1.356	0.118
3H5: 72–74	21.52–21.54	21.53	86.891	52.564	16.359	1.168	8.476	0.025	3.384	2.192	1.242	1.339	0.141
3H6: 67–69	22.97–22.99	22.98	86.586	52.514	16.659	1.151	8.545	0.022	3.250	1.836	1.133	1.350	0.126
4H2: 67–69	25.58–25.60	25.59	86.877	52.382	16.437	1.153	8.582	0.027	3.460	2.128	1.236	1.344	0.128
4H3: 65–67	27.06–27.08	27.07	86.429	52.044	16.609	1.131	8.709	0.025	3.172	2.093	1.137	1.354	0.155
4H4: 66–69	28.57–28.60	28.585	86.487	52.075	16.799	1.141	8.621	0.023	3.224	1.994	1.110	1.364	0.136

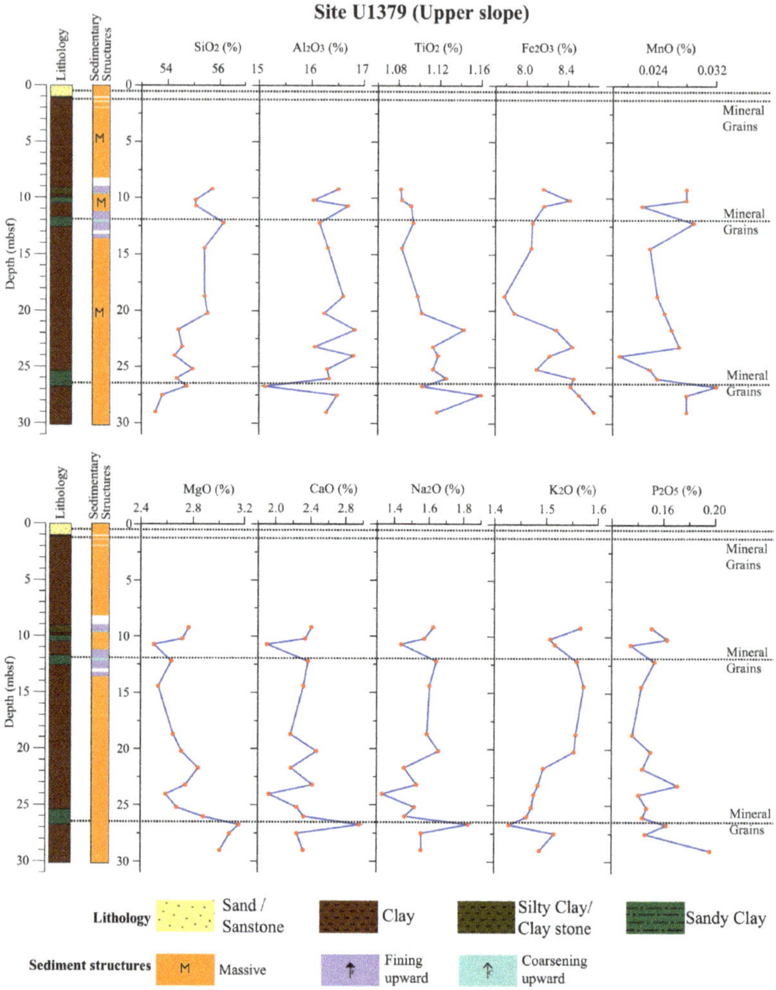

Fig. 7 Downhole variation of concentration of major element content for the Site U1379. Also plotted are shipboard described lithology and sedimentary structures

3.2.1 Sites U1379 and U1378

Major elemental composition has been analyzed for cores from upper slope (Site U1379) and middle slope (Site U1378), Figs. 7 and 8, respectively. Upper slope sediments show relatively higher SiO_2 concentration (average 55 wt%) than the lower slope (52 wt%). SiO_2 concentrations decrease at a subsurface depth of 20.21 mbsf (Fig. 7) whereas the middle slope core shows highest concentration at about 18 mbsf with the overlying and underlying sediments having lower concentrations (Fig. 8). Al_2O_3 content shows no variation between the upper and middle slope cores, but the concentration gradually decreases in deeper core. The TiO_2 concentration is

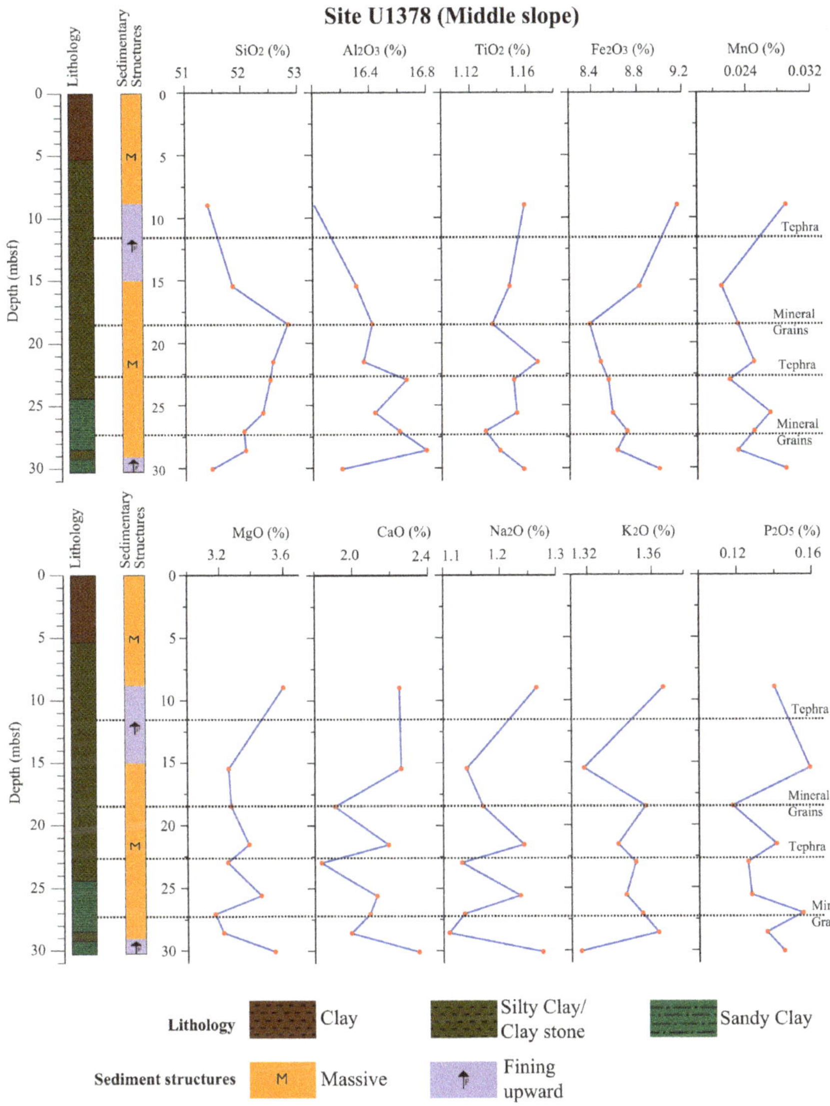

Fig. 8 Downhole variation of concentration of major element content for the Site U1378. Shipboard lithology and sedimentary structures are plotted for comparison

higher in lower slope than upper slope. The Fe_2O_3 is higher in lower slope than upper slope, with both showing a trough between 15 and 25 mbsf depth. MnO concentration are nearly similar in both cores and shows similar trend with Fe_2O_3. MgO is higher in middle slope than upper slope may be there is an influence of oceanic crust (basaltic weathering). CaO content is nearly similar in the two cores, however, the concentrations of important rock forming alkalies K, Na and Mg show upper to middle slope contrast (Fig. 9). Silica versus total alkalies (TAS) diagrams are commonly used in petrochemical studies of hard rocks to distinguish the type of rocks. But, a modified TAS (see Kurian et al. 2013) is employed (Fig. 10) to determine the provenance of sediments. The geochemical data of two cores fall distinctly in two groups, falling mainly in the intermediate field, with mid-slope sediments falling at the basic to intermediate boundary (Fig. 10). Both silica and alkaline elements are low in middle slope regions possibly driven by size-sorting with finer sediments possibly with higher smectite content reaching the middle slope region. This is also supported by distinctly higher Mg content in middle slope compared to the upper slope region

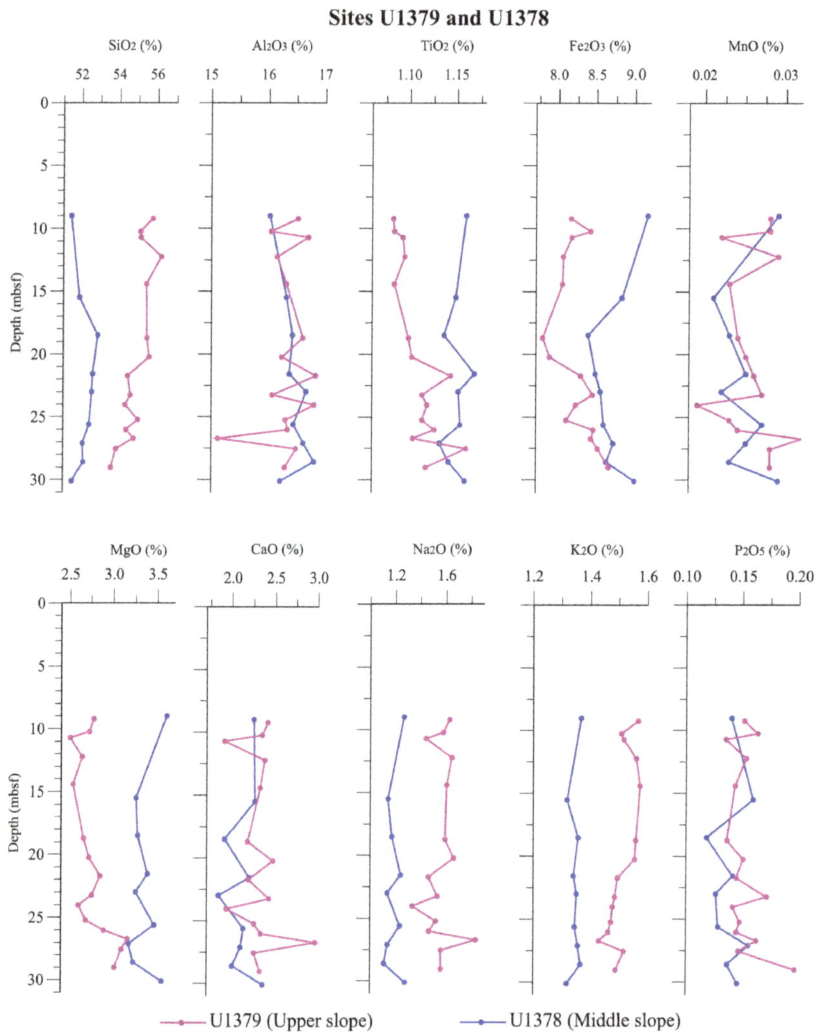

Fig. 9 Major element chemistry for the sites U1378 and U1379 plotted together

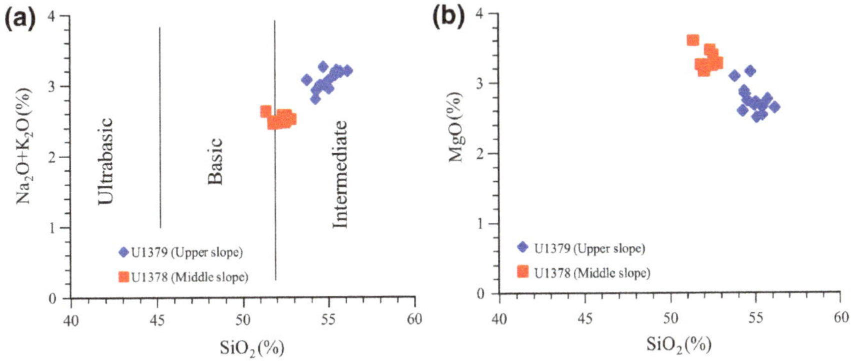

Fig. 10 **a** A modified TAS diagram showing a distinct grouping of two sites with most of the sediments falling between basic to intermediate source rock composition; **b** silica versus Mg variation showing higher Mg and Si in middle slope

(Fig. 10) suggests the presence of smectite. Smectites can be a good sink for Mg and the Mg in sediments can co-vary with smectite/montmorillonite content (Nath et al. 1989). Owing to their smaller grain size (e.g. Thamban et al. 2002), smectites tend to be transported to farther areas than other coarser-sized clay minerals (e.g., Kurian et al. 2008). A-CN-K plots (see Nath et al. 2000; Kurian et al. 2013) help evaluate the provenance as well as the degree of weathering. A-CN-K plot of two sites studied here (Fig. 11) display the grouping of each site, with a progressive transition between two sites. Though, the sampling resolution is sparse, changes in concentrations of key elements follow the visual changes of presence of tephra and

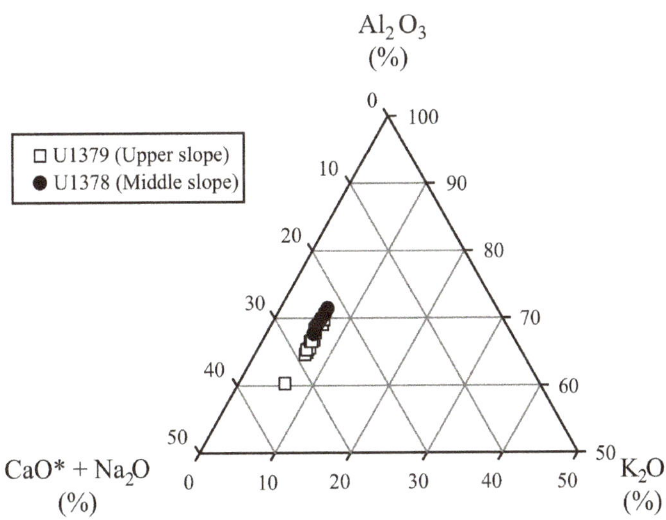

Fig. 11 A-CN-K diagram showing the compositional fields of two sites

mineral grains (Figs. 7 and 8). Distinct change in major element at about 12 and 25–26 kyr BP in upper slope (U1379) coincide with changes noticed in clay mineral content and also characterized by the presence of visible coarser mineral grains at these depths (Figs. 7 and 12) which have been interpreted to represent periods of episodic sedimentation possibly through as turbiditic currents.

3.3 Carbon and Nitrogen Isotopes

A total number of 60 sediment samples together from sites U1379 and U1378 were analyzed for carbon and nitrogen content and their isotope studies and these data are presented in Tables 5 and 6, respectively. The downhole distribution of OC, TN, $\delta^{13}C_{org}$ and $\delta^{15}N$ parameters for the cores from sites U1379 and U1378 are shown in Figs. 13 and 14, respectively. These parameters are plotted together in Fig. 15 for correlation and with respect to age in Fig. 16.

To assess the relative proportions of terrestrial and marine organic carbon present in the sediments a simple $\delta^{13}C$-based two end-member mixing model is used. The model used here is originally proposed by Calder and Parker (1968), modified later (Schultz and Calder 1976); and has been adopted by several workers in the past (e.g., Hu et al. 2006; Schlunz et al. 1999; Ramaswamy et al. 2008). The equation used is as follows.

$$\mathrm{TC}\%_{0} = \delta^{13}C_{\mathrm{marine}} - \delta^{13}C_{\mathrm{org}}/\delta^{13}C_{\mathrm{marine}} - \delta^{13}C_{\mathrm{terrestrial}} \times 100$$

where

TC is the terrestrial carbon
$\delta^{13}C_{org}$ is the measured $\delta^{13}C$ of a given sample
$\delta^{13}C_{marine}$ is the marine end member $= -20.5‰$
$\delta^{13}C_{terrestrial}$ is the terrestrial end member $= -25.1‰$.

The end member values are taken from Ramaswamy et al. (2008).

3.3.1 Site U1379

The $\delta^{13}C_{org}$ ratio varied from −21.32 to −25.45‰ throughout the core. The sediments near surface were distinctly enriched in $\delta^{13}C_{org}$ (~−21.32‰; Fig. 13; Table 5). But a depleting trend of sedimentary $\delta^{13}C_{org}$ with depth down to 3.4 mbsf ($\delta^{13}C_{org}$ ~ −25.45‰; Fig. 13; Table 5) was noticed. The sediments deeper than 3.4 mbsf again showed enriching trend throughout rest of the core. The OC content was very low (<1%) and similar throughout the core, except the top section (0–3.4 mbsf), where the sediments are enriched (>2%). The nitrogen content was below detection limit except for the topmost section. $\delta^{13}C_{org}$ shows excursions at 9–12 mbsf and follows the lithological changes and sedimentary structures. The top 1 m sediment has higher

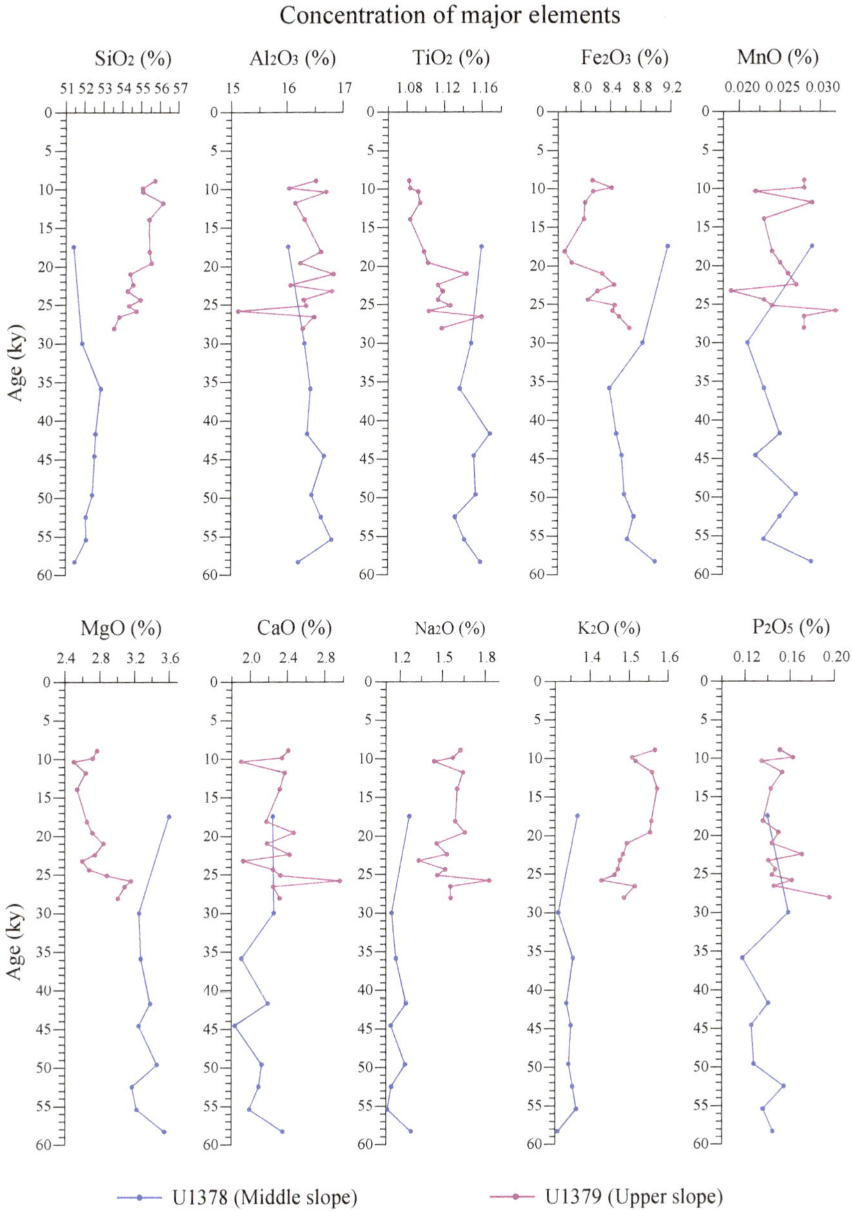

Fig. 12 Major element chemistry for the sites U1378 and U1379 plotted together as a function of age

Table 5 Organic carbon and nitrogen content, their isotope ratios ($\delta^{13}C_{org}$ and $\delta^{15}N$) and the estimated terrestrial carbon (%TC) for the Site U1379 (upper slope)

Sample ID	Depth (mbsf)	Mean depth (mbsf)	$\delta^{13}C_{org}$	$\delta^{15}N$	%OC	%N	Molar C/N ratio	%TC
1H1: 40–43	0.40–0.43	0.42	−21.316	5.537	1.996	0.225	10.3	17.739
1H1: 107–109	1.07–1.09	1.08	−22.008	–	1.026	–	–	32.783
1H1: 126–128	1.26–1.28	1.26	−24.496	–	0.757	–	–	86.870
1H2: 24–26	1.74–1.76	1.75	−24.528	–	0.686	–	–	87.565
1H2: 40–43	1.90–1.93	1.92	−23.882	–	0.581	–	–	73.522
1H3: 40–43	3.40–3.43	3.42	−25.447	–	0.896	–	–	107.543
1H4: 40–43	4.90–4.93	4.92	−24.841	–	0.689	–	–	94.370
1H5: 40–43	6.40–6.43	6.42	−24.377	–	0.755	–	–	84.283
2H1: 100–102	7.70–7.72	7.71	−24.855	–	0.79	–	–	94.674
2H2: 100–102	9.20–9.22	9.21	−23.597	–	0.499	–	–	67.326
2H3: 50–52	10.20–10.22	10.21	−24.301	–	0.633	–	–	82.630
2H3: 100–102	10.70–10.72	10.71	−25.121	–	0.814	–	–	100.457
2H4: 30–32	11.50–11.52	11.51	−25.224	–	0.942	–	–	102.696
2H4: 100–102	12.20–12.22	12.21	−23.758	–	0.504	–	–	70.826
2H5: 30–32	13.00–13.02	13.01	−25.017	–	0.766	–	–	98.196
2H6: 20–22	14.40–14.42	14.41	−23.344	–	0.717	–	–	61.826
3H1: 100–102	17.20–17.22	17.21	−23.846	–	0.667	–	–	72.739
3H2: 100–102	18.70–18.72	18.71	−23.566	–	0.746	–	–	66.652
3H3: 100–102	20.20–20.22	20.21	−23.040	–	0.664	–	–	55.217

(continued)

Table 5 (continued)

Sample ID	Depth (mbsf)	Mean depth (mbsf)	$\delta^{13}C_{org}$	$\delta^{15}N$	%OC	%N	Molar C/N ratio	%TC
3H4: 100–102	21.70–21.72	21.71	−23.107	–	0.662	–	–	56.674
3H5: 100–102	23.20–23.22	23.21	−23.163	–	0.654	–	–	57.891
3H6: 30–32	24.00–24.02	24.01	−23.939	–	0.805	–	–	74.761
4H1: 100–102	25.20–25.22	25.21	−23.534	–	0.73	–	–	65.957
4H2: 30–32	26.00–26.02	26.01	−23.695	–	0.777	–	–	69.457
4H2: 100–102	26.70–26.72	26.71	−23.452	–	0.676	–	–	64.174
4H3: 30–32	27.50–27.52	27.51	−23.385	–	0.777	–	–	62.717
4H3: 116–118	28.36–28.38	28.37	−23.164	–	0.821	–	–	57.913
4H4: 30–32	29.00–29.02	29.01	−22.509	–	0.878	–	–	43.674

C_{org} and enriched isotopic values which grade down significantly to low C_{org} and depleted isotopic values (Fig. 13) in sediments below that where a change in texture and presence of visible mineral grains were noted in onboard lithologs.

Source of OM:

The wide variation of $\delta^{13}C_{org}$ from this core suggests a variation in contribution from different sources. The $\delta^{13}C_{org}$ enriched surface sediments indicate greater present day oceanic contribution for organic matter, which is consistent with low TC content (17.7–32.7%; Table 5 and Fig. 16). The depleted $\delta^{13}C_{org}$ values in subsurface sediments imply a greater contribution from $\delta^{13}C_{org}$ depleted terrestrial source. Depleted values at about 3–4, 7–8, 12 and 13 mbsf (Fig. 13) indicate episodes of increased terrestrial OM deposition or improved preservation. The 12 and 13 mbsf depleted layers sandwich a sandy layer with fining upward sequence and enriched layer (Fig. 13) suggesting a depositional change. At these depths, the end-member calculations show that the organic matter in these sediments is completely terrestrial in nature (Table 5 and Fig. 13). Overall, the TC content in this site is distinctly higher than in U1378, consistent with the proximity of this site to land. Age-wise estimates of TC (Fig. 16) show a distinct deglacial increase, following low values during LGM, suggesting an increased run-off related terrestrial carbon input. Enriched $\delta^{13}C_{org}$ values from 30 to ~15 mbsf (Fig. 13) possibly indicate increased glacial productivity in the overlying waters.

Table 6 Organic carbon and nitrogen content, their isotopic ratios ($\delta^{13}C_{org}$ and $\delta^{15}N$) and estimated terrestrial carbon (%TC) for the Site U1378 (middle slope)

Sample ID	Depth (mbsf)	Mean depth (mbsf)	$\delta^{13}C_{org}$	$\delta^{15}N$	%OC	%N	Molar C/N ratio	%TC
1H1: 68–70	0.68–0.70	0.69	−21.554	6.064	1.669	0.447	4.4	22.913
1H2: 68–70	2.18–2.20	2.19	−21.407	6.684	2.291	0.245	10.9	19.717
1H3: 68–70	3.68–3.70	3.69	−21.486	6.547	1.875	0.195	11.2	21.435
2H1: 68–70	5.98–6.00	5.99	−21.852	7.837	1.843	0.190	11.3	29.391
2H2: 68–70	7. 48–7.50	7.49	−22.105	8.481	1.563	0.161	11.3	34.891
2H3: 68–70	8.98–9.00	8.99	−22.102	6.028	1.743	0.174	11.7	34.826
2H3: 96–98	9.26–9.28	9.27	−22.756	5.316	1.525	0.154	11.6	49.043
2H4: 32–34	10.12–10.14	10.13	−24.260	–	0.744	–	–	81.739
2H4: 39–41	10.19–10.21	10.20	−22.502	6.737	1.342	0.138	11.3	43.522
2H4: 50–52	10.30–10.32	10.31	−22.543	6.838	1.474	0.167	10.3	44.413
2H4: 68–70	10.48–10.50	10.49	−22.245	5.635	1.573	0.191	9.6	37.935
2H4: 99–101	10.79–10.81	10.80	−23.326	5.978	1.164	0.178	7.6	61.435
2H4: 108–110	10.88–10.90	10.89	−22.967	5.696	1.244	0.133	10.9	53.630
2H4: 118–120	10.98–11.00	10.99	−23.086	–	1.185	–	–	56.217
2H5: 40–42	11.70–11.72	11.71	−23.086	–	1.185	–	–	56.217
2H5: 68–70	11.98–12.00	11.99	−22.442	4.048	1.401	0.15	10.9	42.217
2H5: 81–83	12.11–12.13	12.12	−24.282	–	0.546	–	–	82.217
2H5: 109–111	12.39–12.41	12.40	−24.463	–	0.672	–	–	86.152
2H6: 48–50	13.28–13.30	13.29	−23.542	–	0.742	–	–	66.130

(continued)

Table 6 (continued)

Sample ID	Depth (mbsf)	Mean depth (mbsf)	$\delta^{13}C_{org}$	$\delta^{15}N$	%OC	%N	Molar C/N ratio	%TC
2H6: 68–70	13.48–13.50	13.49	−22.551	6.851	1.515	0.144	12.3	44.587
3H1: 65–67	15.45–15.47	15.46	−22.223	6.036	1.678	0.159	12.3	37.457
3H2: 68–70	16.98–17.00	16.99	−22.080	4.374	1.185	0.128	10.8	34.348
3H3: 68–70	18.48–18.50	18.49	−21.735	7.538	1.488	0.150	11.6	26.848
3H3: 116–118	18.96–18.98	18.97	−21.891	4.118	1.638	0.168	11.4	30.239
3H4: 68–70	19.98–20.00	19.99	−21.852	4.754	1.469	0.151	11.3	29.391
3H5: 72–74	21.52–21.54	21.53	−22.154	5.807	1.471	0.151	11.4	35.957
3H6: 67–69	22.97–22.99	22.98	−22.256	5.287	1.453	0.147	11.5	38.174
4H2: 67–69	25.58–25.60	25.59	−21.490	3.700	1.427	0.166	10.0	21.522
4H3: 65–67	27.06–27.08	27.07	−21.734	3.958	1.559	0.177	10.3	26.826
4H3: 119–121	27.60–27.62	27.61	−22.512	0.575	1.302	0.149	10.2	43.739
4H4: 66–69	28.57–28.60	28.58	−22.250	1.018	1.546	0.189	9.5	38.043
4H5: 66–69	30.07–30.10	30.08	−22.401	2.845	1.555	0.198	9.2	41.326

3.3.2 Site U1378

The C_{org} content is 2.3 near surface which is much higher than 0.7% at 10 mbsf depth. Following this, the downcore C_{org} content was similar (1.3–1.5%) throughout the core except at 17 mbsf depth. In general, the $C_{org}$is higher in middle slope than the upper slope (Fig. 14). The nitrogen content varied from 0.13 to 0.45% throughout the core. In general, the surface sediments are rich in nitrogen (~0.4%) compared to the deeper sediments (<0.15%—Figs. 13 and 14). The $\delta^{13}C_{org}$ in this core varied in a narrow range (−21.44 to −22.5‰) except at depth between 9.3 and 13.3 mbsf, where it ranged between −22.25 and −24.46‰ (Fig. 14). The $\delta^{15}N$ in this core varied from 0.5 near bottom to 8.5‰, with enriched values down to about 27 mbsf reflecting probably the denitrification in the overlying waters. C, N content and their

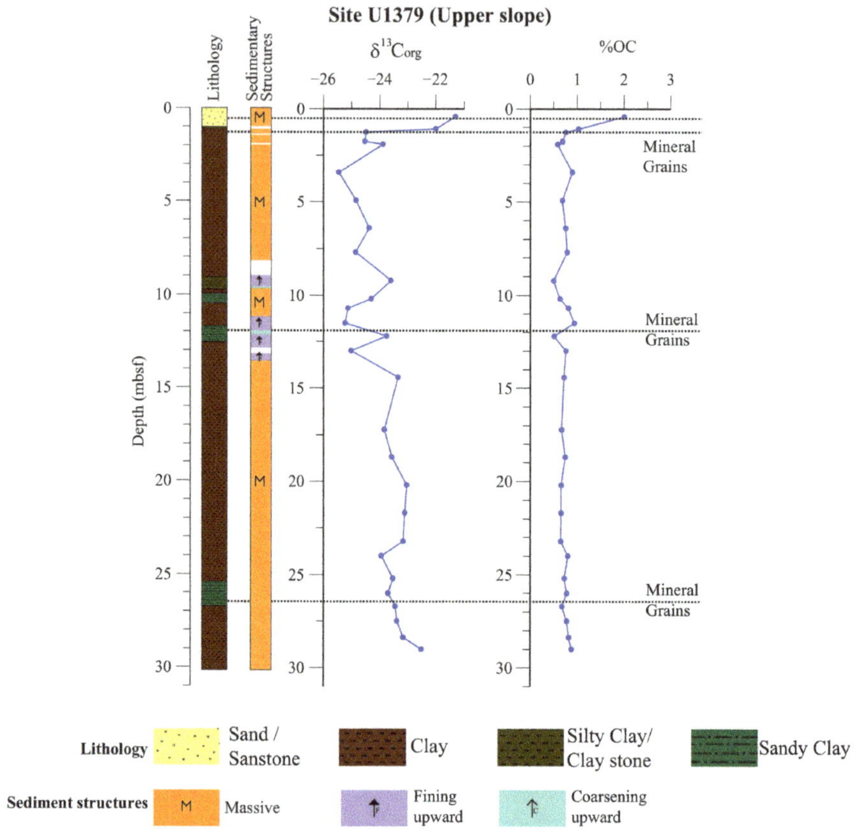

Fig. 13 Downhole variation of organic carbon content and its isotope ratio ($\delta^{13}C_{org}$) for the Site U1379. Also plotted are shipboard described lithology and sedimentary structures. Nitrogen was below detection for this site

isotopes were highly variable at the depths (9–15 mbsf) with a fining upward sequence (Fig. 14) which also contained a tephra layer (Fig. 14). However, no discernible change were seen in these parameters at other tephra layer.

Source of OM:

The enriched $\delta^{13}C_{org}$ values nearly throughout the section studied here except that between 9 and 15 mbsf suggest a dominant marine organic matter contribution which is in contrast to shallower core. The dominance of marine OM was also evidenced from higher $\delta^{15}N$ ratio. The sediments at depth between 9.3 and 13.3 mbsf showed a larger variation in source with episodes of depleted C_{org} and isotopic values reflecting dominant OM contribution from terrestrial sources during last glacial period possibly related to run-off or turbidity currents. The TC computed from $\delta^{13}C_{org}$ values is distinctly less than that in upper slope (Fig. 16). In the 60 kyr record covered by this core, the organic matter of marine origin is dominant and the variation with time is

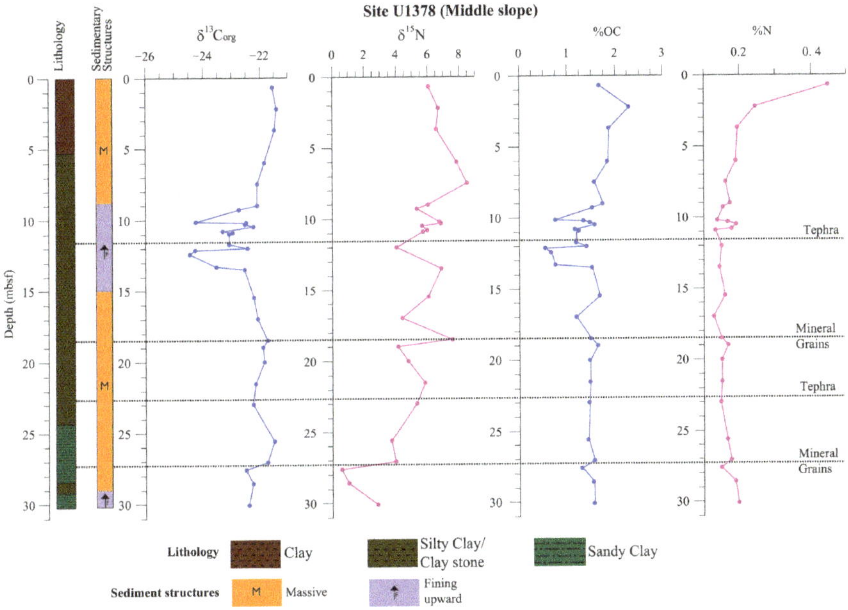

Fig. 14 Downhole variation of organic carbon, nitrogen content and their isotope ratios ($\delta^{13}C_{org}$ and $\delta^{15}N$) for Site U1378 plotted along with the lithology and sedimentary structures

gradual. Except at the LGM where the episodic deposition of TM is seen when the sea level would have been lower and the site would have been closer to the shore, the site has received mainly the organic matter of marine origin. Following this episodic deposition of TM, the TM has reduced since the deglaciation possibly reflecting increased productivity in the overlying waters.

4 Summary and Conclusions

We present the results of clay mineral, major element, carbon, nitrogen content and their isotopes in organic matter for sediment cores retrieved from Costa Rican margin during Integrated Ocean Drilling Program Expedition 334. During this expedition, drilling was carried out at two sites U1379 and U1378 in the upper plate (Caribbean Plate). The Site U1379 is located in the upper slope and the Site U1378 is located in the middle slope. The downhole variation data of clay minerals—smectite, illite and chlorite and their crystallinity are presented in this paper. The clay mineral kaolinite is absent in the analyzed sediment samples. Smectite was found to be the dominant mineral indicating that the source rocks are mainly the young volcanic rocks in central Costa Rica region. Crystallinity and clay mineral content were different in sediment sections showing lithological change such as the presence of tephra, mineral grains

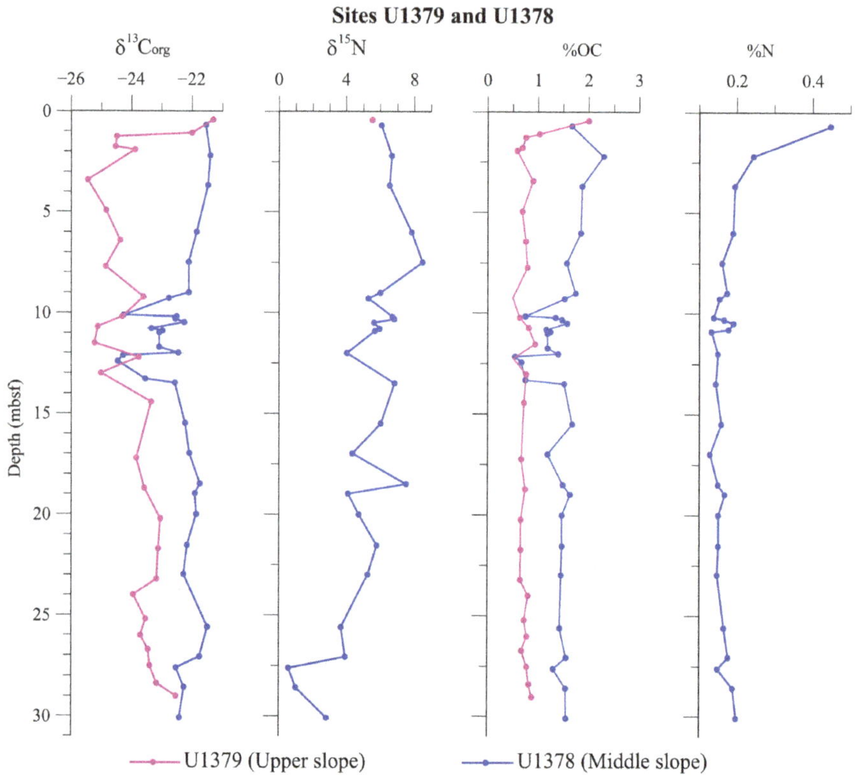

Fig. 15 Downhole variation of organic carbon and nitrogen isotope ratios ($\delta^{13}C_{org}$ and $\delta^{15}N$) for both the sites U1378 and U1379 plotted together for comparison

and with upward fining sequence. Poor crystallinity of clay minerals with fining upward sequence in the middle slope would indicate at least two major episodes of slumping/mass deposition from the shallower areas during the glacial period. Major rock forming elements such as SiO_2, TiO_2, CaO, MgO and Na_2O were distinctly different in the two cores studied here. Geochemical discrimination diagrams allowed the determination of provenance (basic to intermediate in composition) and show the two sites have different composition but within a narrow range mainly controlled by clay mineral sorting during the transport. Major chemical changes were seen at the depths described onboard to have lithological changes. C and N isotopes were also found to be different at depths with lithological changes. While the upper slope contains predominantly terrestrially deposited organic matter, the middle slope with high C_{org} and N content contain mainly marine OM with enriched N isotope indicating denitrification condition. The glacial sediments between 9 and 15 mbsf with fining upward sequence in the core from the middle slope had terrestrially derived carbon. Fining upward sequence, tephra deposition, occurrence of poorly crystalline clay minerals and terrestrially derived carbon in the middle slope core suggests episodes

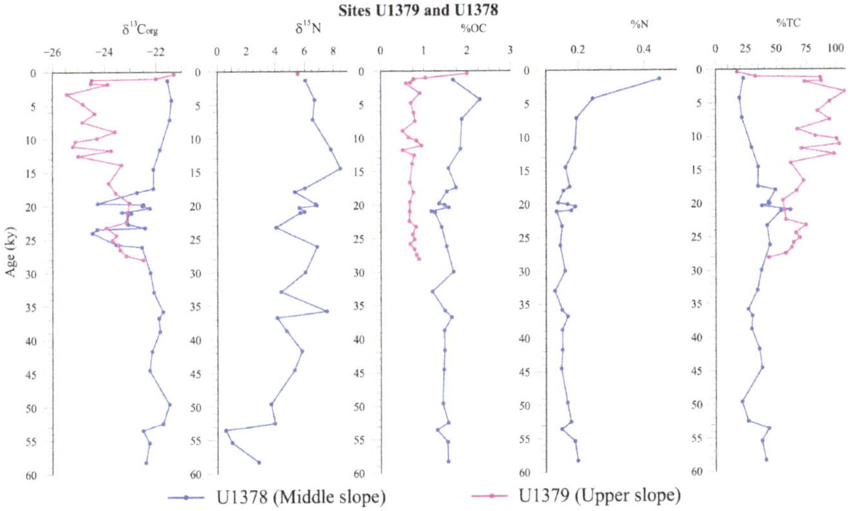

Fig. 16 Organic carbon and nitrogen isotope ratios ($\delta^{13}C_{org}$ and $\delta^{15}N$) for both the sites U1378 and U1379 plotted together as a function of age

of slumping during 25–20 kyr BP. Elemental changes were also noticed in upper slope at about 25 kyr BP reflecting sedimentation changes during glacial period. The slope failure induced events such as turbiditic sedimentation and mass flows are evident from the data presented, mainly during the last glacial period in the Costa Rican margin.

Acknowledgements The present work used samples provided by the Integrated Ocean Drilling Programme (IODP) drilled during IODP Expedition 334. We thank Ministry of Earth Sciences (MoES), Govt. of India and National Centre for Polar and Ocean Research, Goa for the funding support for participation of VY in IODP Expedition 334 and for subsequent onshore data analysis in the form of a project (NCAOR/IODP/17.11/11) under IODP-India Programme. We also thank Director, CSIR-National Institute of Oceanography, Goa and Director, National Centre for Polar and Ocean Research, Goa, for their kind support. VY is grateful to the entire IODP Expedition 334 team for their help and support received onboard. We are thankful to an anonymous reviewer for his encouraging remarks and the Editors for their support. We gratefully acknowledge the technical support received from Gireesh Prabhu for XRD analysis, Caesar Moraes for XRF analysis and Anita Garg for IRMS analysis. Tyson Sebastian has drawn the A-CN-K diagram. All the figures are plotted with Generic Mapping Tools (Wessel and Smith 1995).

References

Calder JA, Parker PL (1968) Stable carbon isotope ratios as indices of petrochemical pollution of aquatic systems. Environ Sci Technol 2:535–539
Chamley H (1989) Clay sedimentology. Springer, New York 623 pp.

Dutch S (2009) Web page: Geology of Costa Rica. https://wwww.uwgb.edu/dutchs/CostaRica2008/CRGeology.HTM. Accessed 09 Jan 2018

Expedition 334 Scientists (2011) Costa Rica Seismogenesis Project (CRISP): sampling and quantifying input to the seismogenic zone and fluid output. IODP Preliminary Report 334. https://doi.org/10.2204/iodp.pr.334.2011

Expedition 334 Scientists (2012a) Expedition 334 summary. In: Vannucchi P, Ujiie K, Stroncik N, Malinverno A, the Expedition 334 Scientists (eds) Proceedings of IODP334, Tokyo. https://doi.org/10.2204/iodp.proc.334.101.2012

Expedition 334 Scientists (2012b) Site U1378. In: Vannucchi P, Ujiie K, Stroncik N, Malinverno A, the Expedition 334 Scientists (eds) Proceedings of IODP334, Tokyo. https://doi.org/10.2204/iodp.proc.334.103.2012

Expedition 334 Scientists (2012c) Site U1379. In: Vannucchi P, Ujiie K, Stroncik N, Malinverno A, the Expedition 334 Scientists (eds) Proceedings of IODP334, Tokyo. https://doi.org/10.2204/iodp.proc.334.104.2012

Hu J, Peng P, Jia G, Mai B, Zhang G (2006) Distribution and sources of organic carbon, nitrogen and their isotopes in sediments of the subtropical Pearl River estuary and adjacent shelf, Southern China. Mar Chem 98:274–285

Islebe GA, Hooghiemstra H (1997) Vegetation and climate history of montane Costa Rica since the last glacial. Quat Sci Rev 16:589–604

Kurian S, Nath BN, Ramaswamy V, Naman D, Gnaneshwar Rao T, Kamesh Raju KA, Selvaraj K, Chen CTA (2008) Possible detrital, diagenetic and hydrothermal sources for Holocene sediments of the Andaman backarc basin. Mar Geol 247:178–193

Kurian S, Nath BN, Chandramohana Kumar N, Nair KKC (2013) Geochemical and isotopic signatures of surficial sediments from the western continental shelf of India: inferring provenance, weathering, and the nature of organic matter. J Sediment Res 83(6):427–442

Lamy F, Hebbeln D, Wefer G (1999) High-resolution marine record of climatic change in mid-latitude Chile during the last 28,000 years based on terrigenous sediment parameters. Quat Res 51:83–93

Nath BN, Rao VP, Becker KP (1989) Geochemical evidence of terrigenous influence in deep-sea sediments up to 8°S in the Central Indian Basin. Mar Geol 87:301–313

Nath BN, Kunzendorf H, Plüger WL (2000) Influence of provenance, weathering, and sedimentary processes on the elemental ratios of the fine-grained fraction of the bedload sediments from the Vembanad Lake and the adjoining continental shelf, southwest coast of India. J Sediment Res Sect A: Sediment Petrol Process 70(5):1081–1094

Ramaswamy V, Gaye B, Shirodkar PV, Rao PS, Chivas AR, Wheeler D, Thwin S (2008) Distribution and sources of organic carbon, nitrogen and their isotopic signatures in sediments from the Ayeyarwady (Irrawaddy) continental shelf, northern Andaman Sea. Mar Chem 111:137–150

Schlunz B, Schneider RR, Muller PJ, Showers WJ, Wefer G (1999) Terrestrial organic carbon accumulation on the Amazon deep sea fan during the last glacial sea level low stand. Chem Geol 159:263–281

Schubert CJ, Nielsen B (2000) Effects of decarbonation treatments on $\delta^{13}C$ values in marine sediments. Mar Chem 72:55–59

Schultz D, Calder JA (1976) Organic carbon $^{13}C/^{12}C$ variations in estuarine sediments. Geochim Cosmochim Acta 40:381–385

Smith WHF, Sandwell DT (1997) Global sea floor topography from satellite altimetry and ship depth soundings. Science 277:1956–1962

Thamban M, Purnachandra Rao V, Schneider RR (2002) Reconstruction of late Quaternary monsoon oscillations based on clay mineral proxies using sediment cores from the western margin of India. Mar Geol 186:527–539

Vannucchi P, Sak PB, Morgan JP, Ohkushi K, Ujiie K, Expedition IODP, IODP Expedition 334 Shipboard Scientists (2013a) Rapid pulses of uplift, subsidence, and subduction erosion offshore Central America: implications for building the rock record of convergent margins. Geology 41:995–998

Vannucchi P, Ujiie K, Stroncik N, the IODP Expedition 334 Scientific Party (2013b) IODP Expedition 334: an investigation of the sedimentary record, fluid flow and state of stress on top of the seismogenic zone of an erosive subduction margin. Sci Drill 15:23–30

von Huene R, Bialas J, Flueh E, Cropp B, Csernok T, Fabel E, Hoffmann J, Emeis K, Holler P, Jeschke G, Leandro MC, PerézFernandéz I, Chavarria SJ, Florez HA, Escobedo ZD, León R, Barrios LO (1995) Morphotectonics of the Pacific convergent margin of Costa Rica. In: Mann P (ed) Geologic and tectonic development of the Caribbean Plate boundary in Southern Central America. Special Paper 295, Geological Society of America, pp 291–308

von Huene R, Ranero CR, Weinrebe W, Hinz K (2000) Quaternary convergent margin tectonics of Costa Rica, segmentation of the Cocos Plate, and Central American volcanism. Tectonics 19:314–334

von Huene R, Ranero CR, Watts P (2004) Tsunamigenic slope failure along the Middle America Trench in two tectonic settings. Mar Geol 203:303–317

Wessel P, Smith WHF (1995) New version of the Generic Mapping Tools released. EOS Trans AGU 76:329

mtCOI Sequence-Based Barcoding of Calanoid Copepods from Lagoon Waters of Lakshadweep, South-west Coast of India

Sanu V. Francis and S. Bijoy Nandan

Abstract Coastal waters of Southwest coast of India including the lagoon waters of Lakshadweep are known to be vital habitat to a wide variety of zooplankton groups by their maximum diversity, and complexity of the trophic organisation. Copepods are the most abundant crustaceans in the world oceans, usually contributing >60% of the total zooplankton abundance. DNA barcode data based on mtCOI sequences for the marine calanoid copepods (111 sequences for 25 species) were documented which was not previously sequenced from the lagoon waters of Lakshadweep, South-eastern Arabian Sea. This work constitutes a powerful tool for identification with minimal errors not only of copepods in the Arabian Sea but in the adjoining marine areas; the results are the first step in building databases of sequences from the south-west coast of India.

Keywords Calanoid copepods · DNA-barcoding · Lagoon waters · mtCOI sequences

1 Introduction

Zooplankton forms an essential component of the pelagic food web in the ocean system. Calanoid copepods are the dominant groups in marine subtropical and tropical waters and exhibit great diversity in morphology and habitats they occupy. Most of the studies in the Arabian Sea have elaborated on the ecology and diversity of calanoids; however, there is a severe lacunae of information on the morpho-taxonomy and molecular systematics of copepods which is a prerequisite for understanding the community ecology of the species and their role in the food web structure. Knowledge on the taxonomy and distribution of calanoid copepods of the coastal waters of the Arabian Sea and lagoon waters of Lakshadweep archipelago are minimal. Copepod crustaceans living in the oceanic realm have received much attention from

S. V. Francis · S. Bijoy Nandan (✉)
Department of Marine Biology, Microbiology and Biochemistry, Cochin University of Science and Technology, Cochin 682016, India
e-mail: bijoynandan@yahoo.co.in

D. K. Pandey et al. (eds.), *Dynamics of the Earth System: Evolution, Processes and Interactions*, Society of Earth Scientists Series,
https://doi.org/10.1007/978-3-030-40659-2_10

molecular biologists, as they are difficult to identify because of their similar morphological characteristics, especially in closely related species. Its great diversity needs accuracy on species identification, because, in some cases, copepods share morphometric and meristic characteristics that tend to confuse the identity of the species.

It is also mentioned that co-occurrence of multiple closely related species makes it difficult to determine their female-male correspondence by morphology (Nishida et al. 2015; Mulyadi et al. 2017; Francis and Nishida 2018; Francis et al. 2018). Hence, molecular techniques are now an essential tool for studies on population structure, phylogeography, and phylogeny of copepods (Sanu et al. 2016). Molecular analysis can provide unambiguous taxonomic discrimination (Bucklin and Wiebe 1998; Lee and Frost 2002; Eyun et al. 2007; Soh et al. 2012), and sequence analysis of mitochondrial genes can reveal intraspecific and interspecific variations (Bucklin et al. 1995; Bucklin and Wiebe 1998; Goetze 2003). Studies were conducted from 2013 to 2016 as part of the Dept. of Biotechnology, Govt. of India funded project on the taxonomy and systematics of pelagic copepods from the lagoon waters of Agatti, Bangaram, Kavaratti, Kalpeni and Minicoy islands of Lakshadweep. The present study documents the molecular systematics of the calanoid copepods from the study area. To document the molecular barcodes, mitochondrial cytochrome C oxidase 1 (mtCOI) sequences were developed for several of the species (for both female and male individuals) and contribute to mainstreaming a barcode library.

2 Materials and Methods

The zooplankton samples were collected from lagoon waters (Fig. 1) by using a plankton net (mesh size 200 μm; mouth area 0.28 m^2). The average depth of the collection site is 6 m with a maximum depth of about 16 m. The net was towed horizontally just below the water surface at a speed of 1 knot for 10 min. For morphological and molecular analysis, the samples were fixed in 4% buffered formalin and 95% Ethyl alcohol, respectively. Formalin preserved specimens were identified to species level using standard keys (Mori 1964; Kasturirangan 1963; Silas and Pillai 1973; Conway et al. 2003; Prusova et al. 2012). The corresponding alcohol-preserved specimens were hydrated in 0.5-mL sterile distilled water for 10–12 h at room temperature before extraction of DNA (Sanu et al. 2016). Genomic DNA was extracted from individual adult copepods using the DNeasy Blood and Tissue Kit (Qiagen) following the spin column protocol. The PCR reaction mixture consisted of 25 μL Master Mix (Takara Clontech EmeraldAmp® GT PCR Master Mix), 1 μL forward primer, 1 μL reverse primer, 8 μL template DNA, and 15 μL distilled deionised H_2O. The amplification primers were LCO-1490 F (5′-GGTCAACAAATCATAAAGATATTGG-3′) and HCO-2198 R (5′-TAAACTTC AGGGTGACCAAAAAATCA-3′) used for amplifying mtCOI gene sequences (Folmer et al. 1994). Amplification was carried out in

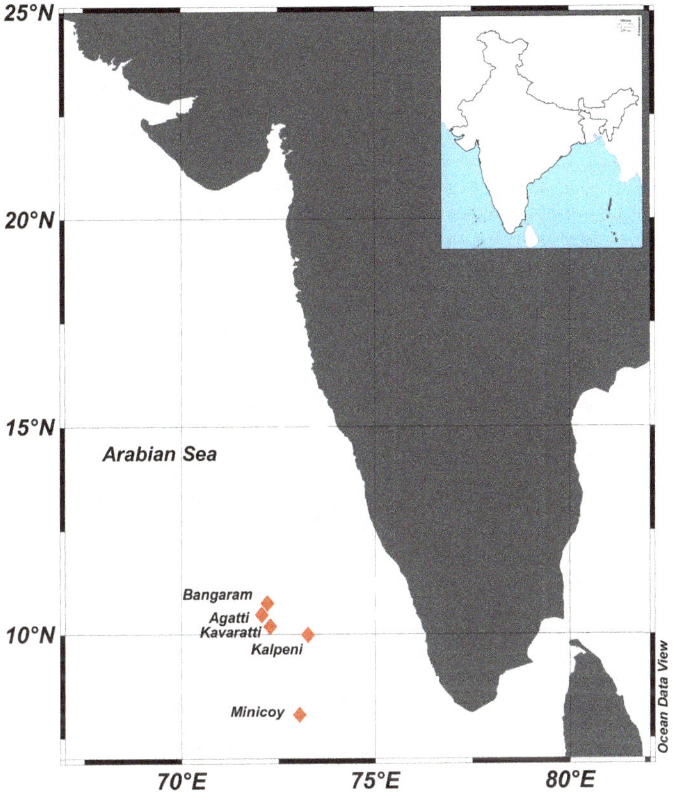

Fig. 1 Sampling locations in Lakshadweep islands during 2013–2016

Agilent technologies thermal cycler (Model no: Sure cycler 8800). The amplification protocol was denaturation at 94 °C for 1 min., annealing at 37 °C for 2 min. and extension at 72 °C for 3 min. The amplification was carried through 40 cycles. Amplified products exhibiting intense bands after agarose gel (1.2%) electrophoresis was purified and sent to SciGenom Labs (SciGenom Labs Pvt, Ltd. Ernakulam, India) for sequencing. Obtained sequences were assembled using BioEdit 7.0.9 (Hall 1999), and alignment was performed using ClustalX (Thompson et al. 1997). Intraspecific pairwise sequence distance and Neighbour-joining (NJ tree) were undertaken using the Kimura 2-parameter model in MEGA5 (Tamura et al. 2011). Bootstrap analysis was performed using 1000 pseudo-replications. Intraspecific Aligned sequences were submitted to the National Center for Biotechnology Information (NCBI) database (www.ncbi.nlm.nih.gov).

3 Results

3.1 Molecular Records and Systematics of the Calanoid Copepods

Mitochondrial COI sequences of 25 copepod species (both female and male individuals) were obtained and submitted in the National Centre for Biotechnology Information (NCBI) database. Among the identified species, mtCOI sequences of *Euchaeta indica* Wolfenden 1905, *Euchirella sp., Candacia catula* Giesbrecht 1888, *C. truncata* Giesbrecht 1888, *Labidocera kroyeri* Brady 1883, *L. acuta* Dana 1849, *L. bataviae* A. Scott 1909, *L. detruncata* Dana 1849, *L. madurae* A. Scott 1909, *L. minuta* Giesbrecht 1889, *Pontella denticauda* A. Scott 1909, *P. spinipes* Giesbrecht 1889, *P. fera* Dana 1849, *P. sinica* Chen and Zhang 1965, *pontellina plumata* Dana 1849, *Pontellopsis armata* Giesbrecht 1889, *Tortanus sp.,* and *Acartia bispinosa* Carl 1907 were represented the first molecular barcode for the species in the NCBI database (Table 1). Before the present study, mtCOI sequences of *Undinula vulgaris* Dana 1849, *Canthoclanus pauper* Giesbrecht 1888, *Subeucalanus subcrassus* Giesbrecht 1888, *Euchaeta concinna* Dana 1849, *Centropages furcatus* Dana 1849, *Temora discaudata* Giesbrecht 1889 and *Calanopia thompsoni* A. Scott 1909 were reported from other marine and coastal environments (Table 2). A total of 111 sequences were developed during the study period from which more than 99% had more than 500 base pairs. Most of the calanoid copepod sequences were developed from the lagoon waters of Lakshadweep islands, that belonged to 10 families and 15 genera. Calanidae, Eucalanidae, Euchaetidae, Aetidae, Centropagidae, Temoridae, Candacidae, Pontellidae, Tortanidae and Acartidae were the major families with barcodes developed in which the family Pontellidae comprises 14 species and five genera (*Calanopia, Pontella, Labidocera, Pontellina* and *Pontellopsis*) (Table 1).

The Neighbour-joining tree of similarity using K2P distance shows the clustering of calanoid copepods (Fig. 2). The mtCOI sequence of molluscan species *Mytilopsis sallei* was selected as an outgroup. The average distance (K2P) among the species was 0.5%, and the divergence between genus 25.63% (Table 3); the highest frequency of intraspecific divergence within species was between 0 and 2%, and the variation between genera, all had more than 17%. All the species were clustered according to bootstrap values. Individuals in the same species got arrayed in a single clade with 100% bootstrap values and the average intraspecific divergence 0.5% within the species adequate to confirm the female-male correspondence and the genetic identity of the species from congeners and other copepods species from different families.

Table 1 COI sequences of calanoid species published in the NCBI database with accession numbers (www.ncbi.nlm.nih.gov) accession number for female and male individuals represented by symbols

Species	Accession number	No.
Order Calanoida		
Family: Calanidae		
Undinula vulgaris Dana, 1849	KJ940173♀ KJ940174♂	2
Canthocalanus pauper Giesbrecht, 1888	KX831912♀ KX831913♂	2
Family: Eucalanidae		
Subeucalanus subcrassus Giesbrecht, 1888	KP842700♀ KP842701♀ KP842702♂, KP842703♂	4
Family: Euchaetidae		
Euchaeta concinna Dana, 1849	KP749945♀ KP749946♀ KP749947♀ KP749948♂, KP749949♂	5
Euchaeta indica Wolfenden, 1905[a]	KY421045♀ KY421046♂	2
Family Aetideidae		
Euchirella sp[a].	KT800383♀	1
Family: Centropagidae		
Centropages furcatus Dana, 1849	KJ561888♀	1
Family: Temoridae		
Temora discaudata Giesbrecht, 1889	KJ940171♀ KJ940172♂	2
Family: Candacidae		
Candacia catula Giesbrecht, 1889[a]	KP068662♀KP068663♀ KP068664♀ KP068665♂, KP068666♂	5
Candacia truncata Dana, 1849[a]	KX831914♀ KX831915♂	2
Family: Pontellidae		
Calanopia thompsoni A. Scott, 1909	KP068656♀ KP068657♀ KP068658♂, KP068659♂	4
Labidocera acuta Dana, 1849[a]	KJ940166♀ KJ940167♀ KJ940168♂, KJ940169♂, KJ940170♂	5
Labidocera bataviae A. Scott, 1909[a]	KU881728♀ KU881729♀ KU881730♂, KU881731♂	4
Labidocera detruncata Dana, 1849[a]	KU881732♀ KU881733♂	2
Labidocera kroyeri Brady, 1883[a]	KP068677♀ KP068678♀	2
Labidocera madurae A. Scott, 1909[a]	KJ701546♀ KJ852655♀ KJ852656♀ KJ911917♀ KJ933427♂, KJ933428♂, KJ933429♂.	7
Labidocera minuta Giesbrecht, 1889[a]	KP068667 ♀ KP068668♀ KP06869♀ KP068670♂, KP068671♂	5
Pontella denticauda A. Scott, 1909[a]	KT800377♀ KT800378♀ KT800379♀ KT800380♀ KT800381♀	5

(continued)

Table 1 (continued)

Species	Accession number	No.
Pontella diagonalis sensu Silas and Pillai, 1973[a]	KT282363♂, KT282364♂, KT282365♂, KT282366♂, KT282367♂, KT282368♂, KT282369♂, KT282370♂, KT282371♂, KT282372♂	10
Pontella spinipes Giesbrecht,1889[a]	KT267166♀ KT267167♀ KT267168♀ KT267169♀ KT267170♀ KT186887♀ KT186888♀ KT186889♀ KT186890♀ KT186891♀	10
Pontella fera Dana, 1849[a]	KT186882♀ KT186883♀ KT186884♀ KT186885♀ KT186886♂	5
Pontella sinica Chen and Zhang, 1965[a]	KT 336554♂, KT 336555♂, KT 336556♂, KT 336557♂, KT 336558♂, KT336559♂, KU881726♀	7
Pontellina plumata Dana, 1849[a]	KP068679 ♂, KP068680♂, KP068681♀	3
Pontellopsis armata Giesbrecht, 1889[a]	KT186892♀ KT186893♀ KT186894♂, KT186892♂, KT186896♂	5
Family: Tortanidae		
Tortanus sp[a]. (Francis and Bijoy Nandan 2019)	KP749950♂, KP749951♂, KP749952♂, KP749953♂, KP749954♀ KP749955♀	6
Family: Acartidae		
Acartia bispinosa Carl, 1907[a]	KP068672♀ KP068673♀ KP068674♀ KP068675♂, KP068676♂	5

[a]Primary submissions in the NCBI database

3.2 Barcode Library Developed for the Calanoid Copepods

The present study provides 111 new barcodes from the lagoon waters of Lakshadweep for 25 calanoid copepod species. The developed barcodes increase the geographical coverage of existing records and could use as a comprehensive library of DNA barcode sequences for lineages of interest. These new data were analysed with previously published sequences in the NCBI database. From the family Calanidae Dana 1849, COI barcode developed for *Undinula vulgaris* and *Canthocalanus pauper*. For *U. vulgaris*, the sequence has 636 base pairs length (bp) (KJ940173 and KJ940174), 654 bp for *C. pauper* (KX831912 and KX831913) and 100% similarity with previously submitted data of the species in the NCBI database. The intraspecific divergence was 0% for both the species. From the family Eucalanidae Giesbrecht, 1893 four sequences with 642 bp (KP842700 to KP842703) developed for *Subeucalanus subcrassus*. The sequences blasted in NCBI database with 99% similarity with previously submitted data of the species. The intraspecific divergence of the species ranged from 0 to 0.6% which is adequate to female-male correspondence. From the family Euchaetidae Giesbrecht, 1893 five sequences (KP749945 to KP749949) of 643 bp were developed for *Euchaeta concinna* and two sequences of

Table 2 COI sequences of calanoid species published before the present submissions in the NCBI database with accession numbers (www.ncbi.nlm.nih.gov)

Species	Accession number	Submitted from
Order Calanoida		
Family: Calanidae		
Undinula vulgaris Dana, 1849	KF977370 to KF977373 KF985239 GU171332 to GU171335	China seas USA Atlantic Ocean
Canthocalanus pauper Giesbrecht, 1888	EU856809, EU856810	Chinese coastal waters
Family: Eucalanidae		
Subeucalanus subcrassus Giesbrecht, 1888	KF977345 to KF977352	China seas
Famil: Euchaetidae		
Euchaeta concinna Dana, 1849	KC287636 KF977273, KF977274	USA China seas
Family: Centropagidae		
Centropages furcatus Dana, 1849	DQ889142	UK
Family: Temoridae		
Temora discaudata Giesbrecht, 1889	KC287892	USA
Family: Pontellidae		
Calanopia thompsoni A. Scott, 1909	KF977243 to KF977248	China seas

498 bp (KY421045 and KY421046) developed for *Euchaeta indica*. NCBI- BLAST (Basic Local Alignment Search Tool) result showed that *E. concinna* with 99% and *E. indica* with 97% similarity with the available data in the NCBI. The intraspecific divergence of both the species was 0%.

The 630 bp sequence (KT800383) developed for *Euchirella sp.* under the family Aetideidae Giesbrecht, 1892. The species was morphologically similar to *Euchirella pulchra* Lubbock 1856. However, the NCBI BLAST result showed that the present possessed 95% similarity with *E. splendens* Vervoort 1963 and only 91% with *E. pulchra* which means the present specimen not identical to both the species. Since *E. splendens* is not reported yet from the Indian Ocean, there is a possibility of the species may a new or some other sister species under the genus *Euchirella* recorded from the Indian Ocean. A 480 bp sequence developed (KJ561888) for the species *Centropages furcatus* under the family Centropagidae. BLAST results indicated that the present sequence 99% similar to the previously submitted data of the species from other marine waters.

Two sequences of 617 bp (KJ940171, KJ940172) developed for the species *Temora discaudata*, under the family Temoridae Giesbrecht 1893. The sequence has 0% interspecific divergence, and the BLAST result indicated that the present sequence 100% similar to the previously submitted sequences. Five sequences of 645 bp (KP068662 to KP068666) developed for the species *Candacia catula* and two sequences of 651 bp (KX831914 and KX831915) for *Candacia truncata* under the

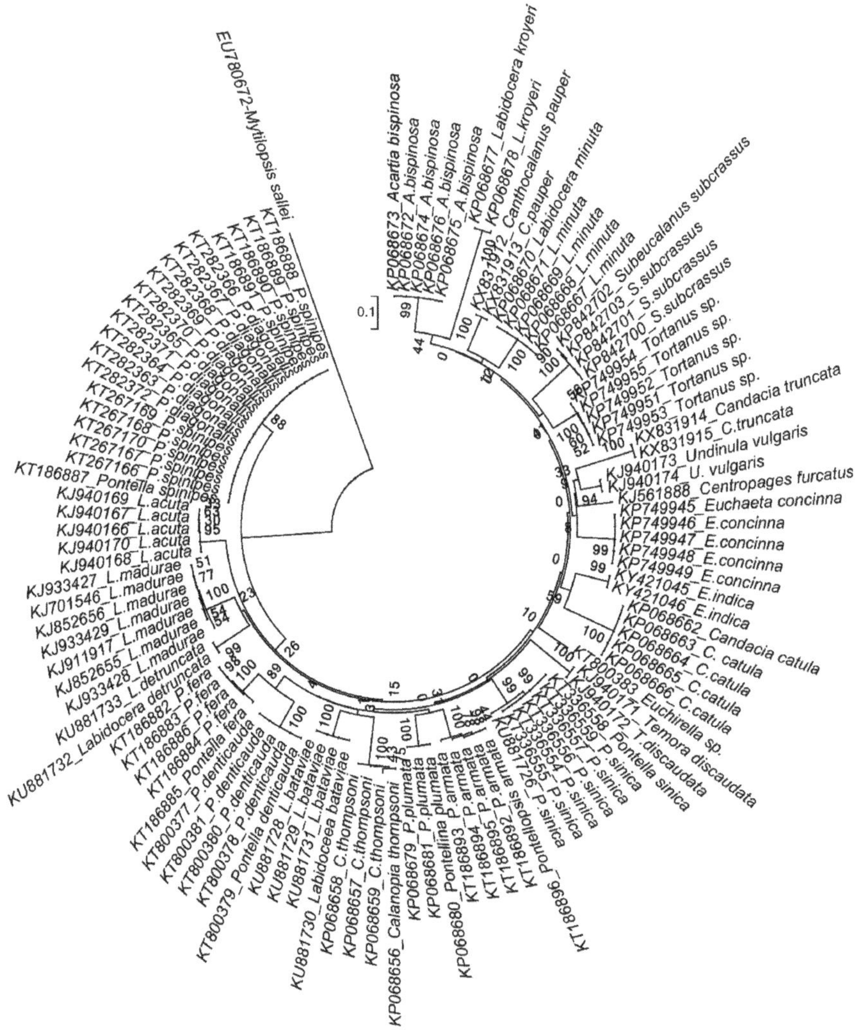

Fig. 2 Neighbour-joining tree for 25 pelagic calanoid copepod species (maximum likelihood, bootstrap = 1000), using K2P distances

Table 3 Genetic divergences (K2P) at different taxonomic levels

Comparison within	Taxa	No. of comparisons	Minimum (%)	Maximum (%)	Mean distance (%)
Species	25	111	0.1	2	0.5
Genus	16	50	17.6	49.2	25.63

family Candacidae Giesbrecht, 1893. The sequences developed for the two species were the first time submission in the NCBI database. The intraspecific distance of the male and female specimen was 0%.

The mtCOI sequences developed for the species under the family Pontellidae *Pontella denticauda, Labidocera acuta, L. bataviae, L. detruncata, L. kroyeri, L. madurae, L. minuta, P. diagonalis, P. spinipes, P. fera, P. sinica, Pontellina plumata, Pontellopsis armata* were primary submission in NCBI database. Four sequences of 627 bp (KP068656 to KP068659) developed for the species *Calanopia thompsoni* has 0.2–0.8% interspecific divergence. For the species, *L. acuta* five sequences of 620 bp (KJ940166 to KJ940170) were generated. The intraspecific divergence within the species was 0–0.3%. Four sequences of 618 bp (KU881728 to KU881731) generated for the species *Labidocera bataviae*. The intraspecific divergence within the species was 0%. Two sequences of 618 bp (KU881732 and KU881733) developed for the species *L. detruncata*. For the species, *L. kroyeri* two sequences of 621 bp (KP068677 and KP068678) were developed. For *L. madurae*, seven sequences developed, and the sequence lengths ranged from 629 to 642 bp. (KJ701546 to KJ933428 and KJ933429). Intraspecific distance of *L. madurae* sequences developed in the present study ranged from 0.2 to 0.9%. For the species *L. minuta*, five sequences of 627 bp developed (KP068667 to KP068671) and the intraspecific distance was 0%.

The developed sequences of female *P. spinipes* and male *P. diagonalis* sensu Silas and Pillai (1973) were submitted to the NCBI database and assigned the following accession numbers: KT186887 to KT186891 and KT267166 to KT267170 for female *P. spinipes*, KT282363 to KT282372 for male *P. diagonalis*. The base pair length for the developed sequences was 639 bp for the male *P. diagonalis*, and 660 bp for the female *P. spinipes*. The female *P. spinipes* and male *P. diagonalis* exhibited 0–0.2% intraspecific sequence divergence. For the species *Pontella denticauda*, five sequences of 618 bp (KT800377 to KT800381) developed and the intraspecific divergence within the species was 0%. Five sequences of 620 bp (KT186882 to KT186886) developed for the species *P. fera*. The intraspecific divergence within this species ranged from 0 to 0.2%. For *P. sinica*, six sequences of 618 bp were developed (KT336554 to KT336559, and KU881726). Intraspecific divergence within the species was 0%. For the species *Pontellina plumata* three sequences of 633 bp were developed (KP068679, KP068680, and KP068681) with 0% intraspecific sequence divergence. Five sequences of 618 bp (KT186892 to KT186896) developed for the species *Pontellopsis armata* showed and intraspecific sequence divergence ranged from 0 to 0.8% in the present study. The developed sequences of female and male *Tortanus sp*. under the genus *Tortanus* Giesbrecht, 1898 were submitted to the NCBI database and assigned the following accession numbers: KP749951 to KP749953 for males and KP749954 and KP749955 for females. The level of intraspecific divergence of *Tortanus* sp. ranged from 0 to 0.5%. Five sequences of 630 bp (KP068672 to KP068676) for the species *Acartia bispinosa* under the family Acartidae Sars 1846 were developed with 0% intraspecific sequence divergence.

4 Discussion

The growing use of DNA barcodes to discriminate and identify marine animal species has included many studies on zooplankton and some studies of planktonic copepods (Bercial et al. 2011). This study presents the results of a comparative analysis of 111 barcode sequences of 25 calanoid species including 85 barcode sequences of 18 calanoid copepod species not included in any previous study. The evaluation comprises Maximum Likelihood and Neighbour Joining automated tree-based and BLAST methods. In the present study, the identification of species based on morphological characteristics was clear for most of the species. The genetic sequence of the species alone does not identify species; a thorough morphological revision is an essential prerequisite to get the correct identification. Morphological identification of calanoid species coupled with the obtained COI gene and its corresponding sequences without any overlap between species. These sequences were the first time attempt and are the start of a library of barcodes of copepods in the Southwest coast of India. The speciation of sequences exhibited in the likelihood analysis showed an intraspecific divergence of 0–0.5%. The Neighbour-joining tree of similarity using K2P distance shows the clustering of calanoid copepods. The average distance (K2P) among the species was 0.5%. Individuals in the same species got arrayed in a single clade with 100% bootstrap values and the average intraspecific divergence 0.5% within the species adequate to confirm the female-male correspondence and the genetic identity of the species from congeners and other copepods species from different families. With regard to mtCOI gene sequence data, which is well accepted and under the threshold value of 4% for calanoid copepods (Bucklin et al. 2010; Sanu et al. 2016; Francis et al. 2018) and is also adequate to confirm the female-male correspondence of the species.

A criticism of metazoan barcoding is the reliance on a single gene, rather than multiple molecular markers. Results obtained from additional genes do not always yield the same results, and caution advised when using only one or few genetic markers. It is possible that there may be better regions for species assignment, and more extended sequences do provide better accuracy and reliability (Roe and Sperling 2007), but the studies by Bercial et al. (2011) proved that even very short COI fragments (<150 bp) show acceptable levels of accuracy for species identification. Another criticism is that if pseudogenes were amplified; this is suspected when two or more sequences are amplified from a single specimen. However, since only a single sequence was amplified from all the samples examined, there is virtually no possibility of pseudogene amplification (Francis et al. 2018). Because of the relatively rapid evolution of the CO1 gene, its variation may help solve taxonomic problems associated with identification, occurrence and female-male correspondence. The molecular barcodes generated in the present study elucidate many taxonomic ambiguities related to the identification and female-male correspondence of the calanoid species (Sanu et al. 2016; Francis and Nishida 2018; Francis et al. 2018; Francis and Bijoy Nandan 2019).

The ability to understand the dynamics of the plankton community depends on the ability to accurately measure the diversity of species and to accurately identify individuals' species morphologically similar. This is an important study made over the lagoon waters of Lakshadweep after the IIOE (1962–65). The present study provides an authentic database for future studies on copepods and related fishery oceanographic studies in the lagoon and coastal waters of the south-west coast of India. The DNA Barcode generated in the present study will be a powerful tool for identification with minimal errors not only of copepods in the Arabian Sea but in the adjoining marine areas.

Acknowledgements This study was funded by the Department of Biotechnology (DBT) Govt. of India [BT/PR4258/AAQ/3/575/2011]. The authors acknowledge for the assistance rendered by the Department of Science and Technology, Lakshadweep. Thanks also to the research fellowship provided to SVF under the University Grants Commission Basic Science Research programme of Govt. of India.

References

Bercial BL, Bradford-Grieve J, Bucklin A (2011) Molecular phylogeny of the calanoida (Crustacea: Copepoda). Mol Phylogenet Evol 59(1):103–113

Bucklin A, Wiebe PH (1998) Low mitochondrial diversity and small effective population sizes of the copepods *Calanus finmarchicus* and *Nannocalanus minor*: possible impact of climatic variation during recent glaciation. J Hered 89:383–392

Bucklin A, Frost BW, Kocher TD (1995) Molecular systematics of six calanus and three Metridia species (Calanoida: Copepoda). Mar Biol 121:655–664

Bucklin A, Hopcroft RR, Kosobokova KN, Lisa MN, Brian DO, Robert MJ, Christopher JS (2010) DNA barcoding of Arctic Ocean holozooplankton for species identification and recognition. Deep Sea Res Pt II 57:40–48

Conway VP, Rowena GW, Joanna HDC, Christopher PG, David B (2003) Guide To the coastal and surface zooplankton of the South-Western Indian Ocean. Mar Biol Ass UK Occasional Publication No. 15

Eyun S, Lee Y, Suh HL, Kim S, Soh HY (2007) Genetic identification and molecular phylogeny of Pseudodiaptomus species (Calanoida, Pseudodiaptomidae) in Korean waters. Zool Sci 24:265–271

Folmer O, Black M, Hoeh W, Lutz R, Vrijenhoek R (1994) DNA primers for amplification of mitochondrial cytochrome c oxidase subunit I from diverse metazoan invertibrates. Mol Mar Biol Biotechnol 3:294–299

Francis SV, Nishida S, Nandan SB (2018) Validation of male *Pontella spinipes* Giesbrecht, 1889 (Copepoda: Calanoida: Pontellidae) based on morphological and mitochondrial COI Gene sequence analysis. Zool Stud 57:16

Francis SV, Nishida S (2018) Amendment of the status of *Pontella sewelli* Heinrich, 1987 as a junior synonym of *P. sinica* Chen and Zhang, 1965, with confirmation of female-male matching by genetic analysis (Copepoda, Calanoida). Crustaceana 91(4):439–449

Francis SV, Bijoy Nandan S (2019) A new species of Tortanus (Atortus) (Copepoda, Calanoida, Tortanidae) from Minicoy Island, southeastern Arabian Sea. Turk J Zool 43:425–436

Goetze E (2003) Cryptic speciation on the high seas; global phylogenetics of the copepod family Eucalanidae. Proc R Soc Lond B 270:2321–2331

Hall TA (1999) BioEdit: a user-friendly biological sequence alignment editor and analysis program for windows 95/98/NT. Nucleic Acids Symp Ser 41:95–98

Kasturirangan LR (1963) Key to the identification of the more common pelagic copepods of the Indian coastal waters. INCOR., Vol 2. New Delhi: CSIR Publications

Lee CE, Frost BW (2002) Morphological stasis in the Eurytemora affinis species complex (Copepoda: Temoridae). Hydrobiologia 480:111–128

Mori T (1964) The pelagic Copepoda from the neighboring waters of Japan. The Soyo company Inc. Japan, pp 90–91

Mulyadi Nishida S, Ohtsuka S (2017) Seven new species of *Tortanus* (*Atortus*) (Copepoda, Calanoida, Tortanidae) from North Sulawesi, Indonesia. Crustaceana 90:77–99

Nishida S, Anandavelu I, Padmavati G (2015) Two new species of *Tortanus* (*Atortus*) (Copepoda, Calanoida, Tortanidae) from the Andaman Islands. Crustaceana 88:216–230

Prusova I, Sharon LS, Elena P (2012) Calanoid copepods of the Arabian Sea region. Academic Publication Board, Sulthan Quaboos University Press, 240 p

Sanu VF, Bijoy Nandan S, Deepak J, Harikrishnan M (2016) Molecular identification and systematic assessment of *Labidocera madurae* A. Scott, 1909 (calanoid copepod) from Lakshadweep Archipelago, southwest coast of India, based on mitochondrial COI gene sequences. Mar Biodiv 46:95–103

Silas EG, Pillai PP (1973) The calanoid copepod family Pontellidae from the Indian Ocean. J mar boil Ass India 15(2):771–858

Soh HY, Kwon SW, Lee W, Yoon YH (2012) A new Pseudodiaptomus (Copepoda, Calanoida) from Korea supported by molecular data. Zootaxa 3368:229–244

Tamura K, Peterson D, Peterson N, Stecher G, Nei M, Kumar S (2011) MEGA5: molecular evolutionary genetics analysis using maximum likelihood, evolutionary distance, and maximum parsimony methods. Mol Biol Evol 28:2731–2739

Thompson JD, Gibson TJ, Plewniak F, Jeanmougin F, Higgins DG (1997) The Clustal_X windows interface: flexible strategies for multiple sequence alignment aided by quality analysis tools. Nucleic Acids Res 25:4876–4882

A Summary of the South China Sea Evolution

Lachit S. Ningthoujam and Dhananjai K. Pandey

Abstract The South China Sea offers a special site for studying continental break-up, sedimentary basin formation, mantle and lithosphere evolution and other land-ocean interactions because of its location and its well-preserved evolutionary records. It behaves as a natural laboratory for understanding of different plate boundaries in a small region. The SCS has got important attention for its international marine transportation and navigation, its rich marine biodiversity and rich oil and gas resources. As the largest marginal sea separating Asia, the largest continent from Pacific, the largest ocean, the SCS plays a key role in land-sea interactions of the Earth system. From the tectonics perspective, there have been several controversies and challenges for understanding the tectonics and evolution of the South China Sea. Here, we present a brief overview on the evolution and formation of the South China Sea.

Keywords South China Sea · Evolution · Marginal sea

1 Introduction

The South China Sea (SCS), which spans from Singapore and the Malacca Strait in the south to the Strait of Taiwan in the north, and from Borneo and the Philippines on the east to Vietnam and south-eastern China on the west, is one of the largest marginal seas in the world. It is located at the convergence of three major plates (the Eurasian, Pacific-Philippine and Indo-Australian plates) along two super-convergent zones (the circum-Pacific and Tethyan zones) (Wang et al. 2000; Li et al. 2013; Zhou and Yao 2009). It is surrounded by passive margins (rifting) in the north and south, an active margin (subduction) in the east, and a transform margin in the west. Hence the evolution of SCS can be thought of as an integrated result of a complex tectonic

L. S. Ningthoujam (✉) · D. K. Pandey
National Centre for Polar and Ocean Research, Vasco da Gama, Goa, India
e-mail: lachit@ncpor.res.in

D. K. Pandey
e-mail: pandey@ncpor.res.in

D. K. Pandey et al. (eds.), *Dynamics of the Earth System: Evolution, Processes and Interactions*, Society of Earth Scientists Series,
https://doi.org/10.1007/978-3-030-40659-2_11

history. Here, the SCS refers a collective feature which includes the deep-sea plain and surrounding slopes and shelves (Fig. 1).

The SCS developed from magma-poor continental margin rifting to subsequent seafloor spreading in the latest Cretaceous to Early Paleocene (Taylor and Hayes 1983; Cullen et al. 2010; Franke 2013) and thus oceanic crust covers the central portion of the basin. Evolution of the SCS is largely believed to be induced by the India-Eurasia collision to the northwest (Morley 2002; Tapponnier et al. 1986), the subduction of the Pacific plate and then the compression from the Philippine sea plate in the east (Li 1994; Zhou et al. 2002), as well as the slab-pull of the proto-South China Sea in the south (Hall 1996; Morley 2002; Taylor and Hayes 1983).

The SCS oceanic basin formed during the early Oligocene to middle Miocene (32–15 Ma) (Li et al. 2015; Taylor and Hayes 1983). Its age is mature enough to experience the entire history of marginal sea development as well as young enough to preserve and exhibit the complete evolutionary records from being distorted (Zhou, and Yao 2009) and moreover it is relatively small so the entire basin is easily accessible. Thus, the basin is considered to be a key site and natural laboratory for studying continental break-up, sedimentary basin formation, mantle and lithosphere evolution, Asian monsoonal climate dynamics and other land-ocean interactions (Li et al. 2012). The lack of salt and Seaward Dipping Reflector (SDRs) and its potential for oil and gas have caught more attention for studying the transition between highly stretched continental and oceanic crust.

Even though studies had been focused primarily on the sea-floor dating and tectonic evolution through modelling (Taylor and Hayes 1980, 1983; Briais et al. 1993; Barckhausen and Roeser 2004), the evolutionary history of the SCS has been in debate for decades (Barckhausen et al. 2014; Briais et al. 1993; Cullen et al. 2010; Franke et al. 2013; Li et al. 2012; Taylor and Hayes 1980, 1983; Zhou et al. 1995). However, in the recent years, with different geophysical data the basin has been well-imaged and scientific drilling expeditions (ODP Leg 184; IODP-349, 367, 368 and 368X) (Wang et al. 2000; Li et al. 2015; Sun et al. 2018 and Childress 2018) have provided crucial clues in constraining the processes of rifting and eventual rupturing of the continental crust during breakup, at this magma-poor rifted margin (drilling locations are shown in Fig. 1). In this paper, we would like to highlight a glimpse of the present scenario of our understanding on the evolution of SCS.

2 Regional Geological Setting

The SCS basin is a 'V'-shaped basin (characteristic of a propagation opening; Courtillot 1982; Taylor et al. 1999) and shows an irregular triangular shape with the major apex towards south-west which reveals a propagating rift (Huchon et al. 2001). The SCS basin has been subdivided into an eastern sub-basin (ESB), northwest sub-basin (NWSB) and southwest sub-basin (SWSB) (Hutchison 1989; Li et al. 2008) (Fig. 1.). The ESB is separated from the SWSB by Zhongnan Fault Zone (Sun et al. 2009) (Figs. 1 and 2). Briais et al. (1993) proposed that the SWSB and the ESB began to

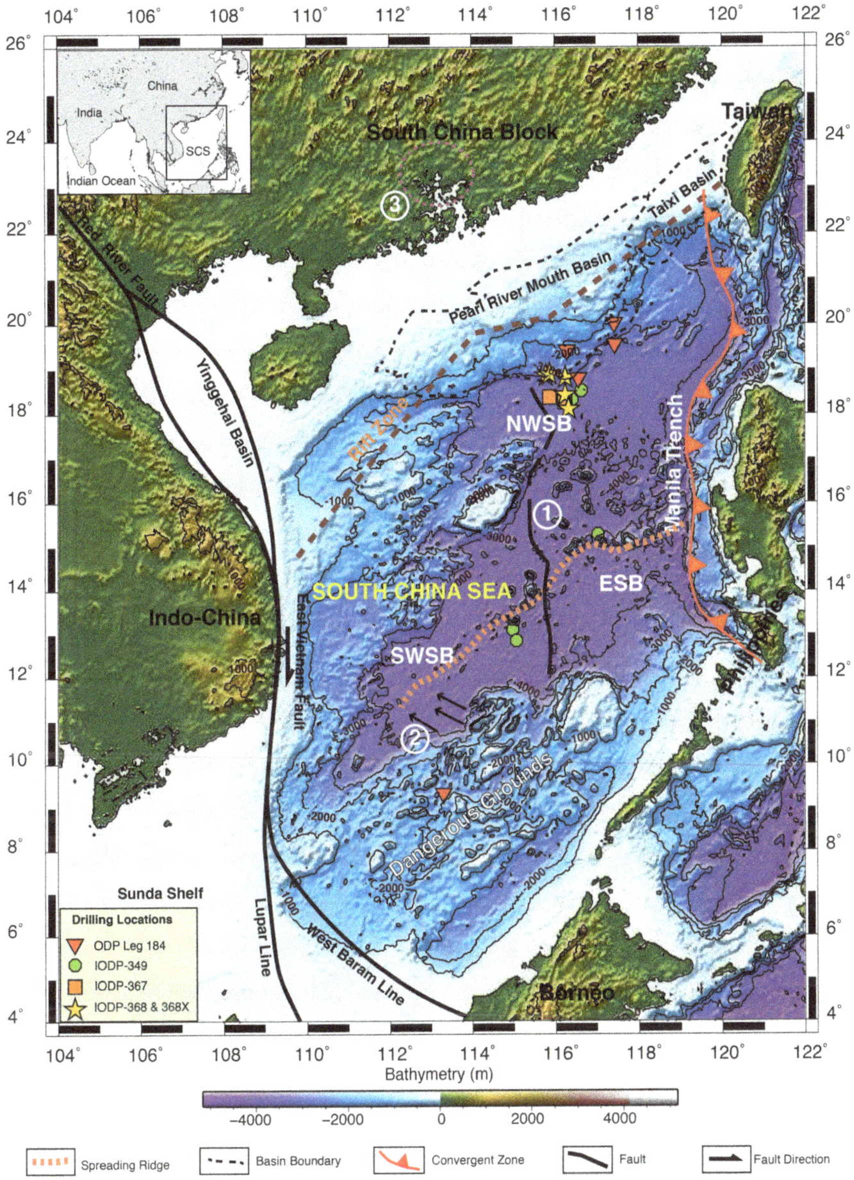

Fig. 1 Map depicts major structures in and around the South China Sea Basin (modified after Sun et al. 2009). Sub-basins in the SCS basin are shown (ESB; East sub-basin, NWSB: North-west sub-basin and SWSB: South-west sub-basin). Location of Zhongnan fault is plotted after Ziying et al. 2019 (indicated as No. 1). In the SWSB, ribbon like structures (indicated as No. 2) are highlighted with arrow-head line segments which implies a decollement (Hayes et al. 1995; Ding et al. 2013). The round shapes (indicated as No. 3) in the northern margin are also highlighted with a dotted circle, these structures are compared to the large granitic bodies cropping out onland in China and Vietnam (Savva et al. 2014). Distinct delineation of the basin is marked by large gradient in the bathymetry

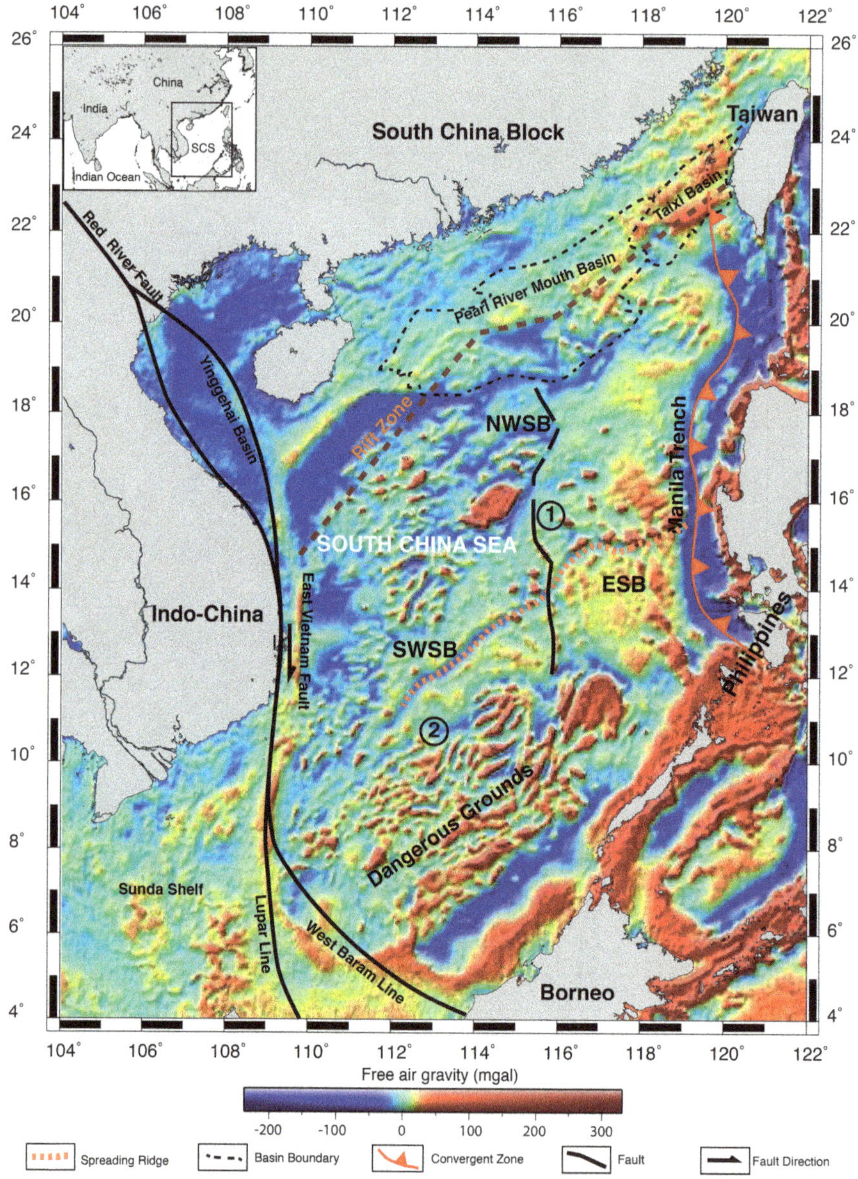

Fig. 2 It shows the free-air gravity anomaly variation in the South China Sea Basin in which major features are highlighted. Zhongnan fault (indicated as No. 1) is distinctly visible. The map reflects high gravity anomalies in the oceanic basin and lows in the surrounding islands (after Yu et al. 2017). The region of ribbon like structures is indicated as No. 2

spread in a N-S orientation at 30 Ma after which the spreading ridge jumped towards south where the spreading episode lasted from 25 to ~16 Ma and the spreading appears to propagate from northeast to southwest.

The conjugate margins of the SCS exhibit various morphological features (Yu et al. 2017) which are shown in Fig. 1. The bottom of the basin is fairly flat with seamounts, sea knolls and depressions of varying sizes. The elongation direction of the seamounts is mainly in E-W (Yu et al. 2017). In the deep area of the SCS, there are more than 50 seamounts that are over 500 m above the seafloor. Compared with those in the central basin, the seamounts in the southwest sub-basin are much smaller and their sails are relatively small.

The south-western part of the SCS shows ribbon like structures (Hayes et al. 1995; Ding et al. 2013) distributed over a wide "Basins and Range-like" province which implies a decollement level in the crust (Gilder et al. 1991; Franke 2013; Pichot et al. 2014) (Fig. 2). The central parts of both the northern and southern margins display a different morphology underlined by large round shapes of around 50–100 km in size. The round shapes in the northern margin may be compared to the large granitic bodies cropping out onland in China and Vietnam (Savva et al. 2014) (Fig. 1). Zhongnan Fault Zone is a striking feature in the middle part of the SCS basin.

Across the South China Sea, the crustal thickness varies significantly (Nissen et al. 1995; Qui et al. 2011; Pichot et al. 2014). The northern margin has an average of 30 km (Nissen et al. 1995) which generally thins towards the South with 10–15 km thick (Qui et al. 2011; Pichot et al. 2014). The coastal zone of the northern margin is marked by a much rougher morphology with elongated flakes composed of granite and Palaeozoic sediments (Shen 2008; Pichot et al. 2014). Basin portions closer to COT boundaries are composed of a thinner crust (between 3 and 5 km on the northern margin and 3 km on the southern side) (Savva et al. 2014).

Satellite gravity images clearly reflect most of the primary tectonic features and many other secondary structural details in the basin (Fig. 2). The gravity map depicts high gravity anomalies in the oceanic basin and lows in the surrounding islands, indicating that the crust of the basin is very thin and characteristic of typical oceanic crust. NE-trending low-intensity anomaly stripes are observed in the high-gravity background of the southwest sub-basin (Yu et al. 2017). The region of dangerous grounds is marked by elongated gravity high and low anomalies.

3 Evolution History

The south-west tapered South China Sea is a backarc basin rifted off the South China Block (Sun 2016) which resulted from the late Cretaceous to Paleogene rifting, late Eocene to middle Miocene seafloor spreading, and post-spreading subduction and closing since the late Miocene. The complex geological evolution of the SCS resulted in a variety of structures across the basin, including the abundant extensional structures and limited magmatic activity that developed in the northern margin of the SCS during and after the rifting and seafloor spreading (Franke et al. 2013). The opening of the SCS was mainly induced by the extrusion of Indochina subcontinent due to Tibet uplift and slab pull due to subduction of the Asian continental crust into the Philippine Sea Plate. Sun et al. (2009) viewed that a Proto-SCS existed before rifting and spreading occurred mainly along the present south-eastern margin although its size and connection with other paleo-oceans are not clear. Based on sea floor spreading magnetic anomaly data from the northeast, the SCS first opened at ~37 Ma, in the late Eocene. IODP Expedition 349 core (Li et al. 2014) shows that initial seafloor spreading started around 33 Ma in the northeastern South China Sea (SCS), but varied slightly by 1–2 Ma along the northern continent-ocean boundary (COB), and the terminal age of seafloor spreading is ~15 Ma in the East sub-basin and ~16 Ma in the Southwest sub-basin. Larsen et al. (2018) also proposed that narrow and fast rift-to-drift transitions along the northern SCS has resulted due to major rifting events within thin lithosphere permitting mantle upwelling, yielding abundant MORB-type melts during final breakup and early seafloor spreading.

The rift stages (in which the basin rifted from the South China Block) that resulted in the formation of the SCS started with an initial uplift of the rift shoulders accompanied by widespread erosion and peneplanation in the Late Cretaceous to Early Paleocene (Taylor and Hayes 1980, 1983; Ru and Pigott 1986; Schlüter et al. 1996; Pubellier et al. 2003). According to Sun et al. (2009), the proto-margins of the SCS experienced three main stages of deformation as the rifting propagated from North to South and then from East to West. By the end of rifting of the NE part of the SCS, spreading began and propagated gradually toward the SW. In the north eastern portion of the SCS, Hsu et al. (2004) and Yeh et al. (2010) interpreted Late Eocene/Early Oligocene oceanic crust and seafloor spreading (37.8–30.1 Ma). However, the nature of the crust there is ambiguous as McIntosh et al. (2013) proposed a hyper-extended crust with a possible mantle exhumation. The timing of seafloor spreading in the central South China Sea has been revised to 32–20.5 Ma (Barckhausen and Roeser 2004) from 32 to 15.5 Ma (Taylor and Hayes 1980; Briais et al. 1993). The post Middle Miocene evolution is considered to be devoid of important tectonic activity except for the Northwest Borneo wedge vertical motions (Sapin et al. 2011).

In the south, rifting delayed until the Eocene-Miocene and often a breakup unconformity of 3–5 Ma is observed underlying strata of 16 Ma and older ages. Since the middle Miocene, the eastern margin has been subducting, subsequently causing the closure of the sea basin and rise of Taiwan in the last 5 Ma (Li et al. 2013).

4 Dynamics of the South China Sea

Researchers have given considerable attention to understand the dynamics and related mechanisms for the evolution of the SCS. Different models have been proposed (Sun 2016) for the evolution of SCS. Of these the models that are widely accepted include (a) Tectonic extrusion model (Tapponnier et al. 1990; Briais et al. 1993); (b) Backarc extension model (Hilde et al. 1977); (c) Two-stage rifting model (Yao 1999); (d) Proto-SCS dragging model (Holloway 1982; Taylor and Hayes 1983; Hall 1996), (e)

Models that involve extension induced by mantle plume (Tamaki and Hoang 1998); and (f) Combinations of proto-SCS pull and extrusion of the Indochina Peninsula, and/or mantle flows (Tamaki 1995; Morley 2002; Zhou et al. 2002; Sun et al. 2006), etc.

4.1 The Tectonic Extrusion Model

This model is based on physical modelling experiments (Tapponnier et al. 1990; Briais et al. 1993) and considered to be the most famous model. The model depicts the major deformation in the Eurasian crust due to the collision between the Indian and Eurasian continents, leading to southward extrusion of the Indo-China Peninsula along the Ailaoshan-Red River sinistral fault. The model was supported by the synchronicity between the strike-slip movement of the fault (Tapponnier et al. 1990) and the spreading of the SCS (Taylor and Hayes 1980, 1983). However, problems emerged with the extrusion model in terms of age discrepancy, compression and extension of the SCS, failing to reflect the real geodynamic conditions and failing to explain the wider spreading in the east of SCS than in the west.

4.2 Backarc Extension Model

This model (Hilde et al. 1977) proposes that the SCS was a back-arc basin rifted from the passive margin of the Eurasian continent. In the model, the subduction direction of the Philippine Sea and the Pacific Plate is not compatible with the extension direction of the SCS even though the subduction of the Philippine Sea Plate was taken as the primary driving force (Karig 1971; Ben-Avraham and Uyeda 1973; Guo et al. 1983). Also, the SCS was assumed as a backarc basin related to subduction of the Neotethys Plate between the Australian and Eurasian continents (Hilde et al. 1977). The model shows that the extension of the SCS started at 100 Ma (Hilde et al. 1977; Stern and Bloomer 1992) without any evidence.

4.3 Two-Stage Rifting Model

Considering geomorphological/geophysical features like water depths, extensional basins surrounding the SCS, and magnetic anomalies in the SWSB of the SCS, this model proposes two extension events in the SCS-the first extension, which occurred in the Late Eocene to Early Oligocene (42–35 Ma), formed the southwest-northeast trending SWSB and the second extension occurred in the Late Oligocene to Early Miocene, resulting in the formation of east–west trending ESB (Yao 1999). Despite the two-stage opening of the SCS predicted by this model is confirmed by recent dating results, the timing is not consistent (Li et al. 2014).

4.4 Proto-SCS Dragging Model

According to this model, a proto-SCS existed south of the current SCS and disappeared through south-eastward subduction beneath the Luzon and Sulu islands in the Late Cretaceous to Paleocene (Holloway 1982; Taylor and Hayes 1983; Lee and Lawver 1994, 1995) which resulted in extension and rifting along the southeast margin of the South China Block and the formation of the SCS (Holloway 1982; Taylor and Hayes 1983; Hall 1996). But, paleomagnetic results suggest that the proto-SCS was very small (Lee and Lawver 1994) to have pulled apart the thick continental lithosphere of South China.

4.5 Models that Involve Extension Induced by Mantle Plume

In this model, a mantle plume is considered as the driving force for the evolution of the SCS (Tamaki and Hoang 1998) which is apparently supported by seismic tomography showing high temperature anomalies beneath the SCS (Huang and Zhao 2006; Zhao 2007). However, absence of large scale magmatism and other geochemical data on some basaltic rocks (Yan et al. 2008; Zou and Fan 2010; Xu et al. 2012; Huang et al. 2013) are against this plume induced model.

4.6 Combined Model

In this model, several events/features like proto-SCS subduction/pull plus mantle flow induced by collision between the Indian and Eurasian continents (Sun et al. 2006); proto-SCS subduction/pull plus southward extrusion of the Indochina Peninsula (Morley 2002); or multiple plate subduction, shearing, and collisions (Tamaki 1995; Zhou et al. 2002) are put together to explain the opening of the SCS. However, neither extrusion nor proto-SCS subduction/pull can explain the formation of the SCS nor can it explain the two-stage extension. As of now, this model is more widely used than the other models.

Remarks
From various perspectives the SCS has been an interesting region for geoscience researchers. Existence of different types of geological features which preserves the complete evolutionary records of the SCS makes it a fascinating area for understanding the geological journey that was taken by a rifted marginal sea which has been

influenced by various plate tectonic processes. Studies with the help of integrated geophysical and geological data would be able to find a conclusive mechanism for the initiation, evolution and formation of the South China Sea.

Acknowledgements This short article is based on the available literatures. We thank the director of NCPOR, Goa. The authors gratefully acknowledge all authors of the research articles which have been referred to compile this article.

References

Barckhausen U, Roeser HA (2004) Seafloor spreading anomalies in the South China Sea revisited. In: Continent-ocean interactions within East Asian marginal seas, vol 149. AGU, Washington, DC, pp 121–125
Barckhausen U, Engels M, Franke D, Ladage S, Pubellier M (2014) Evolution of the South China Sea: revised ages for breakup and seafloor spreading. Mar Pet Geol 58:599–611
Ben-Avraham Z, Uyeda S (1973) The evolution of the China Basin and the mesozoic paleogeography of Borneo. Earth Planet Sci Lett 18(2):365–376
Briais A, Patriat P, Tapponnier P (1993) Updated interpretation of magnetic anomalies and seafloor spreading stages in the South China Sea: implications for the tertiary tectonics of Southeast Asia. J Geophys Res: Solid Earth 98(B4):6299–6328
Childress L (2018) The expedition 368X scientists. Int Ocean Discov Program, Exped 368X Prelim Rep: IODP Publ
Courtillot V (1982) Propagating rifts and continental breakup. Tectonics 1(3):239–250
Cullen A, Reemst P, Henstra G, Gozzard S, Ray A (2010) Rifting of the South China Sea: new perspectives. Pet Geosci 16(3):273–282
Ding W, Franke D, Li J, Steuer S (2013) Seismic stratigraphy and tectonic structure from a composite multi-channel seismic profile across the entire dangerous grounds, South China Sea. Tectonophysics 582:162–176
Franke D (2013) Rifting, lithosphere breakup and volcanism: comparison of magma-poor and volcanic rifted margins. Mar Pet Geol 43:63–87
Franke D, Savva D, Pubellier M, Steuer S, Mouly B, Auxietre JL, Meresse F, Chamot-Rooke N (2013) The final rifting evolution in the South China Sea. Mar Pet Geol 58:704–720
Gilder SA, Keller GR, Luo M, Goodell PC (1991) Timing and spatial distribution of rifting in China. Tectonophysics 197:225e243
Guo LZ, Shi YS, Ma R (1983) On the formation and evolution of the mesozoic-cenozoic active continental-margin and island-arc tectonics of the western Pacific-Ocean. Acta Geol Sin 57(1):11–21
Hall R (1996) Reconstructing cenozoic SE Asia. Geol Soc Lond Spec Publ 106(1):153–184
Hayes DE, Nissen SS, Buhl P, Diebold J, Bochu Y, Zeng W, Chen Y (1995) Throughgoing crustal faults along the northern margin of the South China Sea and their role in crustal extension. J Geophys Res: Solid Earth 100(B11):22435–22446
Hilde TW, Uyeda S, Kroenke L (1977) Evolution of the western Pacific and its margin. Tectonophysics 38(1–2):145–165
Holloway NH (1982) North Palawan block, Philippines—its relation to Asian mainland and role in evolution of South China Sea. AAPG Bull 66(9):1355–1383
Hsu SK, Yeh YC, Doo WB, Tsai CH (2004) New bathymetry and magnetic lineations identifications in the northernmost South China Sea and their tectonic implications. Mar Geophys Res 25(1–2):29–44

Huang J, Zhao D (2006) High-resolution mantle tomography of China and surrounding regions. J Geophys Res: Solid Earth 111(B9)

Huang XL, Niu Y, Xu YG, Ma JL, Qiu HN, Zhong JW (2013) Geochronology and geochemistry of Cenozoic basalts from eastern Guangdong, SE China: constraints on the lithosphere evolution beneath the northern margin of the South China Sea. Contrib Min Pet 165(3):437–455

Huchon P, Nguyen TNH, Chamot-Rooke N (2001) Propagation of continental break-up in the southwestern South China Sea. Geol Soc Lond Spec Publ 187(1):31–50

Hutchison CS (1989) Geological evolution of South-east Asia, vol 13. Clarendon Press, Oxford, p 368

Karig DE (1971) Origin and development of marginal basins in the western Pacific. J Geophys Res 76(11):2542–2561

Larsen HC, Mohn G, Nirrengarten M, Sun Z, Stock J, Jian Z, Klaus A, Alvarez-Zarikian CA, Boaga J, Bowden SA, Briais A, Chen Y, Cukur D, Dadd K, Ding W, Dorais M, Ferré EC, Ferreira F, Furusawa A, Gewecke A, Hinojosa J, Höfig TW, Hsiung KH, Huang B, Huang E, Huang XL, Jiang S, Jin H, Johnson BG, Kurzawski RM, Lei C, Li B, Li L, Li Y, Lin J, Liu C, Liu C, Liu Z, Luna AJ, Lupi C, McCarthy A, Ningthoujam L, Osono N, Peate DW, Persaud P, Qiu N, Robinson C, Satolli S, Sauermilch I, Schindlbeck JC, Skinner S, Straub S, Su X, Su C, Tian L, van der Zwan FM, Wan S, Wu H, Xiang R, Yadav R, Yi L, Yu PS, Zhang C, Zhang J, Zhang Y, Zhao N, Zhong G, Zhong L (2018) Rapid transition from continental breakup to igneous oceanic crust in the South China Sea. Nat Geosci 11(10):782

Lee TY, Lawver LA (1994) Cenozoic plate reconstruction of the South China Sea region. Tectonophysics 235(1–2):149–180

Lee TY, Lawver LA (1995) Cenozoic plate reconstruction of Southeast Asia. Tectonophysics 251(1–4):85–138

Li CF, Li J, Ding W, Franke D, Yao Y, Shi H, Pang X, Cao Y, Lin J, Kulhanek DK, Williams T (2015a) Seismic stratigraphy of the central South China Sea basin and implications for neotectonics. J Geophys Res: Solid Earth 120(3):1377–1399

Li CF, Lin J, Kulhanek DK (2013) South China Sea tectonics: opening of the South China Sea and its implications for southeast Asian tectonics, climates, and deep mantle processes since the late Mesozoic. Int Ocean Discov Program Sci Prospect 349:1–111

Li C-F, Lin J, Kulhanek DK (2015) The Expedition 349 Scientist. Proc Int Ocean Discov Program: IODP Publ 349

Li CF, Wang P, Franke D, Lin J, Tian J (2012) Unlocking the opening processes of the South China Sea. Sci Drill 14:55–59

Li CF, Xu X, Lin J, Sun Z, Zhu J, Yao Y, Zhao X, Liu Q, Kulhanek DK, Wang J, Song T (2014) Ages and magnetic structures of the South China Sea constrained by deep tow magnetic surveys and IODP Expedition 349. Geochem Geophys Geosyst 15(12):4958–4983

Li CF, Zhou Z, Hao H, Chen H, Wang J, Chen B, Wu J (2008) Late Mesozoic tectonic structure and evolution along the present-day northeastern South China Sea continental margin. J Asian Earth Sci 31(4–6):546–561

Li PL (1994) Structural features and oil-and-gas accumulation in Pearl River Mouth basin. Guangdong Geol 9(4):21–28

McIntosh K, van Avendonk H, Lavier L, Lester WR, Eakin D, Wu F, Liu CS, Lee CS (2013) Inversion of a hyper-extended rifted margin in the southern central range of Taiwan. Geology 41(8):871–874

Morley CK (2002) A tectonic model for the Tertiary evolution of strike—slip faults and rift basins in SE Asia. Tectonophysics 347(4):189–215

Nissen SS, Hayes DE, Bochu Y, Zeng W, Chen Y, Nu X (1995) Gravity, heat flow, and seismic constraints on the processes of crustal extension: Northern margin of the South China Sea. J Geophys Res: Solid Earth 100(B11):22447–22483

Pichot T, Delescluse M, Chamot-Rooke N, Pubellier M, Qiu Y, Meresse F, Sun G, Savva D, Wong KP, Watremez L, Auxiètre JL (2014) Deep crustal structure of the conjugate margins of the SW South China Sea from wide-angle refraction seismic data. Mar Pet Geol 58:627–643

Pubellier M, Ego F, Chamot-Rooke N, Rangin C (2003) The building of pericratonic mountain ranges: structural and kinematic constraints applied to GIS-based reconstructions of SE Asia. Bull Société Géologique Fr 174(6):561–584

Qui XL, Zhao MH, Ao W, Lü CC, Hao TY, You QY, Ruan AG, Li JB (2011) OBS survey and crustal structure of the SW sub-basin and Nansha block, South China Sea. Chin J Geophys 54(6):1009–1021

Ru K, Pigott JD (1986) Episodic rifting and subsidence in the South China Sea. AAPG Bull 70(9):1136–1155

Sapin F, Pubellier M, Lahfid A, Janots D, Aubourg C, Ringenbach JC (2011) Onshore record of the subduction of a crustal salient: example of the NW Borneo Wedge. Terra Nova 23(4):232–240

Savva D, Pubellier M, Franke D, Chamot-Rooke N, Meresse F, Steuer S, Auxietre JL (2014) Different expressions of rifting on the South China Sea margins. Mar Pet Geol 58:579–598

Schlüter HU, Hinz K, Block M (1996) Tectono-stratigraphic terranes and detachment faulting of the South China Sea and Sulu Sea. Mar Geol 130(1–2):39–78

Shen W (2008) Post-orogenic extension in the Pearl River Delta region (South China): an integrated morphological, structural, geophysical and thermochronological study. HKU theses Online (HKUTO)

Stern RJ, Bloomer SH (1992) Subduction zone infancy: examples from the Eocene Izu-Bonin-Mariana and Jurassic California arcs. Geol Soc Am Bull 104(12):1621–1636

Sun W (2016) Initiation and evolution of the South China Sea: an overview. Acta Geochim 35(3):215–225

Sun Z, Jian Z, Stock JM, Larsen HC, Klaus A, Alvarez Zarikian CA, (2018) The Expedition 367/368 scientists. Proc Int Ocean Discov Program: IODP Publ 367/368

Sun Z, Zhong Z, Keep M, Zhou D, Cai D, Li X, Wu SH, Jiang J (2009) 3D analogue modeling of the South China Sea: a discussion on breakup pattern. J Asian Earth Sci 34(4):544–556

Sun Z, Zhou D, Zhong Z, Xia B, Qiu X, Zeng Z, Jiang J (2006) Research on the dynamics of the South China Sea opening: evidence from analogue modeling. Sci China Ser D: Earth Sci 49(10):1053

Tamaki K (1995) Opening tectorfic of the Japan Sea. Plenum Press, New York

Tamaki K, Hoang N (1998) Mantle extrusion'A model for dispersed volcanism and DUPAL-like asthenosphere in East Asia and the Western Pacific. Mantle Dyn Plate Interact East Asia 27:67

Tapponnier P, Lacassin R, Leloup PH, Schärer U, Dalai Z, Haiwei W, Jiayou Z (1990) The Ailao-shan/Red river metamorphic belt: tertiary left-lateral shear between Indochina and South China. Nature 343(6257):431

Tapponnier P, Peltzer G, Armijo R (1986) On the mechanics of the collision between India and Asia. Geol Soc Lond Spec Publ 19(1):113–157

Taylor B, Hayes DE (1980) The tectonic evolution of the South China Basin. Tecton Geol Evol Southeast Asian Seas Isl 23:89–104

Taylor B, Hayes DE (1983) Origin and history of the South China Sea basin. Tecton Geol Evol Southeast Asian Seas Isl: Part 2(27):23–56

Taylor B, Goodliffe AM, Martinez F (1999) How continents break up: insights from Papua New Guinea. J Geophys Res: Solid Earth 104(B4):7497–7512

Wang P, Prell WL, Blum P (2000) Proceedings of the Ocean Drilling Program, initial reports, vol 184. http://www.ldeo.columbia.edu

Xu Y, Wei J, Qiu H, Zhang H, Huang X (2012) Opening and evolution of the South China Sea constrained by studies on volcanic rocks: preliminary results and a research design. Chin Sci Bull 57(24):3150–3164

Yan Q, Shi X, Wang K, Bu W, Xiao L (2008) Major element, trace element, and Sr, Nd and Pb isotope studies of cenozoic basalts from the South China Sea. Sci China, Ser D Earth Sci 51(4):550–566

Yao BC (1999) Spreading of the southwest basin of the South China Sea and its tectonic significance. South China Sea Geol 9:20–36

Yeh YC, Sibuet JC, Hsu SK, Liu CS (2010) Tectonic evolution of the Northeastern South China Sea from seismic interpretation. J Geophys Res: Solid Earth 115(B6)

Yu X, Xue C, Shi H, Zhu W, Liu Y, Yin H (2017) Expansion of the South China Sea basin: constraints from magnetic anomaly stripes, sea floor topography, satellite gravity and submarine geothermics. Geosci Front 8(1):151–162

Zhao D (2007) Seismic images under 60 hotspots: search for mantle plumes. Gondwana Res 12(4):335–355

Zhou D, Yao B (2009) Tectonics and sedimentary basins of the South China Sea: Challenges and progresses. J Earth Sci 20(1):1–12

Zhou D, Chen HZ, Wu SM, Yu HS (2002) Opening of the South China Sea by dextral splitting of the East Asian continental margin. Acta Geol Sin 76(2):180–190

Zhou D, Ru K, Chen HZ (1995) Kinematics of cenozoic extension on the South China Sea continental margin and its implications for the tectonic evolution of the region. Tectonophysics 251(1–4):161–177

Ziying X, Jun W, Hongfang G, Guihua S, Meijing S, Xin N, Rongwei Z (2019) Research progress of the zhongnan-Lile Liyue fault zone in the South China Sea basin. J Trop Ocean 38(2):86–94

Zou H, Fan Q (2010) U-Th isotopes in Hainan basalts: implications for sub-asthenospheric origin of EM2 mantle endmember and the dynamics of melting beneath Hainan Island. Lithos 116(1–2):145–152

Microbial Community Profile of Deep-Sea Sediment from Eastern Arabian Sea (IODP 355)

A. Ganesh Kumar, R. Vijaya Raghavan, G. Dharani and M. A. Atmanand

Abstract Deep-sea sediment core collected from the Laxmi Basin in the eastern Arabian Sea during IODP 355 was analyzed for bacterial diversity using next generation V3-V4 of the16S rRNA gene amplicon sequencing. A total of 523,427 high quality 16S rRNA gene sequences obtained clustered in to 66,290 operational taxonomic units (OTU). Majority of these sequences were affiliated with *Proteobacteria* followed by *Firmicutes*, *Chloroflexi*, *Actinobacteria* and *Bacteroides*. The presence of well-known denitrifying bacterial groups, such as *Micrococcus*, *Paracoccus*, *Pseudomonas*, *Lactobacillus*, *Herptosiphon*, *Nitroscoccus*, *Nitrococcus* and *Nitrospira* sp. in our sequencing data suggests that bacterial community from the deep-sea sediment of the Laxmi Basin harbor certain specialized bacterial groups.

Keywords Subsurface sediments · Nitrogen cycle · Denitrification · Biogeochemical cycles

1 Introduction

Bacteria and archaea inhabiting subsurface sea bed accounts for more than half of the total microbial cells in the oceans (Starnawski et al. 2016). Despite their high relative abundance in deep sea subsurface sediments diversity and metabolic functions of these microorganisms are relatively unknown (Zhang et al. 2018). Therefore, studies are required to help understand the role of complex native microbial communities in biogeochemical processes and their adaptation strategies in such extreme environment. Deep sea sub-surface sediments are characterized by unique geochemical signatures of carbon, sulfur, iron and manganese cycles (Jorgensen et al. 2012). Due to restricted energy substrates, the growth of microbial communities decreases with sediment depth. As a consequence, the subsurface microorganisms persist over prolonged time scales with much slower metabolism (Bradley et al. 2018). Interestingly,

A. Ganesh Kumar (✉) · R. Vijaya Raghavan · G. Dharani · M. A. Atmanand
Marine Biotechnology Division, National Institute of Ocean Technology, Pallikaranai, Chennai 600100, India
e-mail: microganesh@niot.res.in

D. K. Pandey et al. (eds.), *Dynamics of the Earth System: Evolution, Processes and Interactions*, Society of Earth Scientists Series,
https://doi.org/10.1007/978-3-030-40659-2_12

hundreds of meters below the sea floor sediments are populated by microbes and the buried organic matter, which is the most important carbon source for sediment microbial communities (Jorgensen et al. 2019). The microorganisms in marine sediments play a major role in nitrogen transformation and thereby balance global nitrogen cycle (Zhao et al. 2019). Earlier studies on molecular ecology of Arabian Sea sediments focused on general prokaryotic, ammonium oxidation (Luke et al. 2016), sulphate reducing and methane cycling populations (Fernandes et al. 2018). The eastern Arabian Sea has a unique and persistent oxygen minimum zone (OMZ). Recent study on OMZ highlighted the importance of alternative pathways of respiration by microorganisms (Luke et al. 2016). The culture independent approach of bacterial diversity analysis of Arabian Sea sediment showed seven phyla which include *Proteobacteria, Bacteriodetes, Actinobacteria, Firmicutes, Chloroflexi, Deferribacter and Acidobacterium* (Harisree et al. 2017). The study in OMZ sediments of the eastern Arabian Sea demonstrated the presence of phylum *Proteobacteria,Planctomycetesa* and *Chloroflexi* as dominant groups (Divya et al. 2011). However, none of these studies recorded the presence of well known denitrifying bacterial groups in their bacterial community profiles. Thus, our data from the Laxmi Basin may provide some useful in-site about denitrifying bacterial groups in the sedimentary bacterial community.

2 Materials and Methods

2.1 Deep Sea Sediment Cores IODP

Sediment samples were collected in May 2015 on-board the Joides Resolution, during the IODP Expedition 355. Sediment cores were taken from the Laxmi Basin U355 (16° 34.2851′ N, 68° 50.3454′ E; 3645 m water depth). All cores for microbial analysis were drilled with full microbiological contamination checks and subsequently whole round cores (WRC) were aseptically cut during on-board, capped and stored. For molecular analysis, the upper 10 cm of the sediment sample was immediately frozen, transported and stored at −80 °C until processed.

2.2 DNA Extraction

Genomic DNA was extracted from 0.5 g sediment sample by homogenization using the FastDNA™ Spin Kit for soil (MP Biomedicals, US) according to the manufacturer's instruction. The extracted total genomic DNA of samples was pooled together into one concentrated sample. The quality and concentration of extracted DNA was analyzed using a nano-spectrometer (DeNovix DS-11, USA).

2.3 16S rRNA Gene Sequencing

The metagenome libraries of the V3-V4 hypervariable region of the 16S rDNA were prepared using V3-V4 region-specific proprietary primers (Takahashi et al. 2014) at Genotypic Technology Pvt. Ltd., Bangalore, India. Briefly, 20 ng of genomic DNA was amplified for 26 cycles using KAPA HiFi HotStart PCR Kit (KAPA Biosystems Inc., Boston, MA USA). The forward and reverse primer concentration is kept at 0.2 μM each. The amplicons were analyzed on 1.2% agarose gel. 1 μl of round 1 PCR amplicons were used for Indexing PCR. The round 1 PCR amplicons were amplified for 10 cycles to add Illumina sequencing barcoded adaptors (Nextera XT v2 Index Kit, Illumina, U.S.A.). Round 2 PCR amplicons were analyzed on 1.2% agarose gel. Sequencing of the V3-V4 hypervariable region was carried out in Illumina NextSeq500 platform.

3 Quality Control, Taxonomic Assignment, Diversity Estimation and Total Cell Counts

The pre-processing of raw Illumina reads was carried out using Fast QC (Kleiner et al. 2017). The forward and reverse reads were merged using FLASH (Wang et al. 2018). The quality, length and percentage of ambiguous bases were set at 30, 100 and 10 respectively. Usually sequence similarity should be >Q30 for a good illummina read. Length of read illumina NextSeq is 150 bp since our read length is 100 we allow ambiguous base upto 10% for data processing. Removal of chimeric sequences was done using UCHIME (Wang et al. 2018). The reads that passed the quality filters were used for subsequent downstream analyses. Taxonomic assignments were carried out using MG-RAST version 4.0, using Greengenes database. Determination of Operational Taxonomic Units (OTUs) at 97, 95 and 90% similarity level and calculation of diversity parameters were carried out using QIIME 2 pipeline. Enumeration of total cells in sediment sample was done using acridine orange direct count method.

4 Result and Discussion

4.1 Total Cell Counts

Quantification of direct cell counts using acridine orange revealed very high cell counts, exceeding 10^6 cells/g of sediment at the surface and the cell number decreased to 10^5 cells/g sediment at a depth of 10 cm. Acridine orange was used based on the optimal properties for rapid enumeration and assessment of bacterial colonization (Cardinale et al. 2008). The cell count indicates high prokaryotic cell densities in the surface sediment. This shows that the top layer of sediment provides more food source for the microorganisms than the bottom layer. High cell count indicated the active participation of microbial community that responds to the fluctuating marine environments (Zhang et al. 2017).

4.2 Taxonomic Assignment and Microbial Diversity

More than 0.7 million reads were obtained from high throughput sequencing of sediment sample. The paired end merging of forward and reverse reads followed by quality control of dataset eliminated 27% of reads, and 0.5 million reads (Table 1) were available for downstream analysis. Taxonomic assignment of pre-processed reads indicated that *Proteobacteria,Firmicutes*, and *Chloroflexii* were found to be the dominant phylum. *Proteobacteria* contributed to 32% of relative abundance of the studied sample, while *Firmicutes* and *Chloroflexii* represented 29 and 14% respectively. *Actinobacteria* was found to represent 2% among the bacterial community. *Clostridia,Chloroflexi* and *Bacilli* were the dominant classes with 16, 11 and 14% relative abundance respectively. Significant abundance was also observed for *Deltaproteobacteria, Gammaproteobacteria, Actinobacteria, Thermotogae, Nitrospira* classes as well. At the genus level, *Herpetosiphon* (14%), *Clostridium* (10%) and *Lactobacillus* (10%) were found to be dominant denitrifiers (Rosch et al. 2002; Lee and Reichenbach 2006). More than 1% representation was also observed from members of *Desulfovibrio, Thermodesulfobium, Nitrococcus, Thermotoga, Peptoniphilus* genera as well (Fig. 1). The metagenome analysis provides conclusive evidence of the genomic potential for an active nitrogen cycle in the Arabian Sea (Luke et al. 2016). These microbial richness and diversity of denitrifiers can be related to the availability of the total organic carbon and nitrogen. This also suggests that changes in biogeochemical environment of sediments may have significant impact on the total microbial communities. In addition, this also can be correlated to the variation in community composition at various geological time scales (Orsi et al. 2017). An interesting observation in the relative sequence abundance is the percentage of unclassified reads. Greengenes database revealed a complex microbial community with high unclassified reads, 15% at the phylum level, 35 and 41% at class and genus level respectively, this increase the possibility that this sample could well be a potential source of various novel microorganisms. This is in agreement with findings by Harisree et al. (2017) where 10.47% of the sequences did not categorize into any taxonomic hierarchy. These research data are associated with unclassified microbial communities with unknown ecological importance. The number of OTU's predicted at 97, 95 and 90% were found to be 66,290, 24,425, 3852 respectively. This indicates the rich bacterial diversity associated with the studied sample which is further substantiated by the various diversity indices such as Chao, Shannon and Simpson (Table 1). The taxonomic distribution also suggested that more than 78% of the phylotypes had a representation which is less than 0.01% of the total bacterial community. This construes the abundance of rare microbial phylotypes which would require further

Table 1 Table representing the number of reads obtained after sequencing, number of sequences filtered through the quality control process and the various diversity parameters of the analysed

		Observed OTU's			Chao			Shannon			Simpson		
Raw reads	After pair-end merging and quality	97%	95%	90%	97%	95%	90%	97%	95%	90%	97%	95%	90%
719,369	523,427	66,290	24,425	3852	281,785.1	49,167.03	5808.035	11.00877	9.89062	6.63121	0.992921	0.987747	0.942872

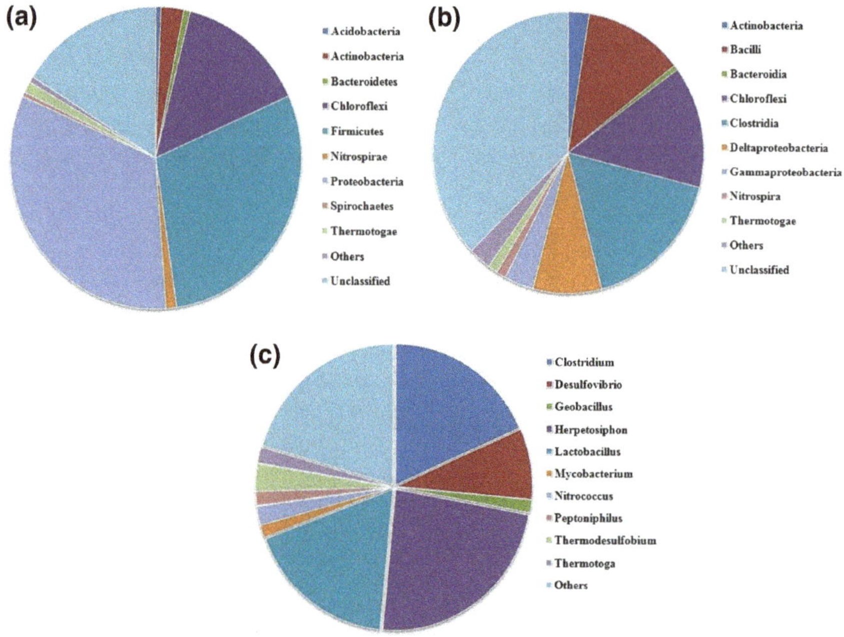

Fig. 1 Taxonomic classification. **a** Phylum, **b** class and **c** genera level based on 16S rRNA sequencing

deep sequencing in order to understand the role of such rare microbiota in biogeochemical processes. In conclusion, this study provides the total microbial diversity and distribution of denitrifiers in the Arabian Sea sub-surface sediments.

Acknowledgements We thank the Secretary, Ministry of Earth Sciences (MoES), Govt. of India and the Director, National Institute of Ocean Technology for their support and encouragement. We also thank the Co-Chief Scientist Dr. Dhananjai Pandey, Dr. Peter Clift, and Staff Scientist Dr. Denise Kulhanek and the IODP Expedition 355 Shipboard Scientific Party for successful completion of sampling and analysis.

References

Bradley JA, Amend JP, LaRowe DE (2018) Bioenergetic controls on microbial ecophysiology in marine sediments. Front Microbiol 9:1–180

Cardinale M, Castro JV, Muller H, Berg G, Grube M (2008) In situ analysis of the bacterial community associated with the reindeer lichen Cladonia arbuscula reveals predominance of Alphaproteobacteria. FEMS Microbiol Ecol 66:63–71

Divya B, Parvathi A, LokaBharathi PA, Nair S (2011) 16S rRNA-based bacterial diversity in the organic-rich sediments underlying oxygen-deficient waters of the eastern Arabian Sea. World J Microbiol Biotechnol 27:2821–2833

Fernandes S, Mazumdar A, Bhattacharya S, Peketi A, Mapder T, Roy R, Carvalho MA, Roy C, Mahalakshmi P, Da Silva R, Rao PLS, Banik SK, Ghosh W (2018) Enhanced carbon-sulfur cycling in the sediments of Arabian Sea oxygen minimum zone center. Sci Rep 8(1): 8665

Harisree PN, Rinu MP, Helvin V, Sarita GB (2017) 16S rDNA-based bacterial diversity analysis of Arabian Sea sediments: A metagenomic approach. Ecol Genet Genomics 3–5:47–51

Jorgensen SL, Hannisdal B, Lanzen A, Baumberger T, Flesland K, Fonseca R, Ovreas L, Steen IH, Thorseth IH, Pedersen RB, Schleper C (2012) Correlating microbial community profiles with geochemical data in highly stratified sediments from the Arctic Mid-Ocean Ridge. Proc Natl Acad Sci USA 109(42):E2846–E2855.

Jorgensen BB, Findlay AJ, Pellerin A (2019) The biogeochemical sulphur cycle of marine sediments. Front Microbiol 10:849

Kleiner M, Thorson E, Sharp CE, Dong X, Liu D, Li C, Strous M (2017) Assessing species biomass contributions in microbial communities via metaproteomics. Nat Commun 8(1):1558

Lee N, Reichenbach H (2006) The Genus *Herpetosiphon*. In: Dworkin M, Falkow S, Rosenberg E, Schleifer KH, Stackebrandt E (eds) The prokaryotes. Springer, New York, NY

Luke C, Speth DR, Kox MAR, Villanueva L, Jetten MSM (2016) Metagenomic analysis of nitrogen and methane cycling in the Arabian Sea oxygen minimum zone. Peer J 4:e1924

Orsi WD, Coolen MJL, Wuchter C, He L, More KD, Irigoien X, Chust G, Johnson C, Hemingway JD, Lee M, Galy V, Giosan L (2017) Climate oscillations reflected within the microbiome of Arabian Sea sediments. Sci Rep 7(1):6040

Rosch C, Mergel A, Bothe H (2002) Biodiversity of denitrifying and dinitrogen-fixing bacteria in an acid forest soil. Appl Environ Microbiol 68(8):3318–3329

Starnawskia P, Bataillon T, Ettema TJ, Jochuma LM, Schreiber L, Chen X, Lever MA, Polz MF, Jorgensen BB, Schramm A, Kjeldsen KU (2016) Microbial community assembly and evolution in subseafloor sediment. PNAS 114(11):2940–2945

Takahashi S, Tomita J, Nishioka K, Hisada T, Nishijima M (2014) Development of a prokaryotic universal primer for simultaneous analysis of bacteria and archaea using next-generation sequencing. PLoS ONE 9(8):e105592

Wang S, Yu M, Wei J, Huang M, Shi X, Chen H (2018) Microbial community composition and diversity in the Indian ocean deep sea Rey-rich muds. PLoS One 13(12)

Zhang Y, Liang P, Xie X, Dai X, Liu H, Zhang C, JiKao S, Jiao N (2017) Succession of bacterial community structure and potential significance along a sediment core from site U1433 of IODP expedition 349, South China Sea. Mar Geol 394, 125–132

Zhang X, Xu W, Liu Y, Cai M, Luo Z, Li M (2018) Metagenomics reveals microbial diversity and metabolic potentials of seawater and surface sediment from a hadal biosphere at the Yap Trench. Front Microbiol 9:2402

Zhao R, Hannisdal B, Mogollon JM, Jorgensen SL (2019) Nitrifier abundance and diversity peak at deep redox transition zones. Sci Rep 9(1):8633

CPSIA information can be obtained
at www.ICGtesting.com
Printed in the USA
LVHW081042200420
654119LV00002B/93